Microbiology of
Food Fermentations

other AVI books

Amerine, Berg and Cruess	Technology of Wine Making, Second Edition
Brink and Kritchevsky	Dairy Lipids and Lipid Metabolism
Charm	Fundamentals of Food Engineering, Second Edition
De Man	Phosphates in Food Processing
Desrosier	The Technology of Food Preservation, Third Edition
Desrosier and Desrosier	Economics of New Food Product Development
Goldblith and Joslyn	Milestones in Nutrition
Graham	Safety of Foods
Greig	Economics of Food Processing
Hall, Farrall and Rippen	Encyclopedia of Food Engineering
Hall and Hedrick	Drying of Milk and Milk Products
Henderson	The Fluid-milk Industry, Third Edition
Inglett	Corn: Culture, Processing, Products
Joslyn and Heid	Food Processing Operations, Vols. 1, 2, and 3
Kazarian	Work Analysis and Design
Kramer and Twigg	Quality Control for the Food Industry, Third Edition, Vol. 1
Levie	Meat Handbook, Third Edition
Margen	Progress in Human Nutrition
Matz	Bakery Technology and Engineering
Matz	Cereal Technology
Matz	Cookie and Cracker Technology
National Canners Association	Laboratory Manual for Food Canners and Processors, Vols. 1 and 2
Pomeranz and Meloan	Food Analysis
Pomeranz and Shellenberger	Bread Science and Technology
Potter	Food Science
Sacharow and Griffin	Food Packaging
Schultz	Food Enzymes
Schultz and Anglemier	Proteins and Their Reactions
Schultz, Cain and Wrolstad	Carbohydrates and Their Roles
Schultz, Day and Libbey	Chemistry and Their Roles
Schultz, Day and Sinnhuber	Lipids and Their Oxidation
Sivetz and Foote	Coffee Processing Technology, Vols. 1 and 2
Sperti	Probiotics
Thorner and Herzberg	Food Beverage Service Handbook
Tressler and Joslyn	Fruit and Vegetable Juice Processing, Second Edition
Tressler, Van Arsdel and Copley	Freezing Preservation of Foods, Fourth Edition, Vols. 1, 2, 3, and 4
Wall and Ross	Sorghum Production and Utilization
Webb and Johnson	Fundamentals of Dairy Chemistry
Webb and Whittier	Byproducts from Milk, Second Edition
Weiser	Practical Food Microbiology and Technology, Second Edition
Weiss	Food Oils and Their Uses
Woolrich	Handbook of Refrigerating Engineering, Fourth Edition, Vols. 1 and 2
Woolrich and Hallowell	Cold and Freezer Storage Manual

Microbiology of Food Fermentations

by CARL S. PEDERSON, Ph.D.

Professor Emeritus of Bacteriology,
Cornell University and New York State
Agricultural Experiment Station,
Geneva, New York

WESTPORT, CONNECTICUT

THE AVI PUBLISHING COMPANY, INC.

1971

Library of Congress Catalog Card Number: 77-165214
ISBN-0-8705-104-3

Printed in the United States of America
BY MACK PRINTING COMPANY, EASTON, PENNSYLVANIA

There were doubts in the mind of my niece, Marcia, when her eighth grade teacher told the class that all bacteria were harmful and should be avoided. Marcia was certain that her uncle was studying certain bacteria that produce desirable changes in food products.

At various periods in history, the world has been described as chaotic, and maybe rightfully so; but if true, that chaos is man-made. When von Leeuwenhoek (1632–1723) first saw bacteria with his perfectly-formed lenses, he implied chaos, and this has often been applied since; however, we live in an orderly world. All life, plant or animal, has a role or function. As our concept of life reveals itself by advancing knowledge, it is observed that the universe is orderly. The various blocks of knowledge fit together. The blocks of knowledge of our lower forms of life fit into the overall plan. We do not understand all of the functions as yet, but future generations may reveal their nature.

The average individual associates bacteria and other lower forms of life with disease, illnesses, food spoilage, and other undesirable conditions. The implication is that these lower forms of life do not have a useful function. Nothing could be further from the truth. It is not the purpose of this treatise to reveal the necessity for these lower forms of life in their role as digesters of dead plant and animal refuse in order to reincorporate such materials in the soil. Rather, the purpose is to illustrate the role of the lower forms of life (certain bacteria, yeasts, and molds) in protecting some of our foodstuffs from further changes by other microorganisms so that they remain suitable nutrient substances required by mankind.

In order to accomplish this role, some of our specialized microorganisms have adapted to certain unusual conditions. They have adapted their metabolic processes to alter foods slightly so that undesirable microbial and enzymatic changes are precluded, and the food remains essentially an excellent nutrient substance. It is the author's hope that this role may be illustrated in the pages to follow.

Furthermore, it is hoped that this treatise will supplement the excellent textbooks and papers dealing with the preparation of fermented foods and emphasize the interrelationship of various fermentations. There is, moreover, a desire to stimulate further study of some of the methods of fermentation, many of which have been in use for centuries. Fermentation and drying are the oldest methods of food preparation known to mankind. The methods are so common that few people realize that microorganisms have changed their food. The microbiological and resulting physical, chemical, and organoleptic changes that occur are important to mankind.

Fermentation is a method of preparing foods to develop in the foods certain desirable characteristics, flavor, aroma, texture, and keeping quality. Temporary preservation of such foods is most important; however, for prolonged periods of storage other preservation methods are often utilized. The changes in flavor, aroma, and texture are as characteristic and distinctive for each food as are the masterpieces of excellent cooks or chefs. The preparation by fermentation of a typical Swiss cheese, a fine loaf of bread, a tangy sausage, or flavorful sour milk cannot be duplicated by any other known process.

Following the correspondence resulting from the doubts in the mind of the writer's niece, hundreds of notations in regard to fermented foods have been collected. In order to give the readers a brief but broad concept of the microbiological changes that occur in the fermentation of different foods, many of the foods consumed by peoples of the world will be mentioned briefly without resort to the thousands of references pertaining to these foods. Fermented foods have been prepared for centuries. General descriptions of many of them may be found in encyclopedias, standard dictionaries, writings of people who have traveled in various areas of the world, and in the many textbooks in the fields of microbiology and food preservation. There are very few new preparations; however, the knowledge obtained by scientists throughout the world during the past century has broadened knowledge and, in general, caused improvement and standardization of quality of the products.

It is hoped that this treatise will give readers a broader concept of the importance of this method of preparing foods in the feeding of the populations of the world.

The writer wishes to thank Dr. D. K. Tressler for encouraging the assembly of this material, Mrs. Pederson for general assistance, particularly in regard to calling attention to various foods noted during reading, Mrs. Charles Albury for reviewing the manuscript, Mrs. Foster Van Dusen for typing, Mr. Paris Trail and Mrs. Monica Juffs for their valuable assistance with illustrations, and the many friends who have helped and encouraged compilation of these pages.

March 1971   CARL S. PEDERSON

Contents

CHAPTER **PAGE**

1. Fermented Foods of the World............................ 1
2. Microorganisms of Fermented Foods...................... 12
3. Chemical Alterations During Fermentation................. 36
4. Growth of Microorganisms in Foods...................... 45
5. Fermented Milk Products................................ 66
6. Fermented Vegetable Products........................... 108
7. Fermented Sausage..................................... 153
8. Some Cereal Foods..................................... 173
9. Alcoholic Beverages.................................... 199
10. Nutritious Fermented Foods of the Orient................. 231
11. Coffee, Cacao, Vanilla, Tea, Citron, Ginger.............. 247
12. Organic Acids... 260
13. Microbial Products Consumed in Foods................... 269
14. Food Fermentation as an Approach to World Feeding Problems.. 275
    Index... 277

CHAPTER 1

Fermented Foods of the World

INTRODUCTION

When the writer was a young boy, one of his weekly chores was to churn cream into butter. This chore was made more pleasant by the reward of a hot biscuit spread liberally with the creamy, sweet-sour, partially churned butter that was carried up the plunger and settled around the chime during churning. That evening the cool fresh buttermilk was served with supper. Among other pleasant experiences remembered was the occasional hike to the nearby sauerkraut factory where a spicy, juicy dill pickle was given by the manager. A trip to the meat market was made more enjoyable by the generous piece of tangy summer sausage given by the meat cutter to be eaten as the meat was carried home. Some of the other foods served regularly included sourdough buckwheat pancakes, freshly prepared cottage cheese, and a smooth thick curdled milk sprinkled generously with sugar and cinnamon. Among the cheeses purchased regularly, besides American, Cheddar, and Swiss cheeses, was a brown, sweet, grainy whey cheese. This was occasionally made at home by boiling down the whey left from freshly-made cottage cheese. On occasion, when the wind blew from the east, the aroma of cider vinegar from the nearby factory was quite pronounced. Occasionally at an evening party or gathering, a sip of beer from a visitor's glass or a small glass of sweet port wine made going to bed somewhat less difficult.

Many of the readers have had similar experiences. Later it was learned that these were all fermented foods. The sour milk products, sauerkraut, pickles, vinegar, and sausage were fermented by acid-forming bacteria; the beer and wines were fermented by yeasts. Naturally, little thought was given to these phenomena at the time. In fact, at that time there were only a few individuals in the world who had any concept of what occurred during the fermentation process. One may ask: What are fermented foods? Why are they so prepared? Who devised or invented the various processes? Why are they important? What actually happens when milk becomes sour, fruit juice turns to wine, cider becomes vinegar, or meat acquires the tangy flavor associated with summer sausage?

Actually, these are age-old processes many of which originated in that period before history was recorded in written form. It will never be known who first observed that some foods could be prepared and preserved by these methods. If one thinks about the development of man in relation to fulfillment of his food needs, he may obtain some concept of the importance of the fermentation processes.

Prehistoric man consumed his food in the raw state. As he roamed about, he usually had little difficulty in finding sufficient food. Since food, however, is a perishable substance, it is subject to degradation and spoilage by its natural enzymes or by the enzymes elaborated by microorganisms. As man developed he learned to store some of his food for future use.

Fermentation and drying are two of the oldest methods of preparation and preservation of foods known to mankind. Even though the original physical and chemical characteristics of the foods may be altered during fermentation, their nutritive values are usually retained to a great extent. Many of the food preservation practices antidate recorded history. Throughout the centuries fermentation has been one of the most important methods for preserving foods. It still remains one of the most important methods. Relatively few people, however, are aware that the many food products consumed regularly are prepared and/or preserved by fermentation processes.

ORIGIN OF FERMENTATION AS A METHOD OF PREPARING FOOD

The origin of fermentation as a method can be surmised from knowledge of the process. If we go back far enough, we find man a savage, little higher than the more advanced beasts among which he lived. He possessed nothing whatsoever but his bare hands with which to protect himself, satisfy his hunger, and meet other needs. He was without organized speech, unable to build a fire, and had no one to teach him. The earliest man had to learn everything for himself by slow experience and tedious effort. Fortunately, he was endowed with a superior intellect, could remember his experiences and profit by them. He slowly developed methods of transmitting his experiences to his fellow man. In his development he learned that to survive, he must live where he could be comfortable and where he could obtain food.

He learned that the fruit of some trees and certain vegetable materials were wholesome and satisfied his hunger. He observed that the seeds of certain plants could be dried, and that they could be preserved in storage for long periods if they were kept dry. In the same manner some fruits and vegetables could be partially dried to be preserved for short periods. When he learned that certain animals were somewhat docile and could be tamed to his advantage, he made a great stride forward in his struggle for continued existence. These developments of agriculture and animal husbandry were some of the most significant advances in man's history. Some students assume that the transition from food gathering to food production took place some 10,000–15,000 yr ago, undoubtedly in various areas of the Middle East. Obviously, the transition did not occur quickly or simultaneously throughout all sections of the world.

Certain animals could be milked, and their milks could be stored in crude bags, pouches, or clay jars. Changes occurred in the milk, in that the milk usually became sour and thicker in body, and sometimes set as a solid curd. Undoubtedly, man soon observed that when the milk curd was smooth and had a pleasing acid odor, it was edible and wholesome, and could be consumed without distress. However, when the milk was roiled or gassy, the flavor was less appetizing and frequently unacceptable. Nevertheless, such milk was sometimes consumed with distressing effects upon the consumer. The more observant individuals may have noted that these differences could be correlated with methods of handling. When the milk was handled carefully and placed in clean containers, the curd was smooth, pleasing, and appetizing.

Similarly, he observed that the juices of various fruits could be fermented to produce a beverage that possessed certain exhilarating characteristics. Furthermore, certain vegetable materials packed tightly in containers, particularly if flavored with salt, became acid but remained edible and wholesome. Since different fruits and vegetables were indigenous to specific areas of the world, these observations were made at various periods of man's development. Similarly, different animals were domesticated in the various areas of the world. At some time in man's development, he noted that dried seeds could be finely ground, blended with water and salt, and then improved in flavor and quality by heating or baking. Since this flour and water blend would serve as an excellent medium for yeast and bacterial growth, if the dough was not heated immediately, leavening would occur and a light airy baked product would result. This observation is sometimes credited to the ancient Egyptians. Also, it was observed that if the flour and water mixture was thin, an alcoholic beverage was produced. As observations were noted, improvements in methods of handling foods and animals were made. The information was passed on from father to son, from mother to daughter.

EXPLANATION OF MICROBIOLOGICAL CHANGES INCOMPLETE

A partial explanation of the causes and nature of some of these changes had to await the development of scientific knowledge centuries later, particularly in microbiology. Most of the scientific knowledge has been accumulated during the past century. Since some of it has not yet been applied to many fermented foods, the true nature of many products has not yet been elucidated.

One may say these methods of food preservation have been developed primarily through observation of supposedly natural phenomena. It is indeed surprising to those with a basic knowledge of microbiology that these methods could have been developed purely by observation, with no understanding of the nature of the inherent changes that occur. Furthermore, it is the more remarkable when one realizes the diverse nature of the fermented foods: some resulting from lactic acid bacterial fermentation; some, alcoholic fermentation by yeasts; some, mold fermentation; some, acetic acid bacterial fermentation; and many, by combinations of these fermentations. Today a large majority of individuals preparing foods by these processes are still practicing "rule of thumb" methods developed over a period of years by their forebears.

FERMENTED FOODS ARE WELL-KNOWN TO CONSUMERS

Fermented milks, cheeses, and butters are prepared from milks obtained from many species of animals: cows, goats, sheep, mares, yaks, carabaos, camels, reindeer, and others. There are over 400 names applied to the cheeses produced throughout the world. There are many types of butter and many variations of milk preparations. Breads in a great variety of forms are used everywhere. Fermented sausages prepared from the meat of many animals and also fish are consumed throughout the world. Individual and mixed vegetables are preserved as acid products. The juices of many fruits undergo an alcoholic fermentation; and practically every civilization has a characteristic beer or alcoholic beverage prepared from cereals, potatoes, and other starchy foods either alone or in combinations. Vinegars are prepared from many of the alcoholic drinks. Consumption of mold-fermented products is common in the Orient.

FIG. 1.1. THE PREPARATION OF ALL FOODS FOR THIS SANDWICH—ROLL, BUTTER, CHEESE, SAUSAGE, SAUERKRAUT, AND PICKLE RELISH WITH VINEGAR—INVOLVE FERMENTATIONS BY BACTERIA OR YEASTS OR BOTH

Although many of these products are well known to the average consumer, he may associate fermentation only with the production of alcoholic beverages. The lactic acid fermentations resulting from the activity of the lactic acid bacteria are, nevertheless, of much greater importance in feeding the populations of the earth.

Methods of preparing many of these foods have been and are being improved. For example, the average consumer is unaware of the fact that the cream used in making sweet cream butter has actually fermented. The sweet cream is pasteurized to kill undesirable microorganisms, cooled, a selected bacterial culture is added under carefully controlled conditions to ferment the cream, and then it is churned to produce the sweet cream butter. The microbiological, chemical, and physical changes involved will be discussed in subsequent chapters.

It has been stated that the fermentations are the result of growth of bacteria, yeasts, molds, or combinations of these. Stated more precisely, the changes that occur are caused by the enzymes elaborated by these microorganisms. Some foods usually said to be fermented are actually cured by the enzymes naturally inherent in the foods.

It would be helpful to establish a suitable and complete table for classifying the many fermentations; however, it is imperative to understand that many fermented foods result from a combination or sequence of growth of various microorganisms. Many fermented foods are insufficiently studied, and among fermentations investigated, some variation may occur depending upon environmental conditions.

VARIETY OF FOODS CONSUMED THROUGHOUT THE WORLD

What a pleasant experience it would be to travel throughout the world, visiting the various peoples in their homes and sharing with them their native foods. Probably the food served in the homes of the extremely wealthy would not represent the diet of the average home; nor would that of the extremely poor where existence may be difficult. In the average home, the hosts might be hesitant to serve a guest because of the simplicity of their homes and foods. A guest, however, would be shown every courtesy.

In a western Asian home, the host may literally break a piece of the round, flat, somewhat porous bread with the oldest honored female guest as he offers a brief prayer. Following this, all others would then also break off a piece of the bread. The foods would be well prepared. The cooked lamb would likely have been marinated in sour milk before cooking, the vegetables would be eaten with curds, and some pickled products would be served. In many areas of the world, you and your host would be served by your hostess, by her sister, or by the daughters. You would be surprised by the variety and novelty of the foods, and you would know that they are unsparing in their desire to make you welcome and to treat you to an excellent meal. If you should express particular interest in a particular type of food, you might be served several types of that food.

In India you might be served roghan josh, a meat marinated for hours in clarified butter and milk curds. Or you may have ran, a meat which had been soaked in buttermilk for 24 hr before it is roasted in a slow oven. Either would contain spices and each would be accompanied by pickled vegetables, a variety of breads, and, of course, rice. Many of the foods would have a slight to high acidity, and possibly a pungent flavor. Then you would realize fully that many of the foods are prepared or preserved by fermentation processes or cooked with foods so prepared. The tart and distinctive flavor of the breads, pickles, clarified butter, and curds are imparted only by the fermentation process. Of course, the actual foods served would vary, but would typify the region. In India, meat would be prohibited in many homes. Idli, a steamed bread, would be prepared from ground rice and black gram beans; in another area, idli would be prepared with wheat and black gram beans. A somewhat similar steamed rice bread, called puto, would be served in the Philippines. In the many areas there may be dozens of varieties of bread, often served hot by your hostess, and sometimes with a sauce which blends with the tart flavor of the bread.

In a Tibetan home, you may be served a hard, sour, native cheese with yak meat or mutton or a seasoned dumpling warmed in a sweet-sour milk. You may note the absence of cows and then realize that the milk is yak milk; the cheese, however, may have been made from ewe's milk. The tea may have been boiled in slightly sour milk, and you may end your meal with a slightly yeasty barley beer. In Kashgar, the housewife may be occupied with milking the goats and ewes or preparing butter or cheese from the previous day's milk. The battered clay or stone pot must be washed daily before the fresh milk is added. Little do they realize that in spite of thorough washing, the beneficial lactic acid bacteria remain impregnated in the pores of the container and serve as an effective starter for the next batch of milk. Nevertheless, a small amount of the yoghurt from the previous day's milk will be added to ensure a good fermentation. If a daughter is to be married, among her gifts will be some of the yoghurt starter. Kashgar cheese is uncured and similar to our cottage cheese.

In an Arab home, you may be served a waffle-like bread of the type that is often carried by the native and is used in several ways. He may have it wrapped around a piece of cheese and an olive. When he has eaten his milk curds or mutton gravy, he may wipe out the bowl with the next to the last piece of bread, saving the last piece for wiping off his mouth.

In Baktiari, you may have chopped vegetables in buttermilk with a large pancake-like bread and a dish of rice fried crisp in butter. A dish of apricots with a bowl of sour milk may be your dessert.

The majority of foods served in the Ukraine will contain some type of fermented food. Cereals form an important part of the diet of the Ukrainians. You may be served the sourdough rolls, pampooshke. The sour, fermented beets, kvas, may be added to many foods, and the free juice would undoubtedly be incorporated in a soup. The fermented vegetables may be spiced and served with cereals. The women know how to use the fermented vegetables and milk, as well as buttermilk, butter, and cheese in the tasty foods of their many borscht dinners. Kholodnyk, a vegetable dish, will contain sour cream and vinegar, and for special occasions shrimp or lobster. Rozsolnik made with pork kidneys will be flavored with sour pickles and a special Ukrainian garlic sausage. Kapoosnyak is similar to spare ribs or pork with sauerkraut, while begos is another sauerkraut-meat dinner.

Although the same types of foods will appear on the tables of homes in many other areas, they may be prepared or cooked differently. In Yugoslavia, in a version of sauerkraut, kiseo kupus, the whole heads of cabbage either white or red are fermented in a weak salt brine. The fermented leaves are wrapped around ground meat and baked to produce sarma. Podvarak is prepared by baking slices of the fermented cabbage with turkey or goose that is stuffed with a sauerkraut dressing. The whole fermented cabbage may also be served cold as a salad or cooked with pork. The brine from red kraut has a pink wine color and is relished as an appetizer.

In Western Asia, the soured milk drinks may have a different flavor, somewhat alcoholic and acid. This is kumiss, the milk drink prepared from mare's milk fermented by a combination of lactic acid bacteria and yeasts. It is sweeter than yoghurt and is mildly intoxicating because of its alcohol content. It is usually consumed with a dark sour rye bread. Further south another acid-alcoholic milk drink, kefir, is used. It is characterized by kefir grains, a mixture of yeasts, bacteria, and curdled milk which somewhat resembles wet popcorn. These grains which settle to the bottom of the liquid are used for starters for succeeding batches of milk. The sour milk of Syria is similar to yoghurt. Its name, leben, is used in the Syrian Bible for milk.

Across the continent of Asia in the small countries, is found a distinctive form of fermented or pickled vegetables, kimchi. In the Korean home you would sit on a cushion on the floor of a room a step or two above the central entrance way. A low table covered with food would be carried in by your hostess. Prominent among the foods would be kimchi of several kinds because you expressed an interest in this food. You try all kinds, for the housewife is proud of her kimchi. The first is so peppery that it may burn your mouth, and you hurry to the next which is mild and acid and not unlike our own sauerkraut. A third may contain nuts or a little fish. You may note a row of various sized jars on the other side of the central area. These contain kimchi of different kinds, packed when the vegetables were harvested. It is not a joke when the Korean housewife asks her neighbor if she has packed her kimchi, because it is an important part of that region's food supply. In various homes as well as different areas one will find a great variety of types of that food.

The fermented vegetable blends are equally important in the diet of some Chinese. The Chinese may have been the first to ferment vegetables, and there is evidence to show that they were prepared at the time of building the Great Wall of China. Wooden tanks or vats are used in some areas as containers. In some areas it is reported that they use clay jars with a moat around the lip. The jar cover is made with a lip which will fit into the moat, so that when water is placed in the moat, there is an effective seal.

Across the China Sea in the Philippines, a similar fermented vegetable, burong gulay, is prepared but not widely used. Puto or Biñan puto, named for the city where it is made, is more common. It is a pleasing steamed bread or cake made from the finely ground flour from a particularly glutinous variety of rice. Although it is usually eaten soon after steaming, it is also appetizing after rewarming in an oven. When one processor was asked who taught him the process, he said his uncle, who had learned it from his forebears who possibly may have immigrated from China or Malay years ago. The young man appreciated the necessity of carrying out the process exactly as he had been taught years ago. Puto is leavened by the carbon dioxide produced by lactic acid bacteria.

A special cheese, somewhat like cottage cheese, may be purchased in banana leaf containers. Santa Cruz cheese made from carabao milk is so named because it may have originated in Santa Cruz. Whey drained from the previous day's preparation and containing a piece of calf's stomach is added to the fresh milk. The whey is highly acid as a result of the growth of lactic acid-producing bacteria; therefore, within a relatively short time the fresh milk curdles. The curd is transferred to the banana leaf cups. Some operators realize that the whey must be acid in order to obtain good cheese.

In some areas of the Philippines you may still buy burong dalag or burong hipon, fermented blends of rice with fish and rice with shrimp. Fish other than the dalag may be obtained in some areas. The use of this product is declining primarily because it is difficult to prepare and the quality of some preparations is inferior. But it is a high-protein food and the methods of preparation should be studied. The salty fish preparations, patis and bagoong, are similar to the nuoc mom of Viet Nam. They are digested or fermented by the enzymes native to the fish. Nata de piña and nata de coco are tenacious jelly-like preparations made from pineapple or coconut by an acetic acid fermentation. The nata is similar to mother of vinegar, but firmer and more tenacious. As in every other country, several types of alcoholic beverages are prepared in the Philippines. Rice or fruits or even the sap from trees are basic fermentable substances.

It is unnecessary to go to Europe or some other area of the world to find many fermented foods. America is the melting pot of the world. The many foods and eating customs, as well as the peoples, have been blended among the nationality groups. The many cheeses of England, France, Germany, Italy, Switzerland, Scandinavia, and elsewhere may be imported but their counterparts are made here. France alone produces some 60 varieties of cheese. It is true that you would have to be invited to a Syrian home to taste shonkleesh, but the true French Roquefort and other European cheeses can be purchased anywhere. Possibly the price will be somewhat higher than its American counterpart, blue cheese. Sauerkraut is a German word, but there is more sauerkraut produced commercially in New York State than in Germany, and it is used in many more ways. Every type of pickle made in Europe has its counterpart in the United States. Pizza might not be recognized by a newly-arrived Italian, but it contains the same type of fermented sausage and cheese. The Lebanon bologna was introduced by the German immigrants generations ago. The Italian Genoa sausage may not be available everywhere, but thuringer, salami, and many others are. Possibly they are better known as summer sausages. There is at present a resurgence of interest in fermented milk drinks such as yoghurt, leben, kefir, and buttermilk. Olives may appear in several food products. Butter, previously prepared from the highly acid fermented cream, has been replaced in general by the so-called sweet cream butter; however, as previously mentioned, the cream is still fermented by lactic acid bacteria.

Many native foods are made in the homes of newly-arrived immigrants or their first generation offspring. Often, the second generation discards the practices as old-fashioned; unfortunately, some of the finer preparations are, therefore, never made available to the younger generations.

This is merely a small part of the story of fermented foods, foods prepared by methods developed centuries ago, many even predating the beginning of recorded history. The methods have been passed from grandparents to parents, and eventually to daughter and son. With new observations, slight improvements in methods have been made; but the basic methods have remained constant. Microorganisms are the agents. It is a story full of interest; a story of survival of mankind and the ways he utilized, prepared, and preserved his foods in wholesome, appetizing, and nutritious forms.

Many of these food have been modernized. Their quality and character have been improved and standardized by modern methods as a result of research conducted in both academic and industrial laboratories. Others are still prepared essentially as they were centuries ago.

INFLUENCE OF RESEARCH

Research conducted during the past 50–100 yr with bread, butter, cheese, sausage, pickles, olives, sauerkraut, soy sauce, wines, beers, and other foods prepared by fermentation has explained the nature of the desirable as well as undesirable changes that occur during and subsequent to fermentation. The results have formed the basis for improvement and standardization. Spoilages and development of off-flavors, such as rancid butter, gassy cheese, ropy bread, black pickles, pink sauerkraut, acetic wines, sour beer, and other deleterious activities, have been reduced to a minimum. Although the larger industries have their own research staffs today, their studies are influenced by the previous and present research conducted in academic institutions. The smaller industries must still rely on academic research for information.

This research should form a basis for studies applied to the hundreds of fermented foods now being prepared entirely by "rule of thumb" methods throughout the world. There is far too little research conducted with many of the fermented foods. Accumulation of knowledge through research with these lesser-known foods now prepared in homes and in home industries is essential for their continued utilization. Improvement of methods through research, followed by dissemination of this knowledge by extension into homes, must be instituted if the use of some of these nutritious foods is to be continued. Their use should be continued because they are not only nutritious and economical, but they have been acceptable to peoples for centuries.

Such studies must start with academic research concerning the basic principles; in this case, first, with the study of the microorganisms which are responsible for the desirable changes which occur as well as the microorganisms which are responsible for spoilage. These studies must be followed by studies of chemical, physical, and nutritional changes. This has been the course of study of most of our well-known fermented foods. Their acceptability has advanced to the stage of discernment of minor quality differences.

This treatise can only give a brief summary of the research conducted with some of our fermented foods. It is hoped that it may encourage, assist, and guide the research student in studying the many other fermented foods. The voluminous literature, both academic and industrial, involving foods such as cheese, butter, sausage, and others cannot be encompassed in this book.

In the chapters to follow, many apparently unimportant fermented products will be mentioned; they may, however, be extremely important in the nutrition of the peoples in the areas where such products are used. The acceptability by these people may depend upon the quality of the products, and their quality may be a direct reflection of the microbiological changes that occur during and subsequent to fermentation. Very little is known in regard to the fermentation of many of these foods. The need for research is great even with these apparently unimportant products. Research with bread, cheese, butter, beers, wines, and others has been instrumental in attainment of the high standards of quality and their general acceptance and safety. This research must serve as a basis of investigation of microbiological and chemical changes among the lesser-known food products.

BIBLIOGRAPHY

ANON. 1937. Encyclopedia Britannica, 14th Edition. Encyclopedia Britannica, New York, Chicago.
FRAZIER, W. C. 1967. Food Microbiology, 2nd Edition. McGraw-Hill Book Co., New York.

JENSEN, L. B. 1942. Microbiology of Meats. The Garrard Press, Champaign, Ill.

PETERSEN, W. E. 1950. Dairy Science, 2nd Edition. J. B. Lippincott Co., Chicago.

PETERSON, M. S., and TRESSLER, D. K. 1963. Food Technology the World Over. Avi Publishing Co., Westport, Conn.

PRESCOTT, S. C., and DUNN, C. C. 1959. Industrial Microbiology, 3rd Edition. Mc-Graw-Hill Book Co., New York.

TANNER, F. W. 1944. Microbiology of Foods, 2nd Edition. Garrard Press, Champaign, Ill.

CHAPTER 2

Microorganisms of Fermented Foods

The preparation and preservation of foods by fermentation processes are dependent upon the production by certain microorganisms of chemical substances that alter the flavor of the food and are generally inhibitive to the growth of undesirable microorganisms. The simplest example of such action is the inhibition of toxin-producing bacteria by the lactic acid produced in many fermented foods. Although fermentation is one of the most ancient methods of food preservation, an accurate concept of the nature of its changes has had to await the development of the science of microbiology that has occurred during the past 100 yr.

A few years ago a high school biology teacher, trying to impress upon her class the relevancy of microorganisms to illness, stated emphatically that all bacteria were harmful. While her teaching may have been commendable from the public health standpoint, since certain bacteria cause disease, the beneficial effects of the great majority of species of microorganisms far outweigh the deleterious aspects of illness caused by relatively few pathogenic species. It is indeed unfortunate that the great majority of people associate bacteria or "germs" only with disease. Microbiologists and public health officials have unwittingly influenced the average person's fallacious belief that all bacteria are harmful. The spectacular effects of pathogenic microorganisms are far more dramatic than the relatively prosaic fact that microorganisms are extremely important in ridding the earth of dead plant and animal material, or that there are more than 50 industrial processes that are dependent upon microbial growth, or that there are great numbers of foods prepared and/or preserved by fermentation.

FEW SPECIES OF MICROORGANISMS INVOLVED IN FERMENTATION

Fermented foods are the result of the activity of a few species of microorganisms among the thousands of species of bacteria, yeast, and molds known to mankind today. The microorganisms that ferment foods to produce desirable changes can be distinguished from those that are responsible for spoilages, illnesses, and occasionally death.

Many of our lower forms of life—insects, worms, and others, including the all-important microorganisms—play a vital role in the continuous process of reducing dead plant and animal residues to their elements. Microorganisms, in contrast to many other lower forms of life, are catabolic rather than metabolic. They alter organic components of food to obtain energy for their growth. While the processes are complex, ultimately the organic components are reduced to their elements to become a part of the soil. Many types of microorganisms have evolved, each of which may have a specific role in this process. The decomposition of many organic substances may involve a series of distinct changes brought about by various microorganisms.

This catabolic process is the important role of bacteria, yeasts, and molds in the cycle of life. To accomplish this, great numbers of species have evolved and adapted to the many specialized environmental conditions that exist. Some species are very specific in function, others are broader in their adaptation to environment. In the processes of decomposing organic matter, some of the substances produced are disagreeable in odor, flavor, and texture, and some are, in fact, injurious to health. Many of the decomposition products of proteins and fats fall in these two categories. Some bacterial species are parasitic in that they have adapted themselves to grow in weakened living plant or animal tissue.

TYPES OF MICROORGANISMS INVOLVED IN FERMENTATION

The lactic acid-producing bacteria, the acetic acid-producing bacteria, and certain alcohol-producing yeasts are highly specialized. These and certain molds are the species so important in food fermentation processes. The group of lactic acid-producing bacteria play an important role. They carry on essential metabolic biological processes without oxygen by means of a complex series of intramolecular oxidations and reductions. They are sometimes referred to as microaerophilic. Since they do not utilize oxygen, the changes they accomplish do not result in decomposition of the foods to their basic components, such as carbon dioxide, water, simple nitrates, and sulfates. Instead, the most commonly recognized end product of their metabolism is lactic acid derived from sugar. They alter other components to a minor extent, and some species produce other products from sugars.

The ability to convert carbohydrates to lactic acid, acetic acid, alcohol, and carbon dioxide with only minor changes in other food components has made this group of bacteria so important to mankind in the preservation of edible and nutritious food. There is little caloric change in the conversion of carbohydrate to lactic acid and very little loss in total nutritive values. Since they do not utilize oxygen, these bacteria obtain only a small part of the energy available, and, therefore, must ferment a great amount of sugar to supply the energy needed for growth and reproduction. These bacteria may be accused of an inefficient use of their energy source. This inefficiency is fortuitous to mankind. The lactic acid they produce is effective in inhibiting the growth of other bacteria that could decompose the food, making it unfit to eat; therefore, the inefficiency of the lactic acid bacteria preserves and provides for mankind a safe food.

While some yeasts ferment sugars to alcohol and carbon dioxide in the absence of air, they require air for optimal growth. The great majority of species which grow on food surfaces require oxygen for growth. Similarly, the acetic acid bacteria require air to oxidize alcohol to acetic acid. Since molds are aerobic, they require oxygen for growth.

MICROORGANISMS ARE SMALL PLANTS

Bacteria, yeasts, and molds are extremely small plants. They are called microscopic plants because their individual cells can only be seen with the aid of a microscope; a massive growth of billions of these microorganisms can be seen with the naked eye. The velvety white, green, black, and various-colored fuzzy mold growths sometimes present on breads, jellies, fruits, meats, and other foods consist of billions of cells in mycelial strands with their colored, resistant spores. Each cell is capable of individual growth and reproduction. The cloudiness of fermenting fruit juice is ordinarily caused by the massive growth of yeasts. The occasionally encountered slimy, white growths on the surfaces of many foods and in acid juices may consist of masses of surface-growing or aerobic yeasts. The average individual is acquainted with yeast cakes, a mass of billions of cells held together with a binder. The cloudiness in fermenting pickle brine or similar material is caused by lactic acid-producing bacteria. The slimy "mother of vinegar" associated with apple cider consists of a mass of bacterial cells. These massive growths were associated with fermented foods as well as spoiled foods centuries ago, even though people then had no concept of their true nature. Yeast and mold growths are more obvious than bacterial growths.

OBSERVATION OF PRESENCE OF MICROORGANISMS A RECENT DEVELOPMENT

Possibly Kircher, a Roman Catholic monk in 1671, was the first to see the organisms which we call bacteria. Although he gave no description of them, it seems probable that among the "invisible worms" he claimed to have seen with the microscope, some were bacteria. Shortly afterward in 1683, the Dutchman von Leeuwenhoek, known as the father of microscopy, did see and describe bacteria which he referred to as "animalicules." Other microscopists soon found similar bodies in many substances that they examined. The idea that diseases might be spread by particles too small to be seen by the unaided eye had been suggested much earlier by Fracastoria of Verona in 1546. The germ theory of disease was formulated in 1762 by Plenciz.

It had frequently been observed that moist food would not keep very long. In the years following von Leeuwenhoek, it had been noted that the spoilage of food was accompanied by growth of myriads of the "animalicules" as they were called by von Leeuwenhoek. Early in the 19th Century, Appert (1809) discovered that perishable food when placed in proper containers, subjected to heat, and hermetically sealed would not spoil.

FERMENTATIONS ARE THE RESULT OF ACTIVITY OF MICROORGANISMS

Soon after the middle of the 19th Century, the initial period of present-day microbiology was ushered in by Louis Pasteur. During his early thirties, he became interested in the phenomenon of fermentation that he attributed to biological agencies. This was in opposition to the theories of Liebig. It was a simple matter for Pasteur to confirm the earlier work of Cagnard de Latour and of Schwann, and to show that alcoholic fermentation could take place only in the presence of yeast. Pasteur was not content with this. He soon proved that there were many types of fermentations, each caused by its own specific agent. Until this time there had been no clear proof that there were more than one kind of fermentation. To Liebig, alcoholic fermentation, putrefaction, and acid fermentations were related. The difficulty was that when dealing with such extremely small organisms, no one knew how to separate one kind from the others to find out what one individual kind of organism could do alone.

Pasteur solved this problem in a very simple and ingenious way. He noticed that when any natural substance such as milk or fruit juice underwent fermentation, microorganisms developed in it. In different kinds of fermentation, different kinds of microorganisms appeared. His conclusion was that different kinds of fermentation were caused by distinctly different ferments. To test this theory, he prepared artificial mixtures containing sugars whose fermentations he was then studying. He sterilized these solutions and then inoculated them with a small portion of the sediment from the fermented fruit juice, sour milk, or whatever ferment he was studying. If his artificial culture solution underwent the typical fermentation and exhibited the same kind of microorganism as in the original container, he concluded that it was this organism that caused the fermentation. He found, for instance, that a certain type of microorganism always grew in large numbers in ordinary sour milk and also appeared in his sugar solutions when inoculated with sour milk. If, however, the milk were first boiled, he found that it fermented in a different manner with production of butyric acid instead of lactic acid. He spoke of these ferments as new yeasts, lactic yeasts, and butyric yeasts. He did not fully realize that he was studying organisms belonging to distinctly different groups of living entities.

DEVELOPMENT OF THE SCIENCE OF MICROBIOLOGY

The science of microbiology in its relationship to the many fermentations has advanced considerably during the past century. Each food is distinctive from any other food. There are numerous types of fermentations, some characterized by the activity of a single microorganism, others by the activity of several types. The growing interest in successive years led to wider study, and the discovery of new types or species of microorganisms, many quite specific in their activity. New methods of study have been developed.

APPLICATION OF NAMES TO MICROORGANISMS

The necessity for applying names to species or kinds of microorganisms and suitable names for the groups of interrelated organisms became evident. Characterization and classification of microorganisms became essential. A name given by one person should be mutually recognized everywhere. As far as practical, students throughout the world should refer to microorganisms by the same names. Today, if reference is made to the well-known milk-souring bacterium, *Streptococcus lactis* (Fig. 2.1), bacteriologists all over the world know that the organism is the small, lactic acid-producing, coccoid-shaped bacterium that produces a smooth soft curd in milk with about 0.7% lactic acid.

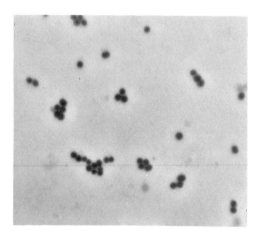

FIG. 2.1. *STREPTOCOCCUS LACTIS*. THE LACTIC ACID-PRODUCING BACTERIUM COMMON TO FERMENTING MILK (MAGNIFIED APPROX. 2500×)

CLASSIFICATION

The purpose of books dealing with classification, such as *Bergey's Manual of Determinative Bacteriology*, is to facilitate finding the accepted name of a microorganism and to offer the first indication of the nature of the organism. The general classification should show the similarities and differences of various types to one another. It must be realized that as knowledge accumulates, new types or species may be recognized and that new information may be made available that may necessitate a change of grouping of species or genera. Classifications are established on the basis of characterization of the organisms.

The early classifications of microorganisms were based primarily upon structural characters, particularly the size and shape of cells. This was a natural development since morphology had been found so useful in drawing up natural classifications of higher plants. Little was known in the early days in regard to physiological characters; therefore, the lactic acid-producing bacteria that are extremely important species in the majority of food fermentations, were widely separated on the basis of morphological differences. Today, although the coccoid forms are separated from the rod-shaped forms on a morphological basis, they are grouped together in one family on the basis of physiological similarities.

Microorganisms are widely distributed in nature; therefore, they are sometimes said to be ubiquitous. Whenever environmental factors are favorable, the microorganisms will grow. Favorable temperature, proper moisture level, and food supply are primary factors that govern growth. A broad classification is essential to a complete understanding of the great variety of changes that occur in the microbial food supply. No attempt will be made to classify all of the microorganisms; rather, the discussion of classification will be confined only to those species that are important in food fermentations. The study of this classification reveals how relatively few species of the great numbers existent are actually involved in food fermentations. This emphasizes the highly specific nature of these and many other organisms.

Microorganisms involved in food fermentations may be grouped as bacteria, yeasts, and molds. They are considered to be members of the plant kingdom in the phylum *Thallophyta*. The *Thallophyta* are defined as undifferentiated masses devoid of root, stem, or leaf. The phylum includes the *Algae* and *Fungi*. The latter include the classes of bacteria, yeasts, and molds. They depend for their food upon organic matter synthesized by other organisms.

The relationship is as follows:

Kingdom: *Plant*
  Phylum: *Thallophyta*
    Subphylum: *Eumycetes*
      Classes: *Schizomycetes* (bacteria)
            Phycomycetes (see Fig. 2.16)
            Ascomycetes (true yeast; see Fig. 2.16 and 2.17)
            Basidiomycetes
            Fungi Imperfecti (asporogenous yeasts: see Fig. 2.16 and 2.17)

In order to present a clearer understanding of the relationships among bacteria, a part of the key from *Bergey's Manual of Determinative Bacteriology, 7th Edition* (Breed *et al.* 1957) will be presented with comments. This is the generally accepted classification among bacteriologists throughout the world. An attempt has been made in this manual to group all bacteria of similar characteristics. No firm rule can be laid down regarding the extent of difference that must exist between species, genera, and families, and further grouping to justify separation of cultures, one from another. Cultures that are considered similar in all respects may be included within a species. For example, when the name *Streptococcus lactis* is used, *lactis* is the species name and is applied to all strains of the lactic acid-producing coccus forms related to the species *Streptococcus lactis*. A family is a group of related genera. An order is a group of related families and a class is a group of related orders.

The class *Schizomycetes,* or fission fungi, is separated into 13 orders, 47 families, and 193 genera in *Bergey's Manual of Determinative Bacteriology* Breed *et al.* (1957).

When food fermentation organisms are discussed, only 2 of the 13 orders are involved, *Pseudomonadeles* and *Eubacteriales.* Because of its importance in food fermentations, the fourth order, *Eubacteriales,* will be discussed first. Order IV *Eubacteriales,* Buchanan 1917 (the true bacteria). Eu-bac-te-ri-a'-les; Greek *eu* meaning true; Greek *bacterium,* a small rod; *ales,* an ending meaning order.

These are simple undifferentiated, rigid cells which are either spherical or straight rods. There are nonmotile as well as motile species. All of the species in certain families are definitely Gram-negative, in other families the majority of species are Gram-positive. Reproduction is by transverse fission.

The order *Eubacteriales* includes 13 families, only 2 of which, *Lactobacillaceae* and *Propionibacteriaceae,* include bacteria that are important in food fermentations.

IV *Eubacteriales* (13 families)
　X *Lactobacillaceae* (2 tribes)
　　I *Streptococceae* (5 genera)
　　　II *Streptococcus* (19 species)
　　　III *Pediococcus* (2 species)
　　　IV *Leuconostoc* (3 species)
　　II *Lactobacilleae* (5 genera)
　　　I *Lactobacillus* (15 species)
　XI *Propionibacteriaceae* (3 genera)
　　I *Propionibacterium* (11 species)

FAMILY LACTOBACILLACEAE

Family X *Lactobacillaceae* Winslow *et al.,* 1917. Lac-to-ba-cil-la'-ce-ae. Latin *lac, lactis* milk; Latin *bacillus,* a little rod; *aceae* denotes family ending.

Long or short rods or cocci that divide like rods in one plane only, producing chains, occasionally tetrads; filamentous as well as so-called false branching forms sometimes occur (Fig. 2.1 through 2.11). Usually nonmotile but may be motile. Gram-positive. Pigment production rare; a few species produce a yellow, orange, red, or rusty brown pigment. Gelatin liquefaction rare. Surface growth on all media is poor or absent (Fig. 2.2). Carbohydrates are essential for good development; they are fermented to lactic acid, some-

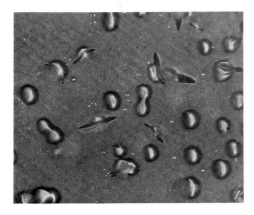

FIG. 2.2. LACTIC ACID BACTERIA GENERALLY DEVELOP ON AGAR PLATES
AS SMALL WHITE ROUND OR LENS-SHAPED COLONIES AMONG THE LARGER
MORE PROFUSE COLONIES OF MANY OTHER SPECIES

times with volatile acid, alcohol, and carbon dioxide as by-products. Nitrites
not produced from nitrates; but among the strict anaerobes there are a few
species that are known to reduce nitrates. Microaerophilic to anaerobic,
found regularly in mouth and intestinal tract of man and other animals, in
food and dairy products, and in fermenting vegetable juices.

This then is a condensed description of the Family X. The descriptions do
not necessarily apply to all genera. The family, as indicated, is divided into
two tribes, *Streptococceae* and *Lactobacilleae.*

TRIBE STREPTOCOCCEAE

Tribe I *Streptococceae* Trevisan, 1889. Cocci occurring singly, in pairs,
and in chains (rarely tetrads). This tribe includes five genera. The first and
fifth, *Diplococcus* and *Peptostreptococcus* are not involved in food fermentations.

Genus II *Streptococcus* Rosenbach, 1884. Strep-to-coc'-cus. Greek ad-
jective *streptus,* pliant; Greek noun *coccus,* a grain or berry.

The species of this genus are of great importance in the dairy industry.
Certain strains are employed as starter cultures in preparing cheeses, butter,
and cultured milk drinks. They develop naturally in ordinary unpasteurized
milk. Some strains are capable of fermenting citric acid with production of
carbon dioxide, acetic acid, and diacetyl when incorporated with a fermentable sugar. Normally, they produce dextro rotatory lactic acid from sugar.
They acidify and curdle milk with a smooth soft curd. Bergey's Manual lists
19 species, the following 4 of these are important in food fermentations.

Species 14. *Streptococcus thermophilus* Orla-Jensen, 1916. This species is
employed as a starter for Swiss cheese. It is easily distinguished from other

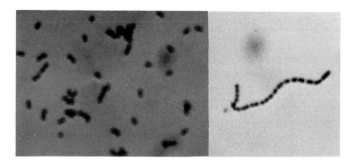

FIG. 2.3. *STREPTOCOCCUS FAECALIS*, THE LACTIC ACID BACTERIUM COMMON
TO FERMENTING VEGETABLE SUBSTANCES

Species of the genus *Streptococcus* frequently grow in chain formation
under adverse environment. (Magnified approx. 2500×)

species of the genus by its high temperature range for growth, 40°–45°C, its
sensitivity to salt, and its inability to ferment maltose.

Species 16. *Streptococcus faecalis* Andrewes and Horder, 1906. This species
(Fig. 2.3) is distinguished from other species of the genus by the wide temperature limits for growth of its strains, their salt tolerance, and their ability
to initiate growth at pH 9.6. These strains are found in the intestines of humans and warm-blooded animals and sometimes in dairy products. They are
the common strains found in vegetable products. There are numerous varieties of this species, based upon physiological reactions.

Species 18. *Streptococcus lactis* (Lister, 1873) Löhnis 1909.[1] This species

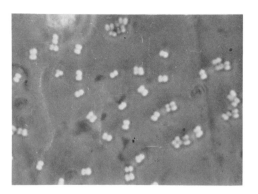

FIG. 2.4. *PEDIOCOCCUS CEREVISIAE*, THE LACTIC ACID BACTERIUM FIRST
ASSOCIATED WITH BEER SPOILAGE BUT NOW KNOWN TO BE COMMON TO
MANY FERMENTING MATERIALS (MAGNIFIED APPROX. 2250×)

[1] This form for naming species is in accordance with *Bergey's Manual of Determinative
Bacteriology, 7th Edition* (Breed *et al.* 1957) and indicates that the species name *lactis* was applied by Lister, but at a later date it was placed in the genus of *Streptococcus* by Löhnis.

(Fig 2.1) is of great economic importance in the dairy industry. Strains of this species are employed as starter cultures in preparing butter, cheese, and milk drinks. The strains that are capable of fermenting citric acid with production of carbon dioxide, acetic acid, and diacetyl are ordinarily selected for starters. They produce a smooth, soft, acid curd in milk.

Species 19. *Streptococcus cremoris* Orla-Jensen, 1919. Strains of this species are also commonly used as starters in the dairy industry. Like *S. lactis* it produces diacetyl, and, although closely related, strains of this species will not grow at 40°C, in a 4% salt broth, or in a medium adjusted to pH 9.2.

Genus III. *Pediococcus* Balcke, 1884, emend. Mees, 1934. Pe-di-o-coc'-cus. Greek noun *pedium,* a plane; Greek noun *coccus,* a berry or sphere.

These are saprophytic organisms found in fermenting vegetables, mashes, meat, beer, and wort. Although they occur as cocci singly, in pairs, and occasionally in short chains, they are more often distinguished by their tetrad grouping. They produce inactive lactic acid, and in larger quantities than streptococci, but most strains do not grow well in milk and seldom curdle it. The following two species are described in Bergey's Manual although many specific names are used in the literature.

Species 1. *Pediococcus cerevisiae* Balcke, 1884. This species (Fig. 2.4) was originally isolated from "sarcina sick" beer, a beer characterized by acid and the "sarcina odor." This odor is due to diacetyl. More recently it has been recognized in various fermenting vegetable juices, in meat for sausage, and in mashes.

Species 2. *Pediococcus acidilactici* Lindner, 1887. This species is closely related to *P. cerevisiae.* It is found in mash and unhopped wort.

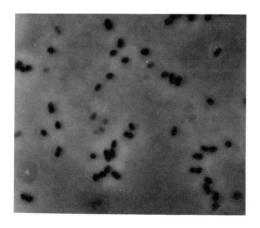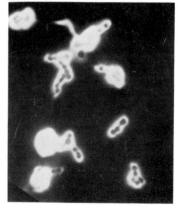

FIG. 2.5. *LEUCONOSTOC MESENTEROIDES,* THE SPECIES OF LACTIC ACID BACTERIA THAT INITIATES MANY VEGETABLE FERMENTATIONS AND IN SUCROSE MEDIA IT PRODUCES A CAPSULAR DEXTRAN (MAGNIFIED APPROX. 2800×)

Genus IV *Leuconostoc* van Tieghem, 1878, emend. Hucker and Pederson, 1930. Leu-co-nos'-toc. Greek *leucos,* clear or light; Latin *nostoc,* algal generic name.

The species of this genus are distinguished from those in the two previously discussed genera in that they are heterofermentative; that is, they produce levorotatory lactic acid, acetic acid, ethyl alcohol, and carbon dioxide from glucose. Fructose is often reduced in part to mannitol. In sucrose solutions many strains grow with a characteristic slime or dextran formation. The cells are normally spherical and, like the streptococci, may occur singly, in pairs,

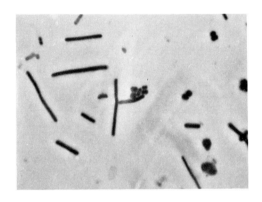

FIG. 2.6. A MIXTURE OF BACTERIAL SPECIES FOUND IN FERMENTING
VEGETABLE MATERIAL

Leuconostoc mesenteroides are the small paired cocci, *Pediococcus cerevisiae* are the tetrad cocci, and *Lactobacillus brevis* and *Lactobacillus plantarum* are the rod-shaped organisms. (Magnified approx. 2250×)

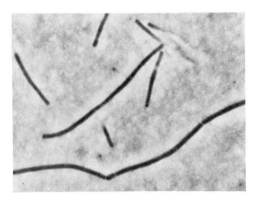

FIG. 2.7. *LACTOBACILLUS BULGARICUS,* THE HIGH ACID-PRODUCING LONG
ROD LACTIC ACID BACTERIUM OF SOUR MILK PREPARATIONS SUCH AS
YOGHURT (MAGNIFIED APPROX. 2250×)

or in short chains. These are saprophytic organisms which were first recognized as nuisance or spoilage organisms in sugar factories where they produced the troublesome, typical dextran slime growth. They are very important in initiating fermentation of vegetables and are found in fruit juices, wines, mashes, and other foods. Of the three species *L. mesenteroides* is most often found among fermenting vegetable material, while *L. citrovorum* is most often found in dairy products.

Species I *Leuconostoc mesenteroides* (Cienkowski, 1878) Van Tieghem, 1878. Strains of this species (Fig. 2.5) were first isolated from slimy sugar solutions in sugar factories, and have been the cause of great financial losses in this industry. The slimy growths on bruised or cut sugar cane and the massive growths on equipment are readily apparent. The strains are now known to be of extreme importance in initiating fermentations of vegetables and other foods.

Species 2. *Leuconostoc dextranicum* (Beijerinck, 1912) Hucker and Pederson, 1930. This species has been associated with dairy starters and vegetables. It might be said to be intermediate between the other recognized species.

Species 3. *Leuconostoc citrovorum* (Hammer, 1920) Hucker and Pederson, 1930. Strains of this species are used as starters in the dairy industry to impart certain desirable characteristics to the products. The strains grow

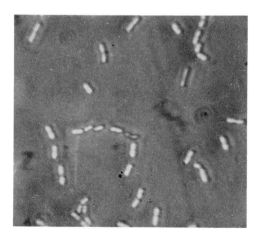

FIG. 2.8. *LACTOBACILLUS PLANTARUM,* THE *LACTOBACILLUS* OF FERMENTING VEGETABLE SUBSTANCES

Lactobacillus casei, the *Lactobacillus* of fermenting milk as well as other lactobacilli, are similar morphologically. The organisms frequently develop in chains of long rods, particularly in adverse environment. (Magnified approx. 2800×)

poorly, if at all, in sucrose solutions, and therefore, do not produce the dextran so commonly associated with *L. mesenteroides*.

TRIBE LACTOBACILLEAE

Tribe II *Lactobacilleae* Winslow *et al.*, 1920. Lac-to-ba-cil'-le-ae. *Lactobacillus*, type genus of the tribe; *eae*, ending indicating tribe.

Bergey's Manual lists five genera in this tribe, but the species of only one genus, *Lactobacillus*, are important in the food fermentation industries.

Genus I *Lactobacillus* Beijerinck, 1901. Lac-to-ba-cil'-lus. Latin *lactis* meaning milk; Latin *bacillus*, a small rod.

These are rod-shaped bacteria that occur singly, in chains, and sometimes filaments (Fig. 2.6 through 2.11). They are usually nonmotile and Grampositive. The genus is divided into two subgenera, the homofermentative and the heterofermentative, comparable to in-growth products from carbohydrates of the cocci genera, *Streptococcus* and *Leuconostoc*. They produce greater quantities of acid than the comparable cocci. These species are saprophytic.

Among the 11 homofermentative species, 7 are primarily associated with milk products. Three of the others and the four heterofermentative species

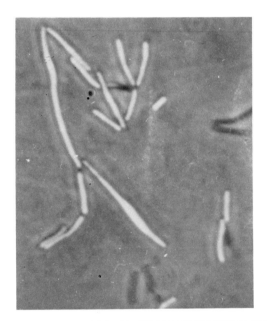

FIG. 2.9.   *LACTOBACILLUS PLANTARUM* (MAGNIFIED APPROX. 4000×)

The long rods in chains are common in acid media or when grown at
elevated temperatures.

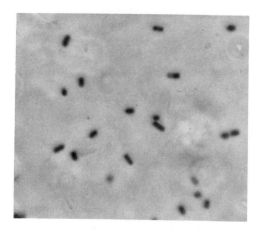

FIG. 2.10. *LACTOBACILLUS BREVIS*, THE MORE COMMON HETEROFER-
MENTATIVE SPECIES OF THE GENUS (MAGNIFIED APPROX. 2800×)

The organisms frequently occur as long rods in chains as do other
heterofermentative species.

are associated with fruit, vegetable, and cereal products. There are numerous other specific names applied to organisms similar in character to those listed. Many of these may be recognized as distinct species after further study.

Homofermentative Species of the Genus *Lactobacillus*

Species 1. *Lactobacillus caucasicus* (Beijerinck, 1889) Beijerinck, 1901 was isolated first from kefir, a fermented milk. The description given by Kern

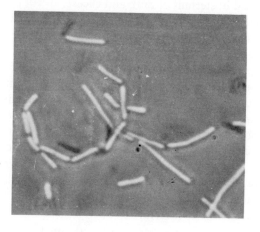

FIG. 2.11. *LACTOBACILLUS BREVIS*, (MAGNIFIED APPROX. 3200×)

The long rods in chains are common in acid media or when grown at
elevated temperatures.

is confused, probably because the organism, a spore former that he isolated, was not the granulated lactobacillus he saw in microscopical preparations of kefir. Because it is the first species named, it is retained.

Species 2. *Lactobacillus lactis* (Orla-Jensen, 1919) Holland, 1920 may be obtained from milk and cheese.

Species 3. *Lactobacillus helviticus* (Orla-Jensen, 1919) Holland, 1920 is similar to the former species but it produces inactive lactic acid instead of the levorotatory lactic acid produced by *L. lactis.*

Species 4. *Lactobacillus acidophilus* (Moro, 1900) Holland, 1920 was originally obtained from the feces of milk-fed infants. It is a low acid-producing strain of the genus *Lactobacillus.* It is the species usually used for implantation of lactic acid bacteria in the intestinal tract.

Species 5. *Lactobacillus bifidus* (Tissier, 1900) Holland, 1920 has been confused with *L. acidophilus.* It is not considered among the food fermenting species.

Species 6. *Lactobacillus bulgaricus* (Luerssen and Kühn, 1907) Holland, 1920 is the high acid-producing lactic acid bacterium of yoghurt (Fig. 2.7).

Species 7. *Lactobacillus thermophilus* Ayers and Johnson, 1924 is the heat-resistant lactobacillus originally isolated from pasteurized milk. Its optimum growth temperature is between 50°C and 62.8°C, and it will not grow below 30°C.

Species 8. *Lactobacillus delbrueckii* (Leichmann, 1896) Beijerinck, 1901 is the high acid-producing species from fermenting vegetable, potato, and grain mashes.

Species 9. *Lactobacillus casei* (Orla-Jensen, 1919) Holland, 1920 is the common lactic acid bacterium of milk and cheese. It will produce more acid than *S. lactis* but will not produce the high acidity attained with *L. bulgaricus.* It produces dextrorotatory lactic acid in contrast to the species *L. plantarum* which produces inactive lactic acid.

Species 10. *Lactobacillus leichmannii* Bergey *et al.*, 1925 has a lower optimum temperature than *L. delbrueckii* but is otherwise similar. It has been isolated from compressed yeast, mashes, and vegetable and dairy products.

Species 11. *Lactobacillus plantarum* (Orla-Jensen, 1919) Holland, 1920 is widely distributed in nature and particularly in fermenting vegetable materials. This species (Fig. 2.8 and 2.9) produces the high acidity attained in such foods. This species is so similar to *L. casei* that it is often difficult to distinguish them. Cultures isolated from milk products will invariably produce dextrorotatory lactic acid while those from vegetable products will invariably produce inactive lactic acid.

Heterofermentative Species of the Genus *Lactobacillus*

Species 12. *Lactobacillus pastorianus* (van Laer, 1892) Bergey *et al.*, 1923 has been isolated from sour beer and distillery yeast. It produces a silky turbidity in unhopped beer. It is the first heterofermentative lactobacillus described.

Species 13. *Lactobacillus buchneri* (Henneberg, 1903) Bergey *et al.*, 1923 has been isolated from sour mash, compressed yeast, and vegetable substances. It may be considered as an intermediate between the next two species.

Species 14. *Lactobacillus brevis* (Orla-Jensen, 1919) Bergey *et al.*, 1934 is the most common heterofermentative species and is widely distributed in plant and animal materials (Fig. 2.10 and 2.11). Many strains are characterized by a marked fermentation of arabinose and usually xylose. Like *Leuconostoc mesenteroides*, previously mentioned, it partially reduces fructose to mannitol, and it may be trained to produce dextran from sucrose. A chromogenic variety has been isolated from the rusty spot condition in cheese.

Species 15. *Lactobacillus fermenti*, Beijerinck, 1901, is quite similar to *L. brevis* but ordinarily has a higher temperature growth range. It is also widely distributed in plant and animal products.

FAMILY PROPIONIBACTERIACEAE

Family XI *Propionibacteriaceae* Delwiche, 1954. Pro-pi-on-i-bac-te-ri-a'-ce-ae. Latin *Propionibacterium* the type genus; *aceae* denotes a family ending. This family includes three genera of irregularly shaped nonmotile, Gram-positive rod-shaped bacteria. Only one, the *Propionibacterium*, is important in foods.

Genus I *Propionibacterium* Orla-Jensen, 1909. There are 11 species listed. All are associated with dairy products, particularly cheese. They ferment lactic acids, carbohydrates, and polyhydroxy alcohols with production of propionic and acetic acids, and carbon dioxide. They are important in Swiss cheese fermentation.

ORDER *PSEUDOMONADALES*, FAMILY *PSEUDOMONADACEAE*

Order I *Pseudomonadales* Orla-Jensen, 1921. This order of rod-shaped bacteria includes only 1 genus of 1 family important in fermented foods, the acetic acid or vinegar-producing bacteria.

Family IV *Pseudomonadaceae* Winslow *et al.*, 1917. Twelve genera.

Genus III *Acetobacter* Beijerinck, 1898. A-ce-to-bac-ter. Latin *acetum*, vinegar; *bacter*, a rod or staff.

They oxidize various organic compounds to organic acids, and are widely distributed in nature where they are particularly abundant in plant materials undergoing alcoholic fermentations. They are important for their role in the production of vinegar. There are seven species listed.

Species 1. *Acetobacter aceti* (Beijerinck, 1898) Beijerinck, 1900.

Species 2. *Acetobacter xylinum* (Brown, 1886) Holland, 1920.

Species 3 to 7 are of lesser importance.

The acetic acid or vinegar bacteria, in contrast to the lactic acid bacteria, are aerobic and oxidizing species whose main activity is the oxidation of alcohol and other carbohydrate substances to acetic acid. The thick, tenacious membrane they form on certain substrates is eaten with enjoyment by the people of some countries.

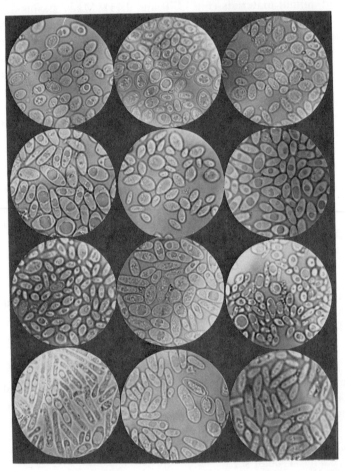

From E. A. Beavens (1940)

FIG. 2.12. MORPHOLOGICAL VARIATIONS AMONG VEGETATIVE CELLS OF STRAINS OF *SACCHAROMYCES CEREVISIAE* VAR. *ELLIPSOIDEUS*

Top, l–r, are cultures No. 341, 248, 206; second row, l–r, are cultures No. 400, 102, 2372; third row, l–r, are cultures No. 331, 266, 427; and bottom row, l–r, are cultures No. 410, 425, and 411.

THE FERMENTATIVE YEASTS

The microscopic differences between bacteria and yeasts were recognized quite early in the history of microbiology. Yeasts (Fig. 2.12 through 2.15), because of their comparatively large size and their method of multiplication by budding, were relatively conspicuous. Knowledge of them progressed rapidly because of their function in fermentation. Fermentation has been utilized by man for centuries. Perhaps no race of man has been unaware of some of the facts regarding fermentation, and all peoples, including some

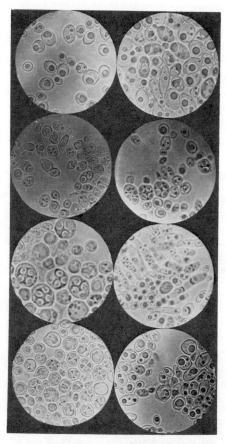

From E. A. Beavens (1940)

FIG. 2.13. VARIATIONS IN SPORULATION AMONG STRAINS OF *SACCHAROMYCES CEREVISIAE* VAR. *ELLIPSOIDEUS*

Top, 1-r, are cultures, No. 213, and 4110; second row, 1-r, are cultures 237 and 305; third row, 1-r, are cultures 102 and 174, and bottom row, are cultures 1343 and 171.

From E. A. Beavens (1940)

FIG. 2.14.   *SACCHAROMYCES CARLSBERGENSIS*, CULTURE NO. 422

primitive tribes, have prepared an alcoholic beverage by fermentation of sweet juices. Leavened and unleavened bread are mentioned in the oldest Hebrew records; leavened bread was that which had been allowed to ferment and therefore became porous and light during baking. The Romans prepared a yeast from grapes, and they artfully placed it in bread dough for the purpose of leavening the dough.

The best known yeasts are those used for producing bread, beer, and wine;

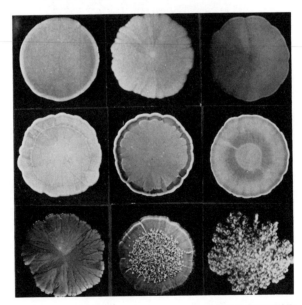

From E. A. Beavens (1940)

FIG. 2.15.   MORPHOLOGICAL VARIATIONS AMONG GIANT COLONIES OF STRAINS OF *SACCHAROMYCES CEREVISIAE* VAR. *ELLIPSOIDEUS*

Top row, 1–r, are cultures No. 161, 2372, and 232; middle row, 1–r, are cultures No. 425, 4134 and 423; and bottom row, 1–r, are cultures No. 417, 265 and 174.

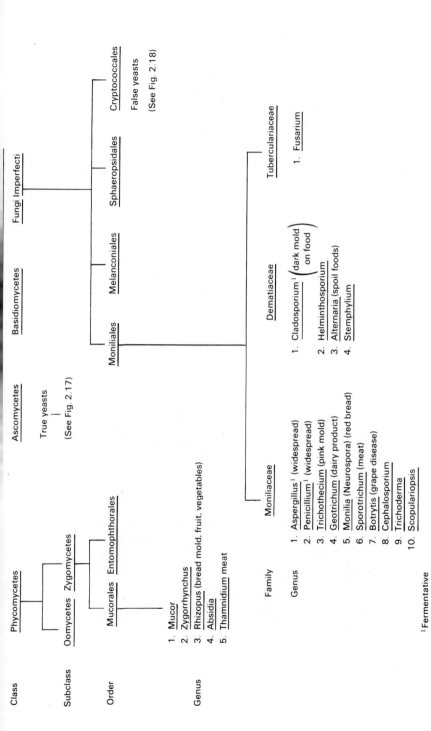

FIG. 2.16. PARTIAL CLASSIFICATION OF *EUMYCETES* TO SHOW RELATIONSHIPS OF MOLDS AND YEASTS OF FOODS

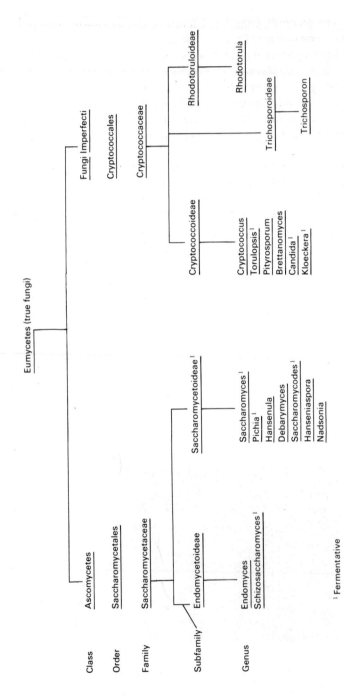

FIG. 2.17. PARTIAL CLASSIFICATION OF CLASSES *ASCOMYCES* AND *FUNGI IMPERFECTI* (TRUE AND FALSE YEASTS)

[1] Fermentative

therefore, yeasts are generally considered fermentative organisms that produce alcohol and carbon dioxide. This concept arises from the fact that nearly all studies have been conducted with bakers' or brewer's yeast, *Saccharomyces cerevisiae* or *S. carlsbergensis*. The food or torula yeast, *Candida utilis*, has been studied to a lesser extent. *Saccharomyces cerevisiae* can shift its metabolism from a fermentative to an oxidative pathway, the latter yielding more energy for cell growth. Many species of yeasts are entirely aerobic, while others are intermediate types with varying respiratory and fermentative metabolism.

The Danish scientist, Emil Christian Hansen, spent some 30 yr studying the morphological and physiological characteristics of yeasts. His recommendation of a systematic relationship was expanded by Guilliermond and later by a series of monographs prepared at the Technical University of Delft by Stelling-Dekker *et al.* and summarized by Lodder and Kreger-Van Rij (1952). Wickerham (1951) introduced certain principles in yeast classification.

The class *Ascomycetes* (Fig. 2.16 and 2.17) is 1 of 4 classes in the phylum *Eumycetes*. These spore-forming yeasts include most of the yeasts of industrial importance and are included in the genus *Saccharomyces*. Many species of the 6 other genera in this subfamily and of the 6 genera of false yeasts in the subfamily *Cryptococcoideae* are important, primarily, as contaminants in fermentations.

Among the numerous species of the genus *Saccharomyces, S. cerevisiae,* the brewing and baking yeast, the variant *S. cerevisiae* var *ellipsoideus* (Fig. 2.12., 2.13, 2.15) used in wine fermentations; and the lager beer yeast, *S. carlsbergensis* (Fig. 2.14) are the most important in industrial food fermentations. Several other species' names have been, and continue to be applied

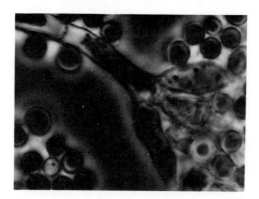

FIG. 2.18. *PENICILLIUM ROQUEFORTI.* A MOLD MYCELIA AND SPORE OF ROQUEFORT AND BLUE CHEESE (MAGNIFIED APPROX. 2000×)

The molds are considerably larger than bacteria.

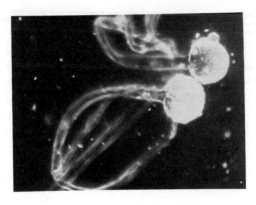

FIG. 2.19. *RHIZOPUS ORYZAE.* A MOLD SPORE SAC AND MYCELIA FROM
THE INDONESIAN FOOD "TEMPEH"

to similar alcohol-producing strains of yeast. Nearly all bakeries, wineries, and breweries use certain selected strains of *Saccharomyces cerevisiae.* A few species of other genera play a role in certain fermentations.

Yeasts play an important role in the food industry. Their enzymes catalyze many chemical reactions, such as the production of carbon dioxide and alcohol so important in leavening the dough of bread and many bread-like products, and in synthesizing certain vitamins of the B group. They are known throughout the world for the production of many alcoholic beverages.

MOLDS

Certain molds are useful in the preparation of several kinds of foods such as mold-ripened cheeses, Roquefort, Camembert, and related types, and many Oriental foods such as soy sauce and tempeh.

The class *Fungi Imperfecti* is divided into four orders (Fig. 2.16). One of these, *Moniliales,* has 3 families, 1 of which, *Moniliaceae,* includes the 2 genera of molds so important in mold-fermented foods, that is, *Aspergillus* and *Penicillium* (Fig. 2.18). There are many other genera of molds associated at times with foods and food fermentations. The species of molds of the genus *Penicillium* share with the species of the genera *Aspergillus, Rhizopus* (Fig. 2.19), and *Mucor* their reputation as "the microscopic weeds." They rot fruit, attack vegetables and meat, injure stored grain, spoil jellies and fruit juices, contaminate storages of fibers, wood, and paper, and often destroy the usefulness of laboratory preparations. The cheese industry has capitalized on their enzyme activities to ripen such cheeses as Roquefort and Camembert. Soy sauce, tempeh, and many other preparations result from fermentations by mold. Scientists have also capitalized on the use of their enzymes for many purposes and the preparation of many chemical substances including a number of antibiotics.

BIBLIOGRAPHY

BEAVENS, E. A. 1940. Morphological and physiological characteristics of strains of yeast related to *Saccharomyces ellipsoideus* Hansen and *Saccharomyces cerevisiae* Hansen. Ph.D. Thesis, Cornell Univ., Ithaca, N.Y.

BREED, R. S., MURRAY, E. G. D., and SMITH, N. R. 1957. Bergey's Manual of Determinative Bacteriology, 7th Edition. Williams & Wilkins Co., Baltimore, Md.

CONN, H. W., and CONN, H. J. 1926. Bacteriology. Williams & Wilkins Co., Baltimore.

FRAZIER, W. C. 1967. Food Microbiology, 2nd Edition. McGraw-Hill Book Co., New York.

LODDER, J., and KREGER-VAN RIJ, N. J. W. 1952. The yeasts, a Taxonomic Study. Interscience Div., John Wiley & Sons, New York.

RAPER, K. B., and FENNELL, D. I. 1965. The Genus *Aspergillus*. Williams & Wilkins Co., Baltimore.

RAPER, K. B., and THOM, C. 1949. A manual of the penicillia. Williams & Wilkins Co., Baltimore.

WICKERHAM, L. J. 1951. Taxonomy of Yeasts. U.S. Dept Agr. Tech. Bull. 1029, 1–56.

CHAPTER 3

Chemical Alterations During Fermentation

Fermentations have aroused the curiosity of philosophers and scientists for centuries. Originally the term was applied to the conversion of grape juice to wine, but it has a much broader connotation today. The detailed information made available within the past century has now been interpreted in terms of enzyme activated reactions that provide energy for the vital processes of living cells and has increased the knowledge of the growth of higher plants and animals.

Microbial physiology and metabolism deal with the activity of living organisms. The elucidation of the many changes that occur has made possible the partial control of many microbial phenomena. Enough is now known about the details of these changes or reactions to permit charting the various reactions, including the roles of enzymes, coenzymes, and minerals involved. It is the purpose of this section to discuss, briefly, biochemical changes as they relate to food fermentations.

The primary purpose of fermentation by microorganisms is the furnishing of energy for their metabolism and growth. It is fortunate for humanity that the metabolism of the lactic acid bacteria is so similar to the metabolic changes that occur in the animal body that humans can use the fermented foods. Furthermore, little loss in nutritive values occurs during such fermentations. The biochemical changes produced in foods through the activities of microorganisms are distinctive manifestations of their existence. Carbohydrates, amino acids and peptones, lipids, vitamins, and minerals are required for growth of lactic acid bacteria. It is essential that they exist in an available form in the food. Yeasts and molds have greater synthetic abilities than the lactic acid bacteria, and therefore, their growth requirements are less stringent; in fact, in some instances substances essential for human nutrition are accumulated in some of their fermentations.

FERMENTATION AS ENZYME-INDUCED REACTIONS

Metabolism is the sum total of many reactions going on in a living cell. The myriad biochemical reactions that collectively sustain life are made possible through the action of highly specialized catalysts, the enzymes. These reactions are of great interest in food fermentations since the intermediate as well as the end products of microbial metabolism play an important role in imparting typical characteristics to a food. Although it has been recognized for centuries that flavor, aroma, texture, and appearance of foods are altered during fermentation and curing, the cause of these alterations was unknown until it has become clarified in part during the past century. This discussion will be concerned with a partial explanation of the reason why fermentation produced by the activity of bacteria and yeasts will occur only in the presence of adequate nutritive substance, without entailing the tremendous detail of reactions that have been elucidated during the past 70 yr.

Although the first basic foundations of microbiology were laid by von Leeuwenhoek, he did not associate the role of the organisms observed with the phenomena of fermentation. The French chemist Lavoisier made quantitative studies of the alcoholic fermentation as early as 1789. Gay-Lussac in 1810 reported the reaction that bears his name, that is, that sugar yields alcohol and carbon dioxide ($C_6H_{12}O_6 \rightarrow 2C_2H_5OH + 2CO_2$). The subsequent work of Müller, Ehrenberg, Cohn, and others led to the demonstration of the plant-like nature of yeasts by Cagnard de La Tour and by Schwann. The theory that fermentation was the result of the activity of living cells was disputed by Liebig, Wöhler, Berzelius, and others. Liebig felt that yeasts were without significance in fermentation. Pasteur's fundamental observations regarding the essentially biological basis of fermentation led not only to a dispute with the Liebig group, but also one with those who believed in the theory of the spontaneous generation of life. Pasteur demonstrated that microorganisms were present in the air, that putrescible material, when heated sufficiently to destroy life, would keep indefinitely, and that fermentations resulted from the activities of living, growing microorganisms. He further demonstrated the specificity of fermentations, that different kinds of fermentations resulted from the activities of various kinds of microorganisms. He further demonstrated that glycerol and succinic acid accompanied the production of alcohol and carbon dioxide, showing that fermentation involved side reactions. Despite Pasteur's observations, the controversy continued.

In 1897, Büchner obtained from yeasts a cell-free extract that was capable of causing a vigorous fermentation of sugar solutions. He demonstrated that the intact cell was not required for fermentation, and that a substance expressed from the yeast cell was the actual agent. He called the substance zymase. Although it was thereby demonstrated that a chemical substance was the agent of fermentation, the essential role of the living cell in elaborating this substance was recognized.

These observations led to further studies that eventually culminated in the well-known Embden-Meyerhof pathway explanation of glycolysis. By 1900 it was established that a variety of chemical entities were produced during the alcoholic fermentation. Harden and Young (1906) proved that inorganic phosphate was essential for yeast growth, and they isolated the intermediate hexose diphosphate. Their studies suggested that a relatively small amount of phosphate could esterify large amounts of sugar. Later Harden and Young (1906) observed that when yeast juice was boiled, its activity was destroyed; however, if such boiled juice was added to freshly prepared yeast extract, the fermenting capacity of the latter was greatly enhanced. This demonstrated the presence of a heat-stable factor that was in some way involved in enzyme activity. Later they demonstrated that fresh yeast juice could be separated into two fractions by ultrafiltration; one fraction was the heat-stable filtrate distinct from the protein-glycogen residue. Neither the filtrate nor the protein residue alone could cause fermentation; but when they were combined, alcoholic fermentation occurred to practically the same extent as with the original extract. The heat-stable, ultrafilterable fraction was later given the name coenzyme, and the protein residue was called apoenzyme.

Following these observations of the nature of enzymes and the complexities of enzyme behavior, numerous studies by many scientists followed. These led to an understanding and clarification of the mechanisms of enzyme action and their role in biochemical fermentations and oxidations. The term, fermentation, has come to refer to the great variety of changes that occur in connection with plant and animal life; the growth and multiplication of cells, the formation of enzymes, and the coordinated activity of the large number of enzymes and specific enzymes by which the organism obtains energy for growth.

ROLE OF VITAMINS IN FERMENTATION

It is an interesting coincidence that the same year that Büchner laid the foundation for modern enzymology, in Java another scientist, Eijkman, discovered that a trace material in rice polishings was capable of preventing the disease, beriberi, in birds. Following Eijkman's observation, many other essential entities, the vitamins, have been discovered. It was many years before it was realized that vitamins were in any way associated with the activity of fermentation enzymes. Since 1930, the mystery of the biochemical relationship of vitamins and minerals to enzyme activity has been at least partly elucidated. Modern vitamin research has contributed the explanation of the biochemical mechanisms of fermentation.

The pathways of research concerning vitamins and enzymes crossed in 1932 when Warburg and Christian obtained from yeasts the so-called yellow enzyme, from which were separated a protein fraction and a highly colored, fluorescent, low molecular weight fraction, a coenzyme.

The pathways of carbohydrate utilization and the roles of enzymes, vitamins, and minerals are far too complex to be included within the scope of this book. With ever increasing research, it becomes increasingly difficult to define the boundaries of carbohydrate metabolism. The several excellent reviews and the various textbooks present detailed reactions illustrating the role of the various entities involved in fermentation. Reference is made in particular to the reviews of Wood (1961), Krampitz (1961), and Anderson and Wood (1969). Insofar as this discussion is concerned, the foregoing information is important in explaining the essential role of vitamins, enzymes, and minerals in lactic acid bacterial fermentation. The overall changes that occur by which glucose is converted to pyruvic acid and thence to lactic acid or to alcohol and carbon dioxide explain the way in which energy is furnished for metabolism in the absence of oxygen. When fermentative microorganisms such as yeasts are placed in a nutrient sugar solution, the sugar enters the cell with the aid of a transport system built into the cell wall or in the cytoplasmic membrane. This may involve a permease. Enzymes necessary for hydrolysis of higher sugars such as sucrose are located in the cell surface or may be extracellular. Glucose is phosphorylated to glucose-6-phosphate with the aid of the enzyme, hexokinase, and adenosine triphosphate (ATP). By a series of reversible chemical transformations catalyzed by enzymes and coenzymes, in which certain minerals play an essential role, the labile pyruvic acid is formed. All of these chemical intermediates have been isolated and their enzymes have been separated. The individual reactions in the pathway from glucose to enol-pyruvic acid are all reversible.

The pyruvate formed is an important intermediate in many yeast and bacterial fermentations. It is a crossroads in metabolism in that it may lead to one or more of several major pathways. Maximum energy would be obtained by its complete oxidation to carbon dioxide and water; this however, is an aerobic oxidation. In a homofermentative lactic acid bacterial fermentation, pyruvate is reduced to lactic acid by removal of hydrogen from pyruvate through activity of the enzyme lactic dehydrogenase. The lactate conversion to propionate involves succinic acid as an intermediate. The pyruvate is decarboxylated in yeast fermentations to yield acetaldehyde that is readily reduced to ethyl alcohol and carbon dioxide.

A minor reaction, but one of great importance to the flavor of foods, involves the release of carbon dioxide from the enol-form of pyruvate and the formation of acetylmethylcarbinol. The acetylmethylcarbinol may be converted readily to diacetyl and 2,3-butylene glycol, products frequently present in fermented foods. Pyruvic acid can be transformed by transaminase to the simple amino acid by introducing an amino group.

By the loss of carbon dioxide and then condensation, oxalacetic acid may be formed leading into a series of reversible reactions involving citric, malic, fumaric, succinic, and other organic acids of the citric acid cycle. The succinic acid, through a complex methyl malonyl reaction, is changed to propionic acid that is so important in certain cheeses of the Swiss cheese type. All of these reactions are catalyzed by various enzyme systems.

COMPLEXITY OF LACTIC ACID BACTERIAL FERMENTATIONS

Many lactic acid fermentations, particularly those involving vegetables, are invariably initiated by strains of *Leuconostoc mesenteroides,* whereby lactic acid, carbon dioxide, ethyl alcohol, and acetic acid are major end products (Fig. 3.1). In the fermentation of fructose, mannitol is produced, and from sucrose, dextrans may be important by-products. These are frequently products of vegetable fermentations; however, they are both subject to subsequent fermentation. The heterofermenters are noted for their ability to ferment the pentose sugars, arabinose and xylose. The heterolactic fermentation differs fundamentally from the homolactic fermentation. Glucose, after conversion to glucose-6-phosphate, is oxidized to 6-phospho-gluconate, an important intermediate. This is converted to the pentose phosphate, ribose-5-phosphate, that in turn splits into a 3-carbon unit finally to yield lactic acid and a 2-carbon unit to yield ethyl alcohol and/or acetic acid.

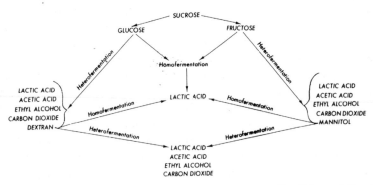

From Pederson and Albury (1969)

FIG. 3.1. MAJOR CHEMICAL SUBSTANCES PRODUCED IN FERMENTATION OF SUCROSE IN A FERMENTATION BY HOMO- AND HETEROFERMENTING LACTIC ACID BACTERIA

In the fermentation of fructose, 3 mol of fructose are reduced to 2 mol of mannitol with 1 mol each of lactic acid, acetic acid, and carbon dioxide. The mannitol, in turn, may be fermented to yield lactic acid, ethyl alcohol, and carbon dioxide. The heterolactic acid bacteria, particularly strains of *Leuconostoc mesenteroides,* have a marked ability to condense the glucose portion of sucrose to form dextran molecules while the fructose portion of the sucrose is fermented to lactic acid, acetic acid, ethyl alcohol, and carbon dioxide. Dextrans are used as food stabilizers.

These are a few of the pathways of sugar metabolism used by yeasts and lactic acid bacteria. These pathways have been elucidated and authenticated by use of various enzymes with the initial carbohydrates and the intermediates. Pathways are presented for the metabolism of proteins and lipids. These are outlined in complex charts such as those distributed by Gilson Medical Electronics and by General Biochemicals.

FOOD FERMENTATIONS

Foods, in general, furnish a satisfactory medium for growth of a variety of microorganisms. Fermentations are enzyme-induced chemical alterations in food; the enzymes involved may be produced by microorganisms or they may be indigenous to the food. All fermentations are complex, but they vary considerably in their degree of complexity.

The character of a fermented food will be determined by the nature and quality of the food itself, the changes that occur as a result of the action of its inherent enzymes, the alterations that occur as a result of the microbial fermentation, and the interactions that occur between the products of these activities and the constituents of the foods. Fermentations by desirable microorganisms impart flavor, bouquet, and texture to the fermented foods. In many lactic acid fermentations, the high acidity, low pH, and low oxidation-reduction potential attained are responsible for inhibition of other organisms and of undesirable chemical changes. In fruit fermentations, ethanol, carbon dioxide, and the absence of oxygen exert a similar effect.

Pure culture fermentations seldom occur naturally. The requirements for growth supplied by natural constituents of food are so similar for both yeasts and many species of lactic acid bacteria that mixed fermentations normally occur. Among the simpler food fermentations are the pure culture fermentations of vegetables induced by homofermentative lactic acid bacteria and the pure culture fermentation of fruit juices by yeasts. The first is characterized primarily by the conversion of simple sugars to lactic acid, while the second is characterized primarily by conversion of sugars to ethanol and carbon dioxide. In both cases, minor amounts of other products are produced and alterations in proteins, lipids, and other constituents are important but much less evident.

Somewhat more complex are those fermentations initiated by heterofermentative lactic acid bacteria and continued by the homofermenters (Fig. 3.1). These are very common to many vegetables (Table 3.1). Lactic acid, acetic acid, ethanol, carbon dioxide, and frequently mannitol and/or dextran are major products of the heterofermentation, while lactic acid is the primary product of homofermentation. The hetero species also utilize pentose sugars more readily than the homos. Again, alterations in lipids, proteins, and other constituents occur but the small amount of these substances limits their influence. The mold fermentations are quite dissimilar and more complex in that they are primarily oxidative and may involve various stages of utilization of the food components. Although fermentations occur naturally, on occasion microorganisms that produce undesirable alterations may develop,

TABLE 3.1

DEVELOPMENT OF ACID AND CHANGE IN BACTERIAL FLORA IN SAUERKRAUT FERMENTATION

					Estimated Number of Each Type × 10^5 per Ml				
Time, Days	Total Acid	pH	Total Plate Count × 10^5 per Ml	Aerobes	Leuco-nostoc mes-enter-oides	Lacto-bacillus brevis	Lacto-bacillus plantarum	Pedio-coccus cere-visiae	Yeasts
5/6	0.15	4.48	1320	1	1319				
1 1/6	0.30	4.23	4400		4400				
1 5/6	0.74	3.93	9650		9150		500		
2 1/6	0.77	4.00	4660		4660				
2 5/6	0.97	3.87	4490		4440	50			
4	1.18	3.67	1250		687	62	313	188	
5	1.19	3.71	970			49	436	145	
6	1.16	3.63	1410			70	1270	70	
8	1.42	3.68	4670			750	3170	250	
10	1.57	3.59	2300			1035	690		
13	1.65	3.59	1200			200	1000		
17	1.61	3.58	546				546		
21	1.78	3.51	251				248		3

particularly before the desirable fermentative species produce by-products that inhibit the undesirable species.

Carbohydrates, in particular the simple sugars, are the most readily available source of energy. Lactic acid is the most obvious product of lactic acid bacterial fermentations, and alcohol and carbon dioxide are the common products of yeast fermentations.

Although sugars furnish the energy for metabolic processes, proteins, lipids, vitamins, nucleic acid, and minerals are essential in synthesis of cell protoplasm. In general, these must be supplied for growth since the fermentative organisms, particularly the bacteria, have relatively poor synthetic abilities. Most of the foods contain a sufficient amount of these substances to permit active growth of fermentative microorganisms. The requirements for these ingredients are apparent in preparations of laboratory media; in fact, the requirements are so definite that assays for these substances in foods may be determined by growth of selected microorganisms.

The fact that changes in protein constituents occur in lactic acid bacterial fermentations has been recognized for many years; however, the changes are minor. Recently, studies of the lipid components have shown that sufficient hydrolysis occurs among the lipids to supply essential components for synthesis of protoplasm (Table 3.2).

The activity of enzymes indigenous to the food cannot be entirely divorced from a discussion of fermentation. Generally, such activity is associated with curing, ripening, and aging rather than fermentation. The curing of meats is usually caused by enzymes inherent to the meat; but it also may be due to

TABLE 3.2

GROSS CHANGES IN THE VARIOUS LIPID FRACTIONS DURING CABBAGE FERMENTATION

Fraction	Mg per 200 Gm dry weight		Increase or Decrease (%)
	Cabbage	Sauerkraut	
Nonesterified fatty acids	86	649	+657
Acetone-soluble lipid fatty acids	526	193	−63
Acetone-insoluble lipid fatty acids	978	504	−49
Unsaponifiable matter	321	183	−43

Source: Vorbeck, et al. (1963).

enzymes produced by growth of molds on the meat surfaces. There is little doubt that meat enzymes are involved in the true lactic acid bacterial fermentation of sausage. The seeds of vegetables are apt to contain a richer supply of enzymes than the vegetable portion. The softening and sloughing of skins of cucumbers are often due to the activity of inherent enzymes. Some of the peculiar changes in lipid components in the sauerkraut fermentation cannot be attributed entirely to the fermentative organisms. The alterations that occur in coffee cherries, the cacao matrix, vanilla beans, tea, citron, and other foods will be discussed later in the chapter dealing with these products, but they also involve complex changes brought about by microbial enzymes and enzymes indigenous to the product.

A series of fermented foods is produced that may be said to represent a progression of fermentations from those characterized primarily by a sugar fermentation to those in which significant changes occur in proteins, lipids, carbohydrates, and other constituents. These range all the way from simple lactic acid bacterial fermentation, dependent upon high acidity and low oxidation-reduction potential, to the cured cheeses that include those requiring the activity of other species of bacteria or molds, such as are essential in Swiss, Limburger, and Roquefort types of cheese, to those products which are essentially mold-fermented foods such as certain Oriental foods. The influence of the growth of one species upon the growth of another is not well known. In like manner, there is very little information in regard to the influence of inherent enzymes upon the growth of microorganisms. It is, however, well known that many fermented foods are improved by aging, and that some of the chemical alterations that occur have been determined.

The numerous pathways in the many fermentations are far too complex to include in this discourse. It is sufficient to emphasize that there are several species of microorganisms capable of producing desirable alterations as well as many that may produce undesirable effects. This fact emphasizes the need for establishment of desirable practices.

The characteristic flavor, aroma, and texture of fermented foods are dependent upon the nature of the food itself, the changes resulting from the activity of microbial enzymes, enzymes indigenous to the food, and the interactions that occur during fermentation and subsequent curing or aging.

BIBLIOGRAPHY

ANDERSON, R. L., and WOOD, W. A. 1969. Carbohydrate metabolism in microorganisms. Ann. Rev. Microbiol. 23, 539-573.

BURROWS, W. 1954. Textbook of Microbiology. W. B. Saunders Co., Philadelphia.

FRAZIER, W. C. 1967. Food Microbiology, 2nd Edition. McGraw-Hill Book Co., New York.

GUNSALUS, I. C., and SHUSTER, C. W. 1961. Energy-yielding metabolism in bacteria. In The Bacteria, Vol. 2, I. C. Gunsalus and R. Y. Stanier (Editors). Academic Press, New York.

HARDEN, A., and YOUNG, W. J. 1906. Alcoholic ferment of yeast-juice. Proc. Roy. Soc. B77, 405-420.

KRAMPITZ, L. O. 1961. Cyclic mechanisms of terminate oxidation. In The Bacteria, Vol. 2, I. C. Gunsalus and R. Y. Stanier (Editors). Academic Press, New York.

PEDERSON, C. S., and ALBURY, M. N. 1969. The sauerkraut fermentation. New York State Agr. Expt. Sta. Bull. 824.

SCHULTZ, H. W. 1960. Food Enzymes. Avi Publishing Co., Westport, Conn.

VORBECK, M. L., MATTICK, L. R., LEE, F. A., and PEDERSON, C. S. 1963. Lipid alterations during fermentation of vegetables by lactic acid bacteria. J. Food Sci. 28, 495-502.

WARBURG, O., and CHRISTIAN, W. 1932. A new oxidation enzyme and its absorption spectrum. Biochem. 254, 438-456. (German)

WOOD, W. A. 1961. Fermentation of carbohydrates and related compounds. In The Bacteria, Vol. 2, I. C. Gunsalus and R. Y. Stanier (Editors). Academic Press, New York.

CHAPTER 4

Growth of Microorganisms in Foods

Foods in the raw state are living substances and even though they may be cut, minced, or ground into fine particles they remain complex biological substances in which natural biological changes, such as enzymatic and chemical reactions, continue to function. These reactions may be modified by moisture, temperature, acidity, and other factors including their blending with other ingredients in the food. This heterogeneous mixture with the continuous changes that are constantly occurring, does not easily provide for simple separation of its basic components for study and analysis. Foods are always subject to the growth of many microorganisms.

Fermentation and drying are the two oldest methods of preserving food. The procedures employed antedate recorded history. Fermentations are complex chemical transformations of organic substances brought about by the catalytic action of enzymes, either native to the substance or elaborated by microorganisms. Fermented foods are foods that have undergone such transformations. Throughout the centuries, fermentation has been and continues to be one of the most important methods for preserving food. Relatively few people are aware of the complex changes that occur during fermentation and the extent to which foods may be altered during fermentation. Fermentation not only involves production of preservative or antibiotic ingredients, notably acids, carbon dioxide, and alcohol, but also results in chemical and physical changes that substantially alter the foods. The production of many chemical substances in addition to those mentioned, alters and improves the flavor of food. Fermentation should be considered a method of preparing food, rather than merely a method of chemical preservation.

The lactic acid-producing bacteria and the alcohol-producing yeasts are, to some extent, curiosities among microorganisms. The lactic acid bacteria are highly specialized and are unable to grow in circumstances that permit growth of many other microorganisms. Why they have developed with such specialized characteristics is somewhat difficult to understand, particularly if their highly specific growth requirements are considered.

It is fortuitous for mankind that certain organisms produce changes in foods that, in turn, prevent growth of many other organisms and, therefore, preserve the food for consumption by man and animals. Furthermore, it is fortunate that the changes wrought by microorganisms are so similar to the changes that occur in assimilation of the food by man. In many cases the flavor, aroma, and texture are improved in addition to being made safer for consumption. The lactic acid bacteria are often described as fastidious in their growth requirements. They are so-called because they are unable to synthesize from simple chemical substances many of the components essential to their cells. They must, instead, utilize the preformed amino acids, vitamins, lipids, and other components contained in the foods for their growth.

Many of our foods contain the ingredients essential for growth of even the most fastidious true lactic acid bacteria, as well as for growth of the fermenting yeasts. The biochemical changes that occur in a food as a result of microbial action are, in their simplest form, a reflection of the changes required by the microorganism to obtain certain components for growth and to obtain energy for the necessary conversions.

Sugars are required for energy; however, since the lactic acid bacteria cannot oxidize the sugars to carbon dioxide and water but rather must convert them to lactic acid as a primary end product, the lactics are extremely wasteful of their energy source. They must utilize large amounts of sugar to obtain sufficient energy for growth and reproduction, and, therefore, they produce large amounts of lactic acid as a by-product or waste. The conversion of sugar to lactic acid is also a primary change occurring in the bodies of higher animals and plants; the lactic acid in the animal system, however, is ultimately oxidized completely to carbon dioxide and water to furnish maximum energy.

Thus, it is indeed fortuitous to mankind that this small group of microorganisms, the true lactic acid bacteria, will grow in many foods, producing minimal changes in their nutritive properties while producing lactic acid and other by-products which prevent the growth of the many other microorganisms.

It is essential that the environment created by the growth of lactic acid bacteria or by fermentative yeasts be retained in the fermented foods in order to prevent surface growth of aerobic yeasts, molds, and bacteria that can utilize and, thereby, destroy lactic acid. However, in some foods, the molds are allowed to develop after the lactic fermentation to produce mold-fermented foods of the nature of Roquefort cheese and some of the mold-fermented foods of the Far East.

The fermentative yeasts produce alcohol and carbon dioxide as primary fermentation products. The growth of many other organisms is suppressed by the alcohol and carbon dioxide produced. The growth of lactic acid bacteria is, however, common in beers and wines, and in some cases their flavor and aroma are characteristically enhanced and desirable. Just as it is essential to prevent growth of molds and yeasts in bacterial fermentations, it is equally essential to prevent growth of the acetic acid bacteria in alcoholic products. This is yet another preservative method, the production of acetic acid.

SIZE OF MICROORGANISMS

Microorganisms, often referred to as microbes meaning tiny living things, are small beyond imagination (Fig. 2.1 through 2.19, Chap. 2). Their smallness is their outstanding property; it explains in part their ubiquity and their ability to alter the character of enormous amounts of organic matter. The microbiologist is so accustomed to using the oil immersion lens for examination of bacteria and yeasts that it rarely occurs to him that he is magnifying the organism by $600-1000\times$. A gram of bacteria on the wet basis may consist of 10^{12} cells. Although a bacterial count of even 10^9 per gram of food is extremely high, it would only constitute $\frac{1}{1000}$ of the weight of the food. Bacteria rarely exceed 1μ in diameter. It would require 25,000 to span an inch, and 3,000,000 could be placed 1 layer thick on a pin head; yet they would not be visible. Yeasts are larger. Bakers' yeast cells, usually oval, vary in size from about $4-6\mu$ to about $5-7\mu$ (Fig. 2.12 and 2.13 in Chap. 2). The size of both bacteria (Fig. 2.5 through 2.11 in Chap. 2) and yeasts varies considerably depending upon nutritional and other environmental conditions. Rapidly growing bacteria and yeasts tend to be smaller than their average size, and yeast buds may be only slightly larger than bacteria. Molds (Fig. 2.18 and 2.19 in Chap. 2) are the largest of this group of microorganisms, and the many species demonstrate considerable variation. Their mycelia may be studied with a low power compound microscope.

In addition to the use of the microscope to study cell morphology, the development of staining procedures has been helpful. More recently the electron microscope has been employed to study the various units of the cell, the cell wall, capsular materials, flagella, cell membrane, and the several protoplasmic substances. The yeast cell is large enough to differentiate its nucleus, vacuoles, granules, and other particles from the cytoplasm. The bacterial cell is too small to distinguish these entities, but the presence of nuclear material is well authenticated. The functions of its various structures have not been clearly elucidated. The cell wall remains rigid even when the cell is plasmolyzed. The cell membrane is less rigid, collapses during plasmolysis, and apparently possesses the power of selective permeability. The protoplasm lies within the membrane. The function of the capsular material surrounding some species, such as *Leuconostoc mesenteroides,* is not clearly defined, but the capsule may be protective in function. The species cited forms a thick dextran capsule when growing in sucrose medium and will produce thick slimy masses of dextran-coated cells on the surface of cut or bruised high sugar-content vegetable material such as sugar cane. The organism is considered microaerophilic and grows poorly on the surfaces of nonsucrose media.

The finer structures of the microbial cell are of considerable interest in relation to its physiology of growth. Bacterial flagella and capsules can be observed with an ordinary microscope using an oil immersion objective. The presence of other structures may be deduced from the physiological changes that occur during growth.

GROWTH

A microbial cell grows by increasing its protoplasm. The rapidity of growth can be explained in part by the extreme smallness of the organism and the selective permeability of the cell membrane. An organism 0.5μ in diameter has an extremely large surface area in relation to its mass. The nutritive substance for growth and the by-products of growth must pass through the cell membrane by diffusion. The high rate of metabolism and the resulting biochemical changes are possible only because of its microscopic size, its large surface area in relation to the protoplasmic mass, and the rapidity of diffusion. Nutrients which exist as molecules too large to be absorbed must be digested to smaller molecules by enzymes excreted by the organism. The synthesis of new protoplasm requires the rearrangement of the chemical substances available. The sum of many hundred chemical reactions in microbial cells accounts for the innumerable varieties of changes in many food products. In many foods the basic required substances are readily available for synthesis of proteins, nucleic acids, lipids, and other constituents of the cell. The cell grows; new protoplasm is differentiated; the bacterial cell divides and the process is repeated. The yeast cell forms buds, the cell constituents divide, and a part passes into the daughter bud.

The measurement of microbial growth is determined by several methods. The actual determination of increase in weight, a method applicable in pure culture study, is applicable in only a few food fermentations. It is quite satisfactory in some cases for determining mold growth. Measurement by increase in turbidity is applicable to fermentation of liquid products. Determination of numbers of cells by microscopic counts fails to distinguish between viable cells capable of reproduction and inactive cells. Use of plating procedures depends upon growth of a cell into a colony on a satisfactory medium; however, not all of the viable cells will necessarily grow to colony size on artificial media, and furthermore, colonies often develop from a pair or groups of microorganisms.

Metabolic activity is frequently used as a criterion of growth of lactic acid bacteria and yeasts in food fermentations. Actual measurement of acids, alcohol, or carbon dioxide are useful measurements of cell activity, but their extent is dependent upon environmental conditions. Determination of activity by determining numbers of microorganisms measured by turbidity or by plating techniques ordinarily yields satisfactory results when applied to food fermentations. Selective media are used sometimes to determine the presence or growth of certain species; however, in a massive fermentation little opportunity is afforded for uniformity, and rates of growth and fermentation may vary considerably.

METHOD OF STUDY OF POPULATIONS

Unlike higher organisms, bacteria and yeasts are too small to identify by morphological characteristics. The only way to identify the species is to isolate the organism and then identify it by use of various physiological tests and by morphological examination. Plating techniques, sometimes using selective media, are helpful in such study. Microorganisms seldom exist as pure cultures in foods (Fig. 2.6, Chap. 2); mixed populations are the rule rather than the exception in nature. The environment will determine to a considerable degree which microorganism will grow and predominate in food fermentations. In general, raw foods exist in an aerobic environment; therefore, the majority of microorganisms present on the foods are aerobic species. When foods are packed into containers of various types, usually after pressing out the juice to cover the food, chopping, shredding, or packing in liquid, air is excluded. Since growth is a response to environment, the aerobic organisms find environment unsuitable, while those species which require little or no air are favored.

FACTORS AFFECTING GROWTH

The rate of growth of microorganisms is extremely important in food fermentations. This is particularly true in reference to comparative rates of growth of organisms responsible for the fermentation (Table 4.1) and the rate of growth of those species which may cause undesirable changes. It is essential that favorable environmental conditions be established in the food to be fermented so that the flora responsible for desired fermentation may mul-

TABLE 4.1

LEUCONOSTOC MESENTEROIDES ORDINARILY HAS A SHORTER LAG PERIOD AND A MORE RAPID RATE OF GROWTH THAN EITHER *LACTOBACILLUS PLANTARUM* OR *LACTOBACILLUS BREVIS*

| Species | Temperature | Percentage of Maximum Acidity Produced by the Species in | | |
		1 Day (%)	2 Days (%)	10 Days (%)
Leuconostoc mesenteroides	10°C (50°F)	3	25	62
Lactobacillus brevis	10°C (50°F)	0	0	10
Lactobacillus plantarum	10°C (50°F)	0	0	18
Leuconostoc mesenteroides	15°C (59°F)	19	47	90
Lactobacillus brevis	15°C (59°F)	0	0	48
Lactobacillus plantarum	15°C (59°F)	0	0	55
Leuconostoc mesenteroides	20°C (68°F)	50	84	100
Lactobacillus brevis	20°C (68°F)	8	18	74
Lactobacillus plantarum	20°C (68°F)	12	31	80
Lactobacillus mesenteroides	25°C (77°F)	77	94	100
Lactobacillus brevis	25°C (77°F)	15	33	91
Lactobacillus plantarum	25°C (77°F)	34	54	92

tiply, ferment the food rapidly, and continue to create an environment unfavorable to undesirable microorganisms. This is accomplished in some fermentations by addition of a starter culture, in others, merely by adjusting
environmental conditions. Different explanations have been suggested to account for varying rates of growth; the rates of growth, however, are primarily
influenced by the presence of nutrients, moisture content, oxygen relationships, degree of acidity, temperature, presence of the microorganisms, their
stage of growth, and other environmental factors.

Moisture

Microorganisms cannot grow in the absence of moisture. The organism ab-
sorbs nutrients only in liquid form through its cell walls and discards waste
materials in the same way. Many microorganisms can survive periods of
drying but will not grow until the water content becomes adequate for the
species. Molds usually require less moisture than yeasts and bacteria. One
may observe mold growth on slightly moist surfaces such as cereal grains,
seeds, or root vegetables in storage. In some mold fermentations, bacterial
contamination is avoided by controlling the moisture content. Yeasts generally require less moisture than bacteria. The moisture content required for
growth of any organism is the quantity required to provide a ready supply of
nutrients for growth.

Moisture requirements are affected by nutrients, temperature, oxygen, and
other factors. The water must be available to the organism; in other words,
it 'must not be combined with various solutes or various hydrophilic colloids.
The available water may be expressed in terms of water activity or the vapor
pressure of the solution divided by the vapor pressure of water. Salt, sugar,
other solutes, or a combination of these makes moisture unavailable and af-
fects vapor pressure. An increasing salt or sugar concentration will, in effect,
dehydrate the medium. The combination of solutes with the presence of acid

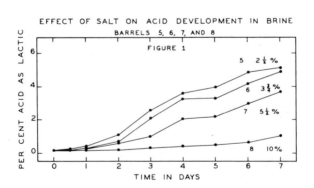

FIG. 4.1. EFFECT OF SALT CONCENTRATION ON DEVELOPMENT OF
ACIDITY IN PICKLE BRINES

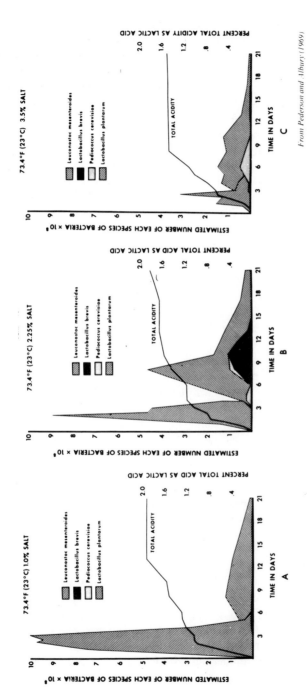

FIG. 4.2. EFFECT OF SALT CONTENT UPON THE DEVELOPMENT OF ACID AND CHANGES IN BACTERIAL FLORA IN SAUERKRAUT AT 23°C

With 1% salt (A), with 2.25% salt (B), and with 3.5% salt (C).

From Pederson and Albury (1969)

or other substances is effective in the preservation of several foods. A great increase in salt may exert a greater osmotic pressure outside the microbial cell and result in cell dehydration. For example, vegetable fermentations will not occur in salt brines much above 40° salòmeter, that is, 10.4% salt (Fig. 4.1). Growth is retarded even below this concentration (Fig. 4.2). Some foods are preserved in high salt-content brines. Although water in its solid state is not available; however, frozen fruit juices sometimes retain sufficient concentrated sugar in water to permit growth of molds.

In general, the water activity or the vapor pressure of the solution divided by the vapor pressure of the solute must be below 0.62 to inhibit mold growth, below 0.78 to stop yeast growth, and below 0.90 to retard growth of bacteria. Certain osmophilic yeasts and bacteria will tolerate a lower available water content.

Acidity

The total acidity or hydrogen ion concentration of a food has a profound effect upon the types of organisms that may grow. Various species differ considerably in acid tolerance. Most molds have a wide tolerance range and may grow in media of pH 2.0–8.5. This exceeds the range of pH developed in food fermentations; therefore, to prevent mold growth their oxygen requirements must be limited. Yeasts are favored by an acid reaction usually in the range of pH 4.0–4.5. They have a wide pH range for growth and will tend to adjust the acidity of their media to a favorable acidity.

The various species of bacteria differ considerably in their pH range for growth. Most bacteria are favored by a reaction near neutrality, a few are favored by an alkaline reaction, and others by an acid reaction. The lactic acid bacteria, acetic acid bacteria, propionic acid bacteria, and a few other specialized species are favored by an acid reaction. The acidity they produce in foods is lower than the pH tolerance of a large majority of other bacteria, including pathogens. It is this characteristic which makes them so valuable in the preservation of foods. The vegetative cells of a few spore-forming bacteria may grow at a pH of 4.0–4.2, but their spores are unable to germinate at this low pH.

The different foods vary in their ability to resist change in pH, that is, in their buffer capacity. The presence of buffering substances, certain salts and proteins, permits fermentation to continue without considerable change in pH. Milks, meats, and seeds have a considerably higher buffer capacity than leafy or root vegetables. Considerable growth and acid production may occur in milk without great change in pH. Green beans and peas are fermentable by lactic acid bacteria, but the pH change is so slow that nonlactic types of bacteria frequently grow and impart undesirable characteristics. The leafy vegetables are readily fermented by lactics and permit a rapid and appreciable decrease in pH.

Oxygen Relationship

Since lactic acid bacteria cannot produce sufficient acid to stop the growth of molds and yeasts in foods, oxygen relationships are very important in the production of lactic acid-fermented foods. The utilization of oxygen is one of the most important factors governing growth and metabolism of microorganisms. Like plants and animals, many microorganisms are strictly aerobic, that is, they require oxygen for growth. Others cannot utilize oxygen in their metabolism, and in some cases, oxygen is toxic to them. These are the anaerobes. Many species of bacteria and yeasts are capable of activity under both aerobic and anaerobic conditions and are referred to as facultative. The lactic acid bacteria may utilize oxygen to a minor extent, but they grow and ferment better at low oxygen tension. They are called microaerophilic. Oxygen is toxic to some strains of lactic acid bacteria. Present knowledge indicates that the absence of oxygen is essential because certain enzymes are not active unless their sulfhydryl groups are in the reduced state. When growing in a food, they will generally lower the oxidation-reduction potential. This is important in retention of color and flavor of many foods and in prevention of growth of other organisms.

Yeasts grow best under aerobic conditions, but the fermentative yeasts can grow under microaerophilic conditions. Production of alcohol and carbon dioxide is an anaerobic process. Molds require oxygen for growth. This is most important in foods fermented by lactic acid bacteria. In some mold fermentations which are followed by bacterial or yeast fermentation, growth of molds can be stopped by packing the food tightly to exclude air. Exposure of food to air results in a rise in the oxidation-reduction potential at the surface; this, in turn, will permit growth of molds and surface-growing yeasts. Oxygen may diffuse for some distance into the food. The oxidizing and reducing substances in foods tend to poise the food at a level favorable for microbial growth.

Fresh or raw foods, fruits, vegetables, meats, and others, support the growth of aerobic organisms at their surface. During active growth, the plant or animal cells are respiring and tend to stabilize the oxidation-reduction potential. Reducing substances, sulfur-containing amino acids, ascorbic acid, and other entities, play an important role in this stabilization. The lactic acid-producing bacteria tend to lower the oxidation-reduction potential and thereby restrict the growth of aerobic microorganisms. In the fermentation of many foods it is necessary that the food be covered in some manner to protect it from air at the surface and from diffusion of oxygen into the food. During the active fermentation of some foods, sufficient carbon dioxide gas is evolved to blanket the surface. After most of the active fermentation is completed, artificial covering must be used such as the plastic covers used on sauerkraut vats. It is interesting that sausage makers realized long ago that ground meat had to be pressed down firmly into the pans for curing. The importance of protecting fermented foods from contact with air has been realized for centuries.

Temperature

Microorganisms differ widely in their optimum, maximum, and minimum temperatures for growth. The temperature at which a food is fermented will greatly influence the rate of fermentation (Fig. 4.3), the species of organisms involved (Fig. 4.4), and the microbiologically-induced changes that occur. Microorganisms that grow best at temperatures between 20° and 40°C are called mesophiles, those that grow above 40°C are called thermophiles, and those that grow below 20°C are called psychrophiles.

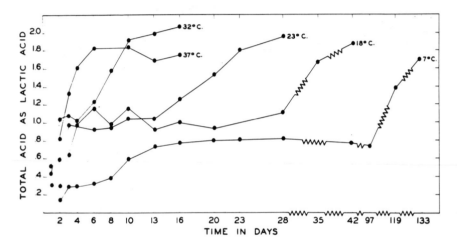

FIG. 4.3. THE EFFECT OF TEMPERATURE OF INCUBATION UPON A VEGETABLE FERMENTATION

Note the rapid initiation of growth at 37°C but eventually the higher acid production at lower temperatures.

The microorganisms involved in most food fermentations are ordinarily classed as mesophiles. The majority of yeasts and molds grow best in the lower mesophilic range, but some may grow slowly even below freezing temperatures. Yeasts used in lager beer fermentations are selected for their ability to grow at the low temperature of 10°C or even lower. Although the fermentation proceeds slowly because the temperature is so low, there is less chance of growth of lactic acid bacteria.

The lactic acid bacteria grow well at temperatures between 20° and 40°C with a general optimum at about 30° to 32°C. A few species of the genus *Lactobacillus* involved in milk and in mash fermentation grow well at 40°C or above. The species *Leuconostoc mesenteroides* has a lower temperature range for growth than other lactics (Fig. 4.4), a fact of great importance in initiating certain vegetable fermentations in cooler climates. It will ferment

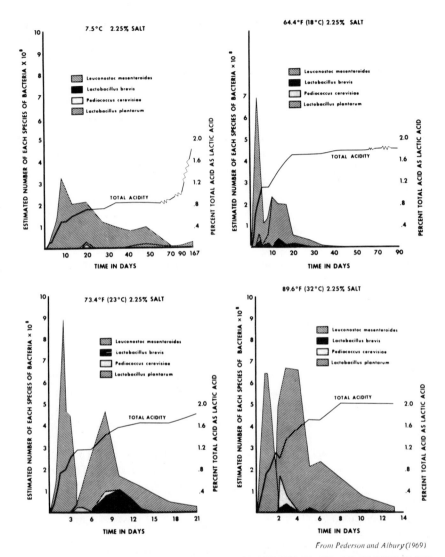

From Pederson and Albury (1969)

FIG. 4.4. THE EFFECT OF TEMPERATURE UPON THE DEVELOPMENT OF ACID AND CHANGES IN BAC-
TERIAL FLORA OF SAUERKRAUT FERMENTED WITH 2.25 PERCENT SALT

A container fermented at 7.5°C was placed in a warmer room after 70 days.

vegetables packed at temperatures as low as 7.5°C producing the acid and
carbon dioxide so essential for inhibition of the truly aerobic psychrophilic
organisms as well as inhibiting oxidative changes. Such fermentations may
remain inactive for months after the initial fermentation is completed. The
product must be warmed before a fermentation by lactobacilli will occur. In
contrast, the starter culture for sausage, a strain of the genus *Pediococcus,*

grows well in the higher mesophilic range and therefore ferments sausage rapidly at the smokehouse or drying temperatures used. On the other hand, certain spoilage organisms of the genera *Pseudomonas, Flavobacterium,* and *Achromobacter* grow at temperatures near the freezing point.

Presence of Microorganisms

The aerobic microorganisms from air, water, soil, containers, sewage, skin surfaces of animals, and organisms from diseased tissues are usually far more numerous on the surfaces of food than are the fermentative species. Fermentation failures often may be attributed to the preponderance of undesirable organisms as well as to the delay in growth of the desirable organisms. This is particularly true with milk products. The aerobic organisms are potential spoilage types. If allowed to develop, they will cause various forms and degrees of off-flavor spoilage, some may cause illness and even death. Fortunately, the large majority of such organisms are usually aerobic and intolerant to the growth substances produced in fermentation and the inhibitives used in preparing fermented foods.

The microorganisms responsible for fermentation of any food are nearly always present, and it is only necessary to establish environmental conditions favorable to their growth.

Natural Inoculation and the Use of Starters

The great majority of fermented foods produced throughout the world still depend upon a natural inoculation with the desired organisms. The microorganisms essential for a satisfactory fermentation may be introduced to a food and gain ascendancy in the fermentation by any one of several ways. Also, conditions may be created in the foods that are not favorable to growth of lactic acid bacteria, yeasts, or other desired organisms. The food may become inoculated from equipment used in preparing the food. The food may be deliberately inoculated with previously fermented or fermenting food. Pure culture starters may be added. Regardless of the method used, environmental conditions must be favorable for the desired microorganism and unfavorable to others.

The creation of conditions favorable for growth of the lactic acid bacteria, the vinegar organism, the alcohol-producing yeast, and other organisms has been practiced unknowingly for centuries. The cutting of vegetables and packing them tightly into a container with salt or salt brine will drive out excess air, essentially creating microaerophilic conditions. The cutting of vegetables releases certain selective inhibitive substances indigenous to the cell protoplasm and at the same time furnishes the lactic acid bacteria with a very satisfactory growth medium. In many fermentations, the addition of salt, nitrates, nitrites, spices, herbs such as hops, garlic and onion, and other substances is effective in retarding growth of competing microorganisms. Conducting fermentations at selected temperatures, such as that used in brewing lager beer, retards growth of unwanted species. The rapid formations of growth products, acids, alcohol, carbon dioxide, and other products, change hydrogen-ion concentration and oxidation-reduction potentials.

For centuries foods have been inoculated by chance contact with containers and equipment. The ancient practice of placing fresh milk in a goatskin bag or a porous bowl previously used for such foods practically assured inoculation with the organism from the previous fermentation. If the milk had a high acid content, the high acid-producing lactobacillus would possibly be the only surviving organism. If the bag contained the kefir granules of curdled milk, lactic acid bacteria and yeasts, an acid-alcoholic fermentation was assured. When clay and wooden containers were first used, the organisms lodged in the cracks or pores, and no amount of scrubbing would remove them. The practice of preparing sausage in certain rooms, always using the same tables, shelves, and containers, assured an inoculation of the meat by organisms lodged in the cracks and crevices. The typical fermentation for cheese such as limburger originally required this form of inoculation. The practice of pre-chewing foods by elders, particularly cereal foods, for preparing infant foods as practiced in many societies, without doubt concomitantly introduces yeasts and lactic acid bacteria.

With the advance of civilization and greater proficiency, the deliberate inoculation of certain raw foods became common practice. The use of pre-soured milk for cheese, butter, and sour milk drinks, the transfer of yeast sediment for inoculating wine and beer, and the setting aside of some of the raw bread dough for the next day's baking were eventual refinements. The borrowing of such starters from neighbors and the inclusion of sour dough, sour milk, and yeast sediment starter in a daughter's gifts at marriage were customary in many societies.

Pure culture starters or inocula are recent innovations that followed the classic studies of Pasteur and others. The numerous studies designed to develop pure cultures of bacteria, yeasts, and molds for use in preparation of cheeses, sour milk drinks, butter, bread, wine, beer, soy bean products, and more recently, sausage have exerted a profound effect on the standardization and improvement of the quality of these food products. Nevertheless, the adaptation of similar principles for inoculation of foods in which a sequence of microbial growth occurs has not been developed so successfully.

Nutrition

The nutritional requirements of the lactic acid bacteria are particularly complex and exacting. Besides an energy source, a variety of essential growth factors must be made available. The yeasts and molds are less exacting in their nutritional requirements. In fact, yeast autolysate contains so many of the necessary growth requirements that when combined with a good protein source, mineral and fermentable carbohydrate, it supplies the necessary requirements for growth of the majority of strains of lactic acid bacteria. It is important to realize that microorganisms in some fermentations die after performing their function, autolyze, and release their protoplasm into the food. This enriches the food from a nutritional point of view. A few lactic acid bacteria require certain growth substances present in foods from which they are isolated, particularly at isolation and for the first few transfers. These may be supplied by the addition of milk, beer, wine, tomato juice, orange juice, and similar foods. Among the lactic acid bacteria, requirements for amino acids, the B vitamins, certain minerals, fatty acids, purines, and pyrimidines are so specific that their presence in foods may be determined quantitatively by microbiological assays using these organisms.

The nutritional qualities of foods have been well elucidated. Many foods contain all of the essential carbohydrates, proteins, vitamins, lipids, minerals, nucleic acids, and other nutrients required for growth of bacteria, yeasts, and molds. Furthermore, microorganisms may invoke mechanisms whereby the various factors essential for growth may be made available to the cell. Although some of the nutritional requirements are infinitesimal, sometimes lactic acid bacteria seem to require supplementary additions.

Carbohydrates.—The microbial cell may be regarded as a machine that manufactures its protoplasm using an energy source supplied by carbohydrates by means of a complex process of respiration. The conversion of simple sugars to lactic acid, alcohol, and other chemical entities furnishes a ready supply of energy for metabolism. The utilization of the available energy is quite incomplete under the anaerobic mechanism by which carbohydrates are utilized. This incomplete utilization requires that a large supply of carbohydrate be available. This anaerobic mechanism of conversion of carbohydrates to obtain energy is quite similar among the lactic acid bacteria and the fermentative yeasts. The unstable intermediate, pyruvic acid, produced by both yeasts and lactics is converted readily to alcohol and carbon dioxide by yeasts and to lactic acid by the lactic acid bacteria. Needless to say, carbohydrates are essential. The various species demonstrate considerable differences in their ability to utilize the different types of carbohydrates, simple sugars, di-, tri-, and polysaccharides, and the glucosides, a character which is used to characterize strains or species.

Proteins.—The enzymes are the proteins on whose surfaces the many metabolic reactions of the cell proceed. The proteins are composed of amino acids linked together through the carboxyl group of one acid and the amino group of another. Each microbial cell has many proteins that are characteristic of the species. The number of amino acids, their order in a chain, and their configuration are the controlling factors determining the nature of the protein. Most of the 25 or more amino acids are found in all proteins. The lactic acid bacteria are often called fastidious, referring in part to their requirements for amino acids. Among the species at least 17 are essential.

A complex relationship exists between proteins, vitamins, carbohydrates, and other nutrients in metabolism. The requirements by a microorganism for certain amino acids are so specific that the organism may be used in assaying for specific amino acids in a particular substance. Even though these requirements are so specific there is little evidence of proteolysis in food fermentations by lactic acid bacteria. Molds, however, are so active that the character of some mold-fermented foods is a reflection of changes in protein constituents and, in particular, in the increase in soluble constituents.

Vitamins.—The vitamins of the B complex are the coenzymes. They may serve as a part of several enzymes that have different functions. The ability or inability of a microorganism to synthesize the vitamin for itself will be reflected in its requirement for this vitamin. If an organism is incapable of synthesizing a vitamin or converting it to the closely related coenzyme, the organism is incapable of growth and fermentation unless the vitamin or related coenzyme is supplied. Many foods are deficient in B vitamins for the nutrition of higher animals. They still may contain a sufficient amount for growth of microorganisms. A few foods are supplemented with or fortified by the addition of vitamins. The growth of certain microorganisms has provided a useful analytical tool for determining the presence and amount of certain vitamins in foods. Also, it has provided a useful tool in discovery of new vitamins and the elucidation of the specific purposes of several vitamins.

Many species of the genus *Saccharomyces* require certain vitamins to supplement other nutrients. *Saccharomyces cerevisiae* is able to concentrate large quantities of thiamine, nicotinic acid, and biotin from growth media and thus form enriched products so that yeasts can be obtained with exceptionally high vitamin contents. A few yeasts can synthesize riboflavin. Baker's yeast and brewer's yeast, the species *Candida utilis* and *Hansenula anomala,* are good sources of thiamine, riboflavin, pantothenic acid, nicotinic acid, pyridoxine, folic acid, biotin, para-aminobenzoic acid, inositol, and choline. Some strains of *Saccharomyces* can produce ergosterol, and that, by irradiation can be converted to vitamin D. Yeast autolysates are of extreme nutritional value in the growth of lactic acid bacteria.

Mineral Elements.—Among the microorganisms there is some variation in requirements. Sodium chloride presumably is required for osmotic regulation, for it does not occur in the known metabolic patterns. Some elements have an important role in the adjustment of hydrogen-ion concentration and oxidation-reduction potential. Species of lactic acid bacteria have an optimal hydrogen-ion concentration in the range of pH 5.5–6.5 and yeast in the range of pH 4.0–6.0. Phosphates are important constituents in several organic substances such as nucleic acids, phosphorylated intermediate metabolites, and coenzymes; therefore, they are active in energy conversion and biosynthesis. Phosphorus has been referred to as the core of metabolism. Sulfur is a constituent of several amino acids of the protein molecule in the vitamin, biotin, and certain sulfur glucosides. Magnesium, potassium, iron, manganese, zinc, copper, probably cobalt, and molybdenum are involved in metabolic patterns. The amounts required are infinitesimal but definite enough so that their presence in foods may be determined by microbiological assay.

Nucleic Acids, Purines, and Pyrimidines.—The nucleus of the cell has aroused the interest of scientists for years. Bacteria are too small for differentiation of the nucleus from the protoplasm. The two major groups of nucleic acids are the ribose nucleic acid (RNA) and the deoxyribose nucleic acid (DNA), both important in metabolism. In the cells they are combined with proteins as nucleoproteins. The nucleic acids consist of the five carbon sugar, ribose; two purines, adenine and guanine; two pyrimidines, cytosine and either uracil or thymine, and four phosphates. The exact proportion of these vary from species to species. When the purine or pyrimidine is combined with sugar, it is called a nucleoside. The nucleoside, combined with a phosphate, is a nucleotide. The purines, pyrimidines, nucleosides, nucleotides, and nucleic acid have distinct physiological roles in the metabolism of microbial cells, the details of which can be obtained from books on microbial physiology. It will suffice to say here that it has been stated that the concept of adenosine triphosphate (ATP) as the driving force in biosynthesis and of the conversion of the oxidative energy to bond energy is undoubtedly one of the most significant achievements of the enzymologist and one of the great unifying principles of biology.

The various lactic acid bacteria are known to require the purines, adenosine and guanine, and the pyrimidines, cytosine, thymine, and uracil.

Lipids.—The lipids are constituents of microorganisms. Fats and the complex phospholipids undergo changes during lactic acid fermentations as characterized by hydrolysis and increase in fatty acids and in choline, that often combines with lactic and acetic acids to form lacteal and acetyl cholines.

The Growth Curve

Whenever microorganisms are present in food and environmental conditions are favorable, growth will occur. The manner in which these organisms develop is especially important in food fermentations. Studies of the growth of microbial cultures offer an opportunity for studying population problems. Growth must occur before fermentation can proceed.

When organisms are counted at regular intervals and the results of these counts are plotted as logarithms of numbers of the organisms, a growth curve is obtained. This curve (Fig. 4.5) is ordinarily divided into phases as follows: A to B—initial stationary or lag phase. During this time the numbers of organisms remain constant or may even decrease slightly. B to C—positive acceleration phase. The phase during which the numbers of organisms are increasing or, in other words, the generation time is decreasing. C to D—log-

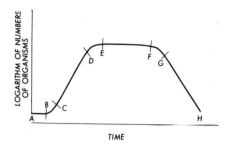

TIME

FIG. 4.5. TYPICAL GROWTH CURVE: A TO B—INI-
TIAL STATIONARY OR LAG PHASE; B TO C—POSI-
TIVE ACCELERATION PHASE; C TO D—LOGARITH-
MIC GROWTH PHASE (LOG PHASE); D TO E—NEG-
ATIVE ACCELERATION PHASE; E TO F—MAXIMUM
STATIONARY PHASE; F TO G—ACCELERATED
DEATH PHASE

arithmic growth phase. Numbers of organisms increase in geometric progression and the generation time is constant. D to E—negative acceleration
phase. During this period the generation time is decreasing but the numbers
of organisms are increasing. E to F—maximum stationary phase. The phase
in which organisms remain constant or old organisms die as rapidly as new
organisms form. F to G—accelerated death phase. The phase during which
the numbers of viable organisms decrease slowly at first but then with increasing rapidity. G to H—death phase or logarithmic death phase. The
period during which a more or less constant rate of death is maintained.

The rate of growth depends upon many environmental factors. It will vary
with different organisms; a bacterial growth curve is, therefore, based upon
the numbers of organisms in a medium at the several stages in development
of the organisms. It is based upon generation time, the time elapsing between
the formation of a cell and its growth and division into two cells. The ability
of an organism to grow and reproduce itself depends upon its ability to form
new protoplasm. Organisms in the lag phase may not reproduce for several
hours after they are introduced into a food. This failure to begin multiplication when introduced into a food has been attributed to the physiological
state of the organism and necessity of adjustment to its new surroundings.
The new nutrients must be absorbed, enzymes may have to diffuse into the
new substrate, and the organism must adjust itself to a new substrate. The
conversion of food into new protoplasm is a period of adjustment during
which the enzymes must adjust their activity. Organisms in their lag phase
may not produce new cells for several hours after inoculation. However,
during the lag phase growth may be occurring, as evidenced by an increase in
the size of bacteria without division of cells. The length of time a culture remains in the lag phase will depend upon the extent of difference between the
new medium and the medium in which it has been growing, upon the tem-

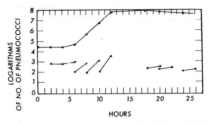

From Chesney (1916)

FIG. 4.6. INFLUENCE OF THE PARENT CULTURE ON
INITIAL GROWTH

perature, and upon the size of the inoculum. If cells are inoculated into similar medium, the culture will continue its rate of growth (Fig. 4.6).

During the second stage, the generation time decreases slowly at first, then at a rapid rate. During the third or logarithmic growth phase, the generation time remains constant, that is, the cells continue increasing at a constant rate. They have become adapted to their new medium. When a rapidly growing starter culture from a food is inoculated in an additional quantity of the same food, it needs no further physiological adjustment. As growth continues, fermentation products, acids, alcohol, carbon dioxide, and others, accumulate and in time act as toxic products retarding the rate of growth and even causing death of some cells. There is evidence that certain growth requirements become depleted. In a lactic fermentation the acids may be neutralized and growth may continue, but ultimately growth will be stopped regardless of neutralization. The physiological state of surviving organisms changes to accommodate to changing environment.

It is an interesting parallelism that the gross morphology of the bacterial cells varies during different phases of growth. During the logarithmic growth phase, cells are multiplying at their most rapid rate and tend to divide more rapidly and to be smaller. They stain uniformly. This is in contrast to cells in the lag phase of growth when they tend to be larger. Bacterial cells of lactic acid bacteria in adverse growing conditions, such as in acid media or growing at temperatures near their maximum, tend to be larger, longer, and to remain in chain formation.

Under ideal conditions, the generation time of some bacteria may be as short as 20 min, for others, it may be several hours. Temperature has a marked effect upon the logarithmic phase and the generation time. In a vegetable fermentation held constant at 7.5°C an estimated generation time for the species *Leuconostoc mesenteroides* was approximately 2 days (Table 4.1). At 18°C a generation time was estimated to be about 2.5 hr. The estimated generation time for the species *Lactobacillus plantarum* at 7.5°C was about 20 days and at 18°C about 18 hr. Metabolism and growth are chemical reactions; therefore, as the temperature is increased, the rate of growth will increase until a temperature is attained where the thermal rate of destruction of the enzymes is compromised with the rate of chemical reaction. For many of the more common bacteria involved in food fermentations this compromise is attained somewhere about 32°C and at a lower temperature for yeasts and molds. The presence of inhibitive substances will lower this point of compromise. A fermentation at 35°C may start more rapidly than at 30°C, but invariably more acid will be produced at the lower temperature. Similarly, every winemaker knows that there will be a higher alcohol production when fermentation occurs at lower ranges of temperature.

Growth Associations in Fermentation

A pure culture of one species of microorganism rarely occurs in nature. The majority of organisms are free living. In food products there may be a mixed population of several species of bacteria, yeasts, and molds, the frequency of any particular population depending upon the physical and chemical environment (Fig. 4.2, 4.3, 4.4, 4.6). Great numbers of species of bacteria find milk an ideal medium for growth. The surfaces of fruits and vegetables harbor many types of microorganisms. Because they are exposed to the air many of them are aerobic; however, microaerophilic types also may be present. When fruit is crushed, the fruit acids deter the growth of many types, and fermenting yeasts usually become the predominant flora. When vegetables are shredded and packed with salt or brine, contact with air is minimized and vegetable cell protoplasm is released; these may offer ideal conditions for growth of lactic acid bacteria.

Foods in general are excellent media for growth of bacteria, yeasts, and molds, and a number of growth associations occur in fermenting foods. Associative growth of yeasts and bacteria occur in bread, wines, milk drinks, and many other foods. Associative growth of molds and bacteria are common in cereal and seed fermentations. Growth associations also occur in spoilage. Organisms are said to be symbiotic when they are mutually helpful or without known effect upon the growth of each other. The growth of lactic acid bacteria and yeasts in the milk drinks, kefir and kumiss, is such a relationship in that the bacteria hydrolyze the lactose, unavailable to the yeasts, to glucose and galactose which the yeasts can then ferment. Some organisms are synergistic in that when growing together they bring about changes which neither can do alone. Other microorganisms are often antagonistic or antibiotic to others. The term metabiotic is used for the relationship in which one organism makes conditions favorable for the growth of another. It is suspected that the growth of *Leuconostoc mesenteroides* in vegetable fermentations creates an environment more suitable for growth of both *Lactobacillus brevis* and *Lactobacillus plantarum*. This is sometimes referred to as a sequential growth. Raw milk supports the growth of many species; however, among the lactics, *Streptococcus lactis* usually initiates fermentation, followed by growth of *Lactobacillus casei* and then *Lactobacillus bulgaricus*. The growth of lactic acid bacteria produces acid unfavorable to many organisms in several fermentations but still permits growth of molds that are responsible for the flavor, aroma, and texture characteristics of several mold-fermented foods such as in the Roquefort and Camembert cheese fermentations.

Sequence of Growth.—The sequence of growth of many species of microorganisms may be considered a natural or orderly sequence required to reduce a complex food to its simple elements. Milk, for example, if left alone may initially be fermented by streptococci and coliforms; fermentation is continued by the acid-tolerant higher acid-producing lactobacilli; followed by film yeasts and molds that utilize the acid and then, in conjunction with proteolytic bacteria, may convert the milk to its simple elements. In cheese and butter factories the fermentation by pure cultures of lactic acid bacteria are allowed to continue only until sufficient acid is produced to separate the curd or to free the butterfat. The vegetable fermentations now illustrate more clearly sequential fermentations and the effects of environment.

Environmental Factors as They Influence Food Fermentations

A discussion of environmental factors as they may influence a specific fermentation may emphasize the importance of the various factors in producing a satisfactory food fermentation. It may be noted that most food products were developed through years of experience in which an occasional individual may have improved the methods. During the past century, the methods of preparation of many of the common foods, bread, cheese, beer, wines, and others, have been standardized and controlled to such an extent that little chance of differences as a result of the effect of environment upon the fermentation is permitted. It should be appreciated that the 400 or more types of cheese, the scores of types of sausage, the many types of bread products, and the various beers and wines were developed by some difference or alteration in condition that resulted in the production of the many types of products. Moisture content, temperature of fermentation, relationship to air, and other factors have caused marked variation in types of products. The contrast between a mild cheddar cheese and an acceptable Roquefort is striking; however, the presence of mold on cheddar cheese is considered spoilage, and conditions are established in manufacture to prevent such growth. The consuming public has become accustomed to the mold-cured Roquefort cheese but does not expect to find mold on cheddar cheese.

The fermentation of vegetables is an ancient process that is presumed to have originated in the Far East. Paw tsay, kimchi, and similar fermented products are consumed in large quantities in the Far East. One Korean when tasting sauerkraut for the first time, called it kimchi of a mild type. The hot kimchi preparations would not be acceptable to most Westerners. As this method of preservation was adopted by Westerners in Europe, the vegetable most common to these people, cabbage, undoubtedly become the prevalent vegetable in the blend. The fermented vegetables of the Far East are blends of many vegetables that vary considerably in amount of each kind used. Also, the sauerkraut produced in various European countries often contains other vegetables, fruits, and spices in addition to cabbage. The fermentation of cabbage to sauerkraut is an old process and presents an example of a fermentation in which effect of environmental factors may be demonstrated.

BIBLIOGRAPHY

BURROWS, W. 1954. Textbook of Microbiology. W. B. Saunders Co., Philadelphia.
CHESNEY, A. 1916. The latent period in the growth of bacteria. J. Exptl. Med. *24*, 387.
FRAZIER, W. C. 1967. Food Microbiology, 2nd Edition. McGraw-Hill Book Co., New York.
GUNSALUS, I. C., and SHUSTER, C. W. 1961. Energy-yielding metabolism in bacteria. *In* The Bacteria, Vol. 2, I. C. Gunsalus, and R. Y. Stanier (Editors). Academic Press, New York.
KRAMPITZ, L. O. 1961. Cyclic mechanisms of terminate oxidation. *In* The Bacteria, Vol. 2, I. C. Gunsalus, and R. Y. Stanier (Editors). Academic Press, New York.
OGINSKY, E. L., and UMBREIT, W. W. 1954. An Introduction to Bacterial Physiology. W. H. Freeman and Co., San Francisco.
PEDERSON, C. S., and Albury, M. N. 1954. The influence of salt and temperature on the microflora of sauerkraut fermentation. Food Technol. *8*, No. 1, 1–5.
PEDERSON, C. S., and ALBURY, M. N. 1969. The sauerkraut fermentation. New York State Agr. Expt. Sta. Bull. *824*.
SARLES, W. B., FRAZIER, W. C., WILSON, J. B., and KNIGHT, S. G. 1956. Microbiology, General and Applied, 2nd Edition. Harper & Row, New York.
SCHULTZ, H. W. 1960. Food Enzymes. Avi Publishing Co., Westport, Conn.
WOOD, W. A. 1961. Fermentation of carbohydrates and related compounds. *In* The Bacteria, Vol. 2, I. C. Gunsalus, and R. Y. Stanier (Editors). Academic Press, New York.

CHAPTER 5

Fermented Milk Products

Milk and milk products have been used as important foods for man since before the dawn of civilization. From historical experience they have long been regarded as the best cornerstone on which to build nourishing diets. It is not known just when animals first became the servant of man nor when milk and milk products were first used. It can be presumed that prehistoric man, while hunting, caught a lactating animal from which he gained food for his family. Primitive peoples had little reason to be concerned with protein foods when they had milk in their diet. They had to be concerned with preservation of milk in its various forms. Evidence of various kinds indicates that the domestication of the cow and the use of her product go back to the time of the early development of the human race. It is known that the cow has served man throughout the ages as a beast of burden, as a source of food, and as an object of worship. Milk and the products of milk have been used for food, for medicines, for sacrificial offerings, and even for cosmetics.

Archaeologists have discovered rock drawings in the Libyan Desert believed to have been made about 9000 B.C. which depict cow worship and the act of milking a cow. Some of the oldest records of man date back to those of Mesopotamia, Egypt, and ancient India. The writings of the Sumerians of Mesopotamia go back to about 6000 B.C. It is apparent from writings, drawings, and friezes that dairying was highly developed by the ancient Sumerians. Their small cows were milked from behind as some African tribes do today.

The ancient Egyptians left evidence in writings, carvings, and in remains in tombs of their interest in dairying. Milk, butter, and cheese were used by Egyptians as early as 3000 B.C. The Egyptians had at least three types of cattle. They milked cows from the side. Like the Sumerians, the Egyptians deified cattle and their god, Ptah, is revealed as a bull.

Records dating back to 2000 B.C. indicate that the people of India raised dairy cattle. The cows were considered holy and had free run of cities and countryside. Even cow urine was considered holy. The ancient Indians evaluated cows on the basis of their milk-producing abilities and probably first used the twirling churn.

The Old Testament has some 40 references to cattle. References to milk, butter, and cheese are frequent, as in Genesis 18:8, "and he took butter and milk;" Judges 5:25, "He asked for water and she gave him milk;" and in Samuel I, Isaiah ordered his son David to take ten cheeses to the captain.

The Greek records go back to about 1550 B.C. and Roman records to 750

B.C. Milk and cheese were important components of the diets of both the Greeks and the Romans, but apparently milk was converted to cheese as the more common food in their diets. Several types of cheese were made during the early Christian era. Obviously, it was known that dry cheese could be preserved better than soured milk. The Greeks obtained much of their milk from goats and the Romans, from sheep.

By the beginning of the Christian era milk and cheese were used for food throughout Europe, but butter was used primarily for ointment purposes. Milk cattle were found all over Europe in the early Christian era, particularly in those areas where sufficient rainfall assured good pasturage for dairy cattle (Fig. 5.1). In drier areas the milk was obtained from sheep and goats, and in extremely wet areas from water buffaloes. From the 16th to the 19th Centuries dairy farming developed progressively. In France, England, and Netherlands cattle were being bred pure, but breeds were not well defined.

From Breed (1939)

FIG. 5.1. AN ALPINE CHEESE FACTORY

In the mountainous areas of several countries, the cattle are pastured in the highlands during summer months, the cows are milked outdoors, and the milk preparations (curdled milk, butter, and cheese) are prepared in small buildings.

Until the 19th Century great changes in the use of milk had not occurred. The market for fluid milk was limited, and even in all except the larger cities, such as London and Paris, people kept the family cow which was pastured nearby. All milk was sold as fresh milk, and at the beginning of the 20th Century it was still measured by the producer from a larger container into whatever receptacle the purchaser provided. In many areas of the world, milk and its products are still produced under conditions comparable to those of centuries ago. Although cows, sheep, and goats furnished the milk in Europe and North America, milk is obtained from many other animals in other areas of the world. The Tatars and Mongols use mares' milk; the Tibetans use yak are mares' milk; the Persians, camels' milk; Tundra folk, reindeer milk;

Peruvians, llamas' and vicunas' milk; some Africans, buffalo and zebra milk; and Far Easterners, buffalo milk. Milk has also been obtained from other animals.

Dairy products consumed in the United States make an important nutritional contribution to our national diet. Although milk is low in ascorbic acid and iron, it is still considered nature's most nearly perfect food. Other foods may contain more of a particular nutrient, but milk is the best balanced nutrient of all foods.

Milk is a complex mixture consisteng of an oil-in-water emulsion stabilized by complex phospholipids and protein absorbed on the surface of fat globules. It contains proteins in colloidal suspension, lactose in solution, numerous minerals, particularly calcium and phosphorus, fat-soluble and water-soluble vitamins, enzymes, and various organic substances.

The composition of milks varies considerably, not only among the species of animals but also between breeds of animals and even from a single animal, depending upon several factors.

GROWTH IN MILK OF MICROORGANISMS

Microorganisms thrive in milk just as plants flourish in a fertile soil. If one were to compare the nutritive requirements of even the most fastidious microorganisms with the composition of milk, one would observe a great similarity. All milks contain proteins, lipids, salts, citrates, the water-soluble vitamins, thiamine, riboflavin, pyridoxine, niacin, pantothenic acid, biotin, folic acid, and other B vitamin factors, vitamins A, C, D, E, and K, enzymes such as lipase, catalase, and phosphatase, and the carbohydrate, lactose—all requirements for growth of the fastidious lactic acid bacteria. Lactose, the unique product of mammary glands, cannot be utilized by all species of microorganisms.

It would be fortunate for consumers if the only species of bacteria that grew in milk were strains of *Streptococcus lactis, S. cremoris, Lactobacillus casei,* and the higher acid-producing species of the genus *Lactobacillus.* Species of the genera *Escherichia, Aerobacter, Micrococcus, Bacillus, Serratia, Pseudomonas, Alcaligenes,* and species of many other genera also find milk an ideal growth medium producing various forms of spoilage: gassiness, unclean flavors, putridity, fruitiness, bitterness, proteolytic conditions, rancidity, ropiness, colored milks, or other forms of spoilage. Various disease-producing organisms may grow also, and serious illnesses and deaths have occurred. Truly, there is little question that milk is about the most completely nutritious food available to mankind, but its very nutritive qualities make it a very desirable medium for growth of microorganisms, and these in the past have caused considerable distress and illness.

Pasteurization will kill the majority of strains of species in these genera, except for spores of spore-forming bacteria and certain thermoduric organisms. If the pasteurized milks are chilled immediately and kept cold, surviving organisms, as well as those that contaminate the milk after pasteurization, have little opportunity to grow. Although pasteurized milk is available throughout many areas of the world, its freedom from microbial growth is brief unless properly cooled and refrigerated after pasteurization; the people of many areas of the world, therefore, must still rely on other methods of preserving the milks as they have had to do for centuries past (Fig. 5.2). Fermentation by lactic acid bacteria has been and still is the best method of retaining the nutritive qualities of milk in spite of the obvious shortcomings of the method. Even fermented milks, however, are subject to mold and yeast growth; however, the casein when separated in cheese, particularly dry cheeses, and butter have better keeping qualities than the fermented milks, an observation that also traces back into very ancient history.

FIG. 5.2. PRIMITIVE CHEESE MAKING

On this goat's hair tent, the Bedouins are drying and curing sour milk
curd cheeses as has been done for centuries.

MICROORGANISMS OF MILK

No discussion of fermented foods would be complete without some discussion of the taxonomy of the microorganisms responsible for fermentation of milk products. There has been more study applied to the microbiology of milk products than to any other fermented food. At times the data may seem to be very confusing. One cannot hope to include a wholly adequate discussion. The many good books on microbiology including those by Foster *et al.* (1957), Hammer (1948), Frazier (1967), Elliker (1949), Orla-Jensen (1919), and, of course, *Bergey's Manual of Determinative Bacteriology,* both the 6th Edition (Breed *et al.* 1948) and the 7th Edition (Breed *et al.* 1957) and *Index Bergeyana* (1966) should be cited. Although it may be said that milk allowed to ferment naturally will demonstrate a sequence of the growth of *Streptococcus lactis, Lactobacillus casei,* and the *Lactobacillus bulgaricus,* this should be considered only in general terms. Preparations of autolyzed extracts of *Streptococcus lactis* or *Streptococcus cremoris* are known to stimulate growth of *Lactobacillus casei* and also *Leuconostoc citrovorum.* Rogers and associates at the U.S. Dept. of Agr. observed that in milk a certain antagonism exists between *Streptococcus lactis* and *Lactobacillus bulgaricus* that seems to be overcome with time.

Streptococcus lactis was first studied by Lister and named *Bacterium lactis* in 1873. In 1909 Löhnis placed the species in the genus *Streptococcus.* A great number of names have been applied to strains or variants of the species (Breed 1928; Breed *et al.* 1948). *Streptococcus cremoris* may be considered a variant, but it is given recognition as a distinct species because of its importance in dairy starters. *Streptococcus cremoris* var. *hollandicus* is a variety first found in the Scandinavian milk drink, taettemjölk. *Streptococcus lactis* var. *diacetilactis* is a citrate fermenting strain; variety *tardus* coagulates milk slowly; variety *maltigenes* produces a malty flavor in milk or cream. *Streptococcus thermophilus* might also be considered a strain of *Streptococcus lactis,* but it is recognized as a species because it is the high temperature-tolerant strain so important in manufacture of Swiss cheese and yoghurt. Whether the organism responsible for the thick and viscous sour milk drinks, taettemjölk and langemjölk, called *Bacterium lactis longi,* is a streptococcus or a lactobacillus is not clear. It should be mentioned that there are a number of species of the genus *Streptococcus* that cause illnesses including septic sore throats, septicemias, and scarlet fever in humans, and mastitis of the cow's udder as well as several other conditions.

In 1920 Hammer described two important species involved in preparation of butter starters, one, *Streptococcus citrovorus,* and the other, *Streptococcus paracitrovorus.* They produce from citric acid the all-important butter flavor, biacetyl or diacetyl. It was later recognized by Hucker and Pederson (1930) working in adjacent laboratories that they were heterofermenters and related to *Leuconostoc mesenteroides* and should be included in the genus *Leuconostoc.* The generic name, *Betacoccus,* was applied by Orla-Jensen to these organisms. This generic name will be found in much European literature.

The genus *Lactobacillus* includes many species important in the fermentation of milk. It is unfortunate in taxonomy that Kern, in describing what he observed in the kefir grains he studied, was not more specific; however, considering the time, 1882, it is surprising that his descriptions were as good as they are. Taxonomically there are at least two species of the genus that occur in kefir grains, the homofermentative lactic acid bacterium described by Beijerinck in 1889 as *Bacillus caucasicus,* now *Lactobacillus caucasicus,* and the heterofermentative lactic acid bacterium described by von Freudenreich 2 yr later as *Bacillus* γ and later in 1904 as *Bacillus casei* γ. It was not realized at these early dates that there were two distinct groups of lactobacilli, one, a homo- and the other, a heterofermenter. Both Beijerinck and von Freudenreich must be credited with some excellent observations in regard to the microbiology of milk. Von Freudenreich recognized several types of lactobacilli as *Bacillus casei* which he distinguished by Greek letters. Later these were renamed by Orla-Jensen (1919) as species in three genera, *Thermobacterium, Streptobacterium,* and *Betabacterium.* These generic names are not generally accepted, and the species will be found in Bergey's Manual in the genus *Lactobacillus. Lactobacillus casei* is the *Bacillus* α of von Freudenreich and *Streptobacterium casei* of Orla-Jensen; *Lactobacillus helviticus* is the *Bacillus* ε of von Freudenreich and *Thermobacterium helviticum* of Orla-Jensen; *Lactobacillus brevis* is the *Bacillus* γ of von Freudenreich and the *Betabacterium breve* of Orla-Jensen; while *Lactobacillus fermenti* is the *Bacillus* δ of von Freudenreich and the *Betabacterium longum* of Orla-Jensen. These species are of importance in milk fermentations.

Lactobacillus acidophilus, Lactobacillus bulgaricus, and *Lactobacillus lactis* are also important milk-fermenting bacteria. Unfortunately there is confusion in the literature in regard to differences between *Lactobacillus casei, Lactobacillus acidophilus,* and *Lactobacillus bulgaricus.* There is little doubt that in many of the studies of the past in which *Lactobacillus bulgaricus* was supposedly used that actually the organism was *Lactobacillus casei.* This has inadvertently led to the use of *Lactobacillus bulgaricus* var. *yoghurti* or even *Lactobacillus yoghurtii* to designate the high acid-producing lactic acid-producing lactobacillus of fermented milk preparations. Furthermore, since Metchnikoff first used *Lactobacillus bulgaricus* in preparation of milk drinks for therapeutic purposes, confusion arose as to the differences between this species and *Lactobacillus acidophilus* and also between the latter species and *Lactobacillus casei. Lactobacillus acidophilus* may be isolated from the feces of milk-fed infants. The species strains produce typical rough colonies and curdle milk with a smooth soft curd. The strains can be implanted in the human intestinal tract. *Lactobacillus casei* may be isolated from great numbers of milk preparations. It and its varieties are commonly found in the mouth. They produce smooth colonies, a firm curd in milk, and usually more acid than the acidophilus species. There is evidence to indicate that they differ only in that they may occur as rough or smooth colonies and can be transformed from one to the other (Pederson

1947). *Lactobacillus acidophilus* produces a mixture of dextro and levo lactic acid while *L. casei* produces dextrorotatory lactic acid.

Curran *et al.* (1933) showed that *Lactobacillus acidophilus* produced inactive lactic acid, formed R or rough or fuzzy colonies, failed to grow at 46°C (114.8°F) or as low as 20°C (68°F), and formed more volatile acid and carbon dioxide than other lactics. A large percentage of the strains that reacted thus were of intestinal origin. Kopeloff and Kopeloff had previously demonstrated that the smooth forms of both *L. acidophilus* and *L. bulgaricus* produced dextrorotatory lactic acid. They suggested that some of the inconsistencies in the literature were due to these differences.

FERMENTED MILK PREPARATIONS

Fermented milk preparations originated in various areas of the world. They were the primitive people's way of preserving milk. A milk fermentation is a process by which a change is produced in milk as a result of the activity of one or more species of microorganisms. The pleasing flavor of a properly fermented milk results largely from the growth of microorganisms and the changes they effect in the milk. The types of fermentation that occur vary from area to area and are governed to a considerable degree by the environment.

There is no record to indicate when man first used milk as a food. Prehistoric man undoubtedly captured one of the more docile animals, possibly a cow, sheep, or goat and obtained milk from the animal to supplement the diet of his offspring. The first time the surplus milk was set aside for later use the milk fermented. Since bacteria are ubiquitous, they were destined to inoculate the milk and ferment it. Ordinarily, a milk fermentation can be caused by any one of several species of bacteria and sometimes several growing at one time or in a series. Possibly the first time this occurred, the milk soured, a curd formed, but the product was still considered edible.

Records indicate that the Sumerians used fermented milk. A bas relief from Babylon from about 5000 yr ago depicts one man milking while another man was pouring milk into a small-mouthed container. Doubtless this milk was inoculated naturally by bacteria present in the minute pores and crevices of the container and fermented quickly. The leben that nourished Abraham in his long life was made from goat's milk. Goats and sheep were used extensively in the dry and rocky pastures of ancient Palestine and Syria. The first book of Samuel refers to delivery of fermented milk and cheese to King David. The milk of mares or yak undoubtedly nourished the Tatars of Genghis Khan.

Nearly every civilization has consumed fermented milk of one type or another. The terms buttermilk, acidophilus milk, yoghurt (Fig. 5.3), and leben are familiar to many people, but in general, few people actually appreciate the fact that these milks are fermented by bacteria and that their charac-

FIG. 5.3. YOGHURT

teristic flavor and texture are the result of this fermentation. Buttermilk is associated with the milk left in churning cream to make butter. Yoghurt is associated in origin with the peoples of the Caucasus and Turkey; leben or leben-raid, with the peoples of Egypt, Syria, Palestine, and other areas of the Arabian peninsula; taettemjölk or tettemelk and filbunke, with the Scandinavians; skyr, with the Icelanders; mazun, with the Armenians; cieddu, with the Italians; dahi, with the Indians; kefir, with the Balkans; kumiss, with the Tatars and Mongolians; kissélo mleka, with the Balkans; and busa, with the natives of Turkestan. However, there are many other fermented milks. The Masai of Kenya weaned babies on fermented milk, and the Bantus, Zulus, and other African groups also prepared such milks.

The products have been prepared from milk obtained from cows, goats, sheep, mares, yak, water buffalo, camels, llamas, vicunas, reindeer, and other animals, even including the leopard.

The early methods of preparation of milk products were based entirely upon chance observation, experience, and environmental conditions. The mothers knew nothing about the fact that microscopic plants prevailed about them and that these microscopic plants entered the milk and caused the fermentation. How could they? The microscope wasn't devised until centuries later, and knowledge of microbiology is now barely a century old. Without doubt, some of the mothers observed that when milk soured with a smooth curd and pleasing acid odor, the milk could be consumed without causing distress or illness. When instead, the milk had an unpleasing odor and was roiled or gassy or otherwise exhibited unsatisfactory character, the infant sometimes became ill after consuming such milk. Some mothers may have observed that conditions in certain containers tended to repeat themselves; a smooth, pleasant curd always came from one container while a poor curd developed in another. How could the mother know that the small plants, forever present, grew in the milk and impregnated themselves in the cracks, pores, and crevices of containers to forever inoculate all milk placed in the container thereafter? Some observant individuals may have associated certain environmental conditions, such as cleanliness, with a smooth, desirable milk. In many societies, religious practices were associated with the animals; the milk, the dairy, and the animals involved were worshipped. Worship centering around the dairy is still practiced among several societies.

Knowledge of microorganisms of natural sequences of species in raw milk or cream are acquisitions of the past century. The types of organisms which develop in raw milk will depend upon the introduction of these organisms in milk. Many species can grow. Some natural antagonisms develop, and some symbiotic and sequential growths occur. Sanitary conditions, prevailing temperatures, and other environmental factors will determine largely the types of organisms that will develop. The many species that produce undesirable changes in milk, species of the genera *Escherichia, Aerobacter, Serratia, Salmonella, Brucella, Shigella, Pseudomonas, Proteus, Bacillus,* and many others find milk a good medium for growth and will grow rapidly if introduced in excessive numbers. All produce undesirable changes, and a few may result in serious illnesses. Fortunately, however, few of them can grow in the acid environment produced by the lactic acid bacteria, even though they may survive the acidity. The lactic acid bacterium that is most likely to grow first in milk will be a strain of *Streptococcus lactis* or closely related variants such as *Streptococcus cremoris, Streptococcus liquifaciens,* or the heat-tolerant *Streptococcus thermophilus.* The species *S. lactis* often outraces other bacteria at temperatures of 15°–25°C (59°–77°F) and soon produces sufficient acid to prevent further growth of acid-intolerant species. The growth of this species is inhibited when an acidity of 0.7–1.0% acid calculated as lactic acid is attained. The amount of acid will depend upon the chemical content of the milk. During the growth of this species, the more acid-tolerant strains of the genus *Lactobacillus* will be developing. *Lactobacillus casei* and *Lactobacillus bulgaricus* or related strains will continue the fermentation. *L. casei* may continue growth until an acidity of 1.5–2.0% acid is attained, and *L. bulgaricus,* until 2.5–3.0% or more acid is produced. The highest acid-producing strains are often referred to as yoghurt strains or *Lactobacillus bulgaricus* var. *yoghurtii.* The degree of acidity attained will vary with the strain, the composition of the milk, the temperature of fermentation, and other factors. Milk is often concentrated for preparation of yoghurt; therefore, if concentrated sufficiently, acidities of 4.0% may be attained. This highly acid condition not only inhibits growth of other bacteria but also causes the death of many. Milk in a container in which the high acid-producing strain is growing may have a great predominance of this high acid-producing strain impregnated in its pores or crevices. Fresh milk added to such a container will be so effectively inoculated with this strain that it may quickly gain ascendancy over all other bacteria present. This principle is practiced in many societies, and even today the addition of small amounts of high-acid milk to otherwise clean containers is common. In warm climates, lactic acid bacterial fermentation may often be initiated by *Streptococcus thermophilus* and continued by *Lactobacillus bulgaricus*. In the colder climates, these species may never initiate growth.

Although other species may survive in the acid milks, the only organisms that will grow readily are strains of yeasts and mold. An extremely high acidity may delay their growth. Species of the genus *Geotrichum* may establish themselves on the surface of such milks and produce the characteristic velvety white growth, utilizing the lactic acid for energy and thereby reducing the acidity. If the acidity is reduced sufficiently, bacteria, including proteolytic types, may develop and decompose the proteins with the typical putrid odor, bitter flavor, and opaque solution. If the growth of these types continues, the organic matter may be decomposed to simple inorganic substances, water, carbon dioxide, ammonia, hydrogen sulfide, and minerals, to become part of soil and again reenter the cycle of new life.

Fermented milks depend upon a lactic acid bacterial fermentation or succession of fermentations for their successful manufacture. Any abnormality in these fermentations will affect the quality of the milk and may even spoil it.

Milk is regarded by many physicians, scientists, and public health officials as one of the most dangerous articles of diet in that it may transmit pathogenic bacteria from diseased animals, personnel engaged in milk handling, or even contaminated utensils. Contamination of cows, sheep, or goats in filthy surroundings or of buffalo wallowing in mud or water flats certainly may introduce many more organisms into milk than when animals are in open pasture land or in clean buildings. Nevertheless, it can be stated that fermented milks, regardless of conditions, have been and still are of extreme importance in the nutrition of the peoples of many societies throughout the world. Pasteurization of milk will decrease the possibility of undesirable changes, but only if the milk is properly cooled and kept cold when it is to be held for any length of time. Pasteurization of milk, followed by fermentation, using a pure culture starter, is a safer procedure for preparing milk for consumption wherever pasteurization equipment is available. Otherwise, many societies must still rely upon the proven methods of preservation by fermentation.

Inoculation

Man has known for centuries that milk becomes sour soon after it is obtained from the animal. He learned that when the milk was handled properly, it formed a smooth uniform curd that could be consumed without causing intestinal distress. Proper handling is always associated with cleanliness, and it may be that religious practices of many societies are thus related. Although milk is usually too acid to permit growth of pathogenic bacteria, sour milk has been incriminated in a few outbreaks of illness. From the results of observations of several workers, there seems little doubt that pathogenic bacteria may survive for a limited time in sour milk. Since sour milk may be a potential source of illness, it is advisable to prepare such milks only from high grade milk that has been pasteurized. The coliforms, tubercle bacilli, and other harmful organisms will thus be eliminated, and safe sour milk preparations can be produced.

Furthermore, from a review of the literature it is quite obvious that the various types of milk vary considerably, not only in different areas but even within a specific area. This has come about by the different methods of handling milk in the different societies. Leben is generally understood to be strictly an acid-fermented milk. Leben, for example, according to one study, resembles kefir in that from it, 3 types of bacteria and 2 yeasts were isolated. Kefir, kumiss, and presumably kuban are acid-alcoholic milks fermented by bacteria and yeasts. Strong kumiss containing as much as 2.3% alcohol and 1.1% acid has been described. Busa of Turkestan has been said to be fermented by the yeast *Saccharomyces busae asiaticae* and the bacterium *Bacterium busa asiaticae* in which as much as 0.78% acid and 7.1% alcohol is produced. Mazun, the Armenian sour milk, has been said to be fermented by yeast and bacteria. Two yeasts of the genus *Saccharomyces* and another of the genus *Torula* have been isolated from taette, which is generally considered as a lactic acid fermentation. Similar descriptions of dadhi of India and cieddu, the Sardinian sour milk, have implicated yeasts. Anyone acquainted with the predilection of yeasts for growing in acid preparations can understand that contaminations are very likely to occur in sour milk and similar preparations such as cottage cheese.

For the manufacture of fermented milks, a starter culture containing the appropriate bacteria is added to the previously pasteurized and cooled milk, and the milk is incubated at a suitable temperature until the desired acidity is attained. Lactic acid is the main product of fermentation, but minor amounts of other fermentation products are always produced, particularly the butter flavor substance, biacetyl. In Europe and America, buttermilk, when properly prepared, has always been a favorite sour milk drink.

The simplest fermented milk preparation is the result of fermentation by *Streptococcus lactis.* This organism produces lactic acid and curdles the milk with a soft creamy curd. More acid than that produced by this species may impart a harsh flavor undesirable to some consumers. If a higher acid product and firm curd is desired, one should prepare a high acid yoghurt. *Streptococcus lactis,* when grown with the *Leuconostoc citrovorum,* lowers the pH so that the aroma-producing species *L. citrovorum* may develop the typical characteristic flavor and aroma.

Buttermilk.—Buttermilk, properly speaking, is the milk remaining after the fat globules in cream are agglomerated by churning and then removed for making butter. Cream for churning contains 30% butterfat and therefore will furnish a large quantity of buttermilk. If the cream is sweet, the buttermilk does not differ materially from ordinary skim milk in nutritional value. It will contain about 0.2–0.5% fat in addition to protein, lactose, ash, and all of the other nutrients common to milk. Nearly all cream is soured to some extent by fermentation to aid in separating the cream from the milk and to impart certain desirable flavor characteristics to the butter. Among these flavor constituents are biacetyl and acetylmethylcarbinol, produced primarily by fermentation of citric acid. Although citric acid occurs in milk in small quantities, some authorities recommend adding citric acid to increase the formation of biacetyl. Physically, the casein is coagulated during fermentation, and the fat globules are enabled to coalesce more readily. The coagulated casein is broken up into small particles as a result of the churning action so that they do not separate readily from the serum or whey. The so-called sweet cream buttermilk has a low acidity of 0.5% or less. Sour cream buttermilk has a higher acidity depending upon the extent of fermentation.

The cream used for butter making in homes and in some creameries is allowed to sour spontaneously. The quality of buttermilk as well as the butter will depend to a large extent upon the nature of the fermentation that has occurred. In years past, housewives who made butter in their homes often were aware of the effect of the methods of handling their cream upon quality of both butter and buttermilk, and where great care was exercised some excellent quality products resulted. Conversely, some inferior quality products were produced from carelessly handled cream.

In many creameries, operators are no longer content to take chances and therefore prefer to have milk or cream delivered to the creamery shortly after milking, pasteurizing the cream and then inoculating the cream with pure culture starters that have been carefully selected for their ability to produce good flavor and aroma. The development of these practices will be discussed in the section on butter making. Practices vary but mixtures of pure cultures of *Streptococcus lactis, Streptococcus cremoris, Leuconostoc citrovorum (Streptococcus citrovorus)*, and *Leuconostoc dextranicum (Streptococcus paracitrovorus)* are generally included.

The market for natural buttermilk is not active for several reasons; the main one is the problem of marketing. This has resulted in the acceptance of a cultured buttermilk, a product which can be produced in quantities needed and under uniform conditions.

Cultured Buttermilk.—It is not necessary to make butter in order to obtain a fermented product with the character of buttermilk. Skim milk, when soured, may have about the same chemical composition as buttermilk. Physically it differs, but if it is thoroughly agitated to break up the curd into finely divided particles that simulate the physical properties of buttermilk, it will be difficult to distinguish it from true buttermilk. At the present time, a large part of the so-called buttermilk is not buttermilk, properly speaking, but is a fermented skim milk of this type. The commercially prepared products are produced by inoculation of milk with the selected culture and fermenting it until from 0.5 to 1.0% acid is formed before it is agitated or churned to break the curd into fine particles. A small amount of butterfat is often added to improve the character, and in some cases, true buttermilk is added to the cultured buttermilk. Traces of salt and flavoring substances are also added at times.

Cultured buttermilk can be prepared in the home, but since many home-makers would find it inconvenient to obtain a commercial starter, they must prepare their own starter. Fairly satisfactory starters have been prepared in homes for many years. To develop a starter it is often suggested that good clean milk be obtained from several sources. About a pint of milk from each source is put into separate clean glass bottles and allowed to stand, covered, 12 to 24 hr in a warm place at 20°–24°C (68°–75°F) until the milk sours and curdles. Considerable variation in aroma and flavor may be observed among these milks. The one that yields the most satisfactory clean aroma and taste with a smooth creamy curd should be the one selected as a starter. If none are satisfactory, this process will have to be repeated. The selected culture should be propagated by adding it to freshly pasteurized and cooled milk. Part of this starter should be retained in the refrigerator for future use. The rest may be used for preparation of the cultured buttermilk. A portion of each buttermilk preparation should be retained to be used as a starter for the next lot of milk. If the flavor is unsatisfactory, it may be necessary to use the original starter or start all over again.

If a more acid product is desired, the milk must be incubated at a higher temperature or for a longer period of time. The species *Lactobacillus casei* is liable to grow under such conditions. To obtain a highly acid product of the nature of yoghurt, the milk will have to be incubated at the high temperature required for growth of *Streptococcus thermophilus* and *Lactobacillus bulgaricus,* that is 40°–45°C (104°–113°F).

Once a good starter is obtained it may be used time and time again by carrying over part of the milk from the previous fermentation. Frequently it will be necessary, however, to revert to the original pure culture starter.

In the preparation of commercial buttermilk where large quantities of milk are to be prepared, it is necessary to build up a large volume of inoculum by preparing a succession of propagations, each about 50 times larger than the preceding one. Since a mixture of pure cultures of bacteria are built up, it is essential that the blend of cultures remains the same; therefore, a good quality skim milk must be used and practically sterilized. Heating the milk for a ½ hr at 180°–190°F is general practice. The milk is cooled to 70°F and inoculated with 2% of the culture and incubated at 70°F. When an acidity of 0.8–0.9% is attained, this milk is inoculated into a larger batch of milk which has been similarly heated and cooled. This may be continued until a large volume of culture is produced. The starter should not be allowed to continue fermentation, since some of the organisms may lose their viability. The final milk for buttermilk should also be heated to 180°–190°F for not longer than 30 min to yield a product with a firm body without imparting a cooked flavor and discoloration of the milk.

In Scandinavian countries, taette, filbunke, and långmjölk were produced in the homes. Ordinarily a culture was carried from day to day, using some of the soured milk from one day's product as a starter for the next lot of milk. A variety, *Streptococcus lactis* var. *hollandicus,* produces a markedly viscous to stringy sour milk and is sometimes used in preparing taette and långmjölk.

Cultured Sour Cream.—Sour cream is used in baking and as a dressing for many foods. A particularly appetizing combination is whipped sour cream blended with dehydrated vegetable soup mix. Sour cream has an aroma somewhat like buttermilk and a rich creamy flavor. Heavy cream is a thick viscous product. It may be fermented in a manner similar to buttermilk, but if agitated too much, the fat globules may have a tendency to agglomerate. The thick viscous creamy product that forms around the chime of a plunger-type churn during churning makes a delicious spread for bread at this stage.

A heavy-bodied, smooth cream may be produced by pasteurizing a light cream at 165°–180°F for 45 sec and homogenizing, followed by fermenting with a commercial starter as described above. The cream has a pleasant, mildly acid flavor and a smooth, custard-like body as a result of the fermentation. It is marketed as sour cream, cultured sour cream, or simply cultured cream and is used in salads, fruits, vegetables, soups, meats, and other foods.

OTHER FERMENTED MILK PREPARATIONS

There are a great number of other fermented milk preparations available, and all have food value equivalent to the milk from which they are produced. Acidophilus milk, Bulgarian buttermilk, yoghurt, kefir, kumiss, and others are available.

Yoghurt, also spelled yoghourt, yogurt, yogourt, yohourti, yourt, jugurt, joughurt, and possibly other ways is a smooth, pleasant, firm, tenacious, custard-like milk with high acidity and often a nut-like flavor. It has been prepared for centuries in various areas of southeast Europe, south Asia, Egypt, and most of the countries around the Mediterranian Sea. The leben, lebeny, or leben-raid of Egypt, Syria, and other countries of the Arabian peninsula, the mazun or matzoon of Armenia, jugurt of Turkey, gioddu of Sardinia, kisselo mléko of the Balkans, mast of Iran, dadhi, dahi, or dahee of India, oxygala of Greece, and others as they are often prepared are also high acid, nonalcoholic, solid curd preparations. It must be realized that there is considerable variation among different preparations of these milks, and they are readily contaminated with yeast, so some preparations may be alcoholic. The important fact is that they are high acid-content milks because they are produced in relatively warm countries where growth of the bacterial species *Streptococcus thermophilus* and *Lactobacillus bulgaricus* is favored. The degree of acidity, the flavor, and the firmness of the curd will depend upon the individual strains of bacteria, the species of animal from which the milk is obtained, the method of preparation, and other factors.

In the cooler mountainous areas of Europe and Asia, the prevailing temperatures may be somewhat similar to those found in the northern temperate climate. Under these conditions, a fermentation by the mesophilic lactics, *Streptococcus lactis* and its variants, *Leuconostoc citrovorum,* and *Lactobacillus casei* occurs. These organisms do not produce a high acidity nor solid, firm curds. *Lactobacillus bulgaricus* may grow slowly, and if the fermentation is allowed to continue for a long time, a high acidity may be attained eventually. The mesophilic organisms mentioned will grow well at 20°–30°C (68°–86°F) and some growth will occur at 10°C (50°F). *Streptococcus thermophilus* and *Lactobacillus bulgaricus* have optimum ranges for growth at about 37°C (98.6°F) and 45°C (113°F) respectively.

The sour milks of Scandinavia are likely to be of the taette, filbunke, långmjölk, or skyr type, unless they are held at high temperature, such as is sometimes done in preparing the skyr of Iceland. For centuries past, the housewife relied upon natural fermentations, existent temperature conditions, inoculation from sour milk containers, or addition of milk from previously fermented milks. Since this remains true in many areas of the world today, the types of milk prevalent are a product of environment regardless of the name applied. Pasteurization of milk, use of pure culture starters, and incubation under controlled conditions have standardized the commercial preparations available today.

Acidophilus Milk

Much of the present-day interest in fermented milks is traceable to the writings of Metchnikoff, a Russian biologist who lived from 1845 to 1911. He had been exposed to the views of Pasteur and was stimulated to an intensive study of bacteria present in the intestinal tract of man. He reasoned that since pathogenic bacteria did not grow well in an acid medium, the creation of an acid condition in the intestines by introduction of lactic acid bacteria would preclude the development of harmful bacteria. He contended that senility resulted from the toxins produced by putrefactive organisms in the intestines. He observed that many Balkan people retained their vigor and lived unusually long lives, and he contended that this was due to the fact that they consumed large quantities of sour milk. He tried to prove that an acid reaction could be established in the intestines by the feeding of a sour milk culture. He selected the organism believed to be *Lactobacillus bulgaricus* to produce the sour milk; *Lactobacillus bulgaricus,* however, cannot be implanted in the intestines. His theories have not been substantiated; his concepts on prolongation of life have never been proved or disproved to the satisfaction of everyone; but his studies stimulated many scientists to study implantation of lactic acid bacteria in the intestinal tract.

Lactobacillus acidophilus, another milk-souring organism isolated from the feces of infants, was described by Moro in 1900 as a normal inhabitant of the intestines. Later work by Rettger and his associates at Yale University demonstrated that this organism was a normal intestinal inhabitant, could be implanted in the intestinal tract, and could alter the flora. The addition of lactose to the diet and daily ingestion of the culture in whey were effective. Whether or not Metchnikoff and Cohendy before him may have had cultures of *Lactobacillus acidophilus* rather than *Lactobacillus bulgaricus* will never be known. The species have been confused with each other for years. Kopeloff contended that acidophilus therapy was a bacterial phenomena and that constipation could be relieved. It has been prescribed by members of the medical profession, and many individuals contend that it has been helpful in controlling their intestinal disorders. The organism can be found in large numbers in the intestinal contents of human infants. It is supplanted by other organisms when the child's diet is varied. It can be established in the intestines of an adult, and those afflicted with alimentary disorders such as constipation and diarrhea are benefited.

Lactobacillus acidophilus does not grow rapidly in milk. To prepare acidophilus milk, an active culture must be obtained from a reliable source, and it must be one that is known to be implantable. This culture is used to inoculate milk previously sterilized in an autoclave or pressure cooker for 20 min at 15 lb pressure and cooled to 37°C (98.6°F). Five percent of tomato juice is sometimes added to milk before sterilization. If a large quantity of milk is to be prepared, the volume of culture must be built up by daily transfers of about 2% of active culture into larger and larger quantities of starter, as previously described for cultured buttermilk. Some of the mother culture should always be retained by storing at about 5°C (59°F) after an acidity of about 0.6% is attained. The milk for consumption should be sterilized and cooled as described above, inoculated with 2% of starter, and incubated at 37°C (98.6°F) until a smooth curd with a creamy consistency and a characteristic acid flavor and aroma is produced. The flavor and aroma are less distinct and pleasant than that of cultured buttermilk fermented by the volatile acid, aroma-producing bateria.

Yoghurt

Yoghurt is one of the oldest fermented milks known. Its use has been increasing rapidly in Europe and America since the advent of use of pure culture starters and standardization of conditions of preparation.

The bacterial starter used in preparation of yoghurt in the Mediterranean countries is a portion of the previously fermented product. This is added to the fresh milk in one way or another. These cultures were the subject of intensive studies by early investigators who found that they contained a variety of bacteria. More emphasis was placed upon the long rod forms of lactic acid bacteria (Fig. 2.7, chap. 2) because they produced more lactic acid, and it was thought they would be more effective in counteracting the putrefactive bacteria in the intestinal tract. Spherical bacteria, the streptococci, were also observed. Various names were applied to the rod-like organisms, but eventually *Lactobacillus bulgaricus* was the name generally applied. Strains of this organism have been used in the starter cultures; however, strains of the mildly acid-producing *Streptococcus thermophilus* have been used in conjunction with the *Lactobacillus bulgaricus* and its variety, *Lactobacillus bulgaricus* var. *yoghurtii*. Some taxonomists prefer to call the species *Lactobacilli yoghurtii* since the name *Lactobacillus bulgaricus* has been used loosely in the past, and many of the strains studied were actually *Lactobacillus casei* strains.

The yoghurt made in the United States for many years was a soft-curd product quite different from the custard-like yoghurt prepared in the Middle East. The organism, supposedly *Lactobacillus bulgaricus,* was used in conjunction with *Streptococcus thermophilus* and fermented at relatively lower temperatures than those prevalent in the Middle East. They resembled the soft-curd product so commonly used in the northern areas of Europe. Methods of preparation vary considerably, but the basic process, fermentation by high acid-producing lactic acid bacteria, is the same. Traditionally, yoghurt was made from milk that had been boiled to evaporate part of the water, thereby concentrating the solids. In some preparations, the volume was reduced as much as ½, and because of the high buffer content, a very high acidity with a very firm curd could be produced. Such milks were sometimes referred to as Bulgarian yoghurt. Today, good quality milk is sterilized and reduced in volume by removing ¼ or more of the water in a vacuum pan and adding as much as 5% milk solids or equivalent quantities of condensed milk. Whole or skim milk or blends of these are used. Of course, milk fat improves the flavor. The concentrated milk is heated at temperatures of from 82°C (179.6°F) to 93°C (199.4°F) for from 30 to 60 min or more and cooled to about 45°C (113°F). Some preparations are homogenized before inoculation and then inoculated with 2% of starter and incubated at 45°C (113°F). The starters are built up in the manner previously described for buttermilk starters. Fermentation is rapid if the starter is active, and a satisfactory acidity may be attained in 3–4 hr. Although an acidity as high as 4% acid may be attained, producers ordinarily cool the milk to about 5°C (41°F) whereupon an acidity of about 0.7% acid or more is attained. A good yoghurt will have a smooth, heavy body and a pleasant acid, almost nut-like flavor. It is commonly eaten with a spoon with other foods and usually flavored with sugar, cinnamon, or other spice, or flavored with any one of a number of flavoring substances.

The milk preparation, skyr, produced in Iceland and now in Denmark is fermented at high temperatures, corresponding to those used in preparation of yoghurt.

Kefir

Two types of milk preparations fermented by a mixture of lactic acid bacteria and yeasts have had a wide acceptance in Eastern Europe and Western Asia, kefir and kumiss. Kefir is made by many tribes under a variety of names such as kephir, kiaphur, kaphir, kefyr, képhir, kéfer, khapon, kepi, and kippe. They all have a similar meaning, i.e., pleasant or agreeable taste. Kefir is characterized by the kefir grains which are small, irregularly shaped, yellowish masses varying in size from a popcorn seed to the popped corn, but more irregular. The grains have been compared in appearance with cauliflower. Actually, the grains consist of irregular masses of bacteria, yeast, and curdled milk. Kefir is used extensively among the peoples of Southern Russia, Turkey, and the Balkans. There is no record or tradition of the origin of these milks, but it is probable that these preparations developed by accident. Mohammedans have declared that the first grains of kephir were put there by Allah. For centuries the people have stored their milk, whether from cow, goat, or sheep, in leather bottles or bags made from the skins of goats (Fig. 5.4). These were hung up in the house during the winter and outside during the summer. Fresh milk was added to the bags when some of the fermented milk was removed, providing a continuous fermentation. Furthermore, there was a continuous addition of microorganisms contained in the

Courtesy of Dr. R. S. Breed

FIG. 5.4. TYPE OF BAG USED FOR CARRYING MILK

fresh milk. The foreign bacteria mingle with the microorganisms present; therefore, it is difficult to designate any single strain or strains as responsible for the fermentations.

The grains themselves are irregular masses of bacteria, yeast, and coagulated protein built up in layers one upon another. When the grains are added to fresh milk they settle to the bottom, and as fermentation proceeds they rise to the surface, carried up with the gas formed. If the fermentation is rapid enough, a layer of the grains in foam will form on the milk; they will settle if shaken or stirred to release the carbon dioxide.

The primary organisms in the grains are lactic acid bacteria and yeasts, but other organisms may be present. It is probable that kefir is produced under different circumstances and that different organisms may be involved. The original description of the biology of kefir by Kern has been a source of confusion in the taxonomy of the lactic acid bacteria. Because of inadequate knowledge of bacteriology in 1882, his descriptions were erroneous in that he called the long slender bacillus, *Dispora caucasica,* because he saw what he believed to be two spores in each of many cells. Whether they were the granules which have been observed frequently since then in lactic acid bacteria, or whether he actually saw spore-forming bacteria will never be known. Seven years later, Beijerinck renamed an organism from kefir grains, *Bacillus caucasicus,* believing it was identical to the high acid-producing bacterium that later was always associated with production of yoghurt. Later, von Freudenreich applied the name *Bacillus caucasicus* to a different lactic acid-producing bacterium. It produced gas as well as lactic acid from lactose. This organism was later called *Betabacterium caucasicum* by Orla-Jensen because it is truly a heterofermenter. It is probable that under different circumstances kefir is produced by different organisms. Many combinations of bacteria and yeast that may form the grains and will produce a lactic acid and alcoholic fermentation may make kefir, but its flavor may be governed by the particular combination of strains. The writer has isolated from kefir grains long rod bacteria resembling *Lactobacillus bulgaricus,* shorter rods resembling *Lactobacillus casei,* heterofermentative lactic acid bacteria resembling *Lactobacillus brevis,* and yeasts. Von Freudenreich applied the name *Saccharomyces kefir* to the yeast, while others have isolated torula yeasts. Kefir may contain about 1% acid and 1% alcohol and may have sufficient carbon dioxide to give it an exhilarating effervescence.

This is not an easy milk to prepare. Kefir grains are not easily obtained in this country. The routine method of preparing sour milk does not promote the growth of grains. Production of this milk drink in goatskin bags by continual addition of fresh milk when fermented milk is removed will promote the formation of grains but is a method which would not be practiced elsewhere than in the Caucasian mountain area where it originated. Method of preparation in the United States would involve preparing the product from pasteurized milk and then collecting the grains to be added to the next batch of kefir. If an alcoholic acid milk is desired, this can be obtained by inducing an alcoholic fermentation in buttermilk by addition of an ordinary bread yeast to buttermilk after addition of sucrose.

Kumiss

Many Asiatic peoples of the vast plains of Western Asia are nomads who travel about seeking pasture for their horses and cattle. They have developed a hardy race of horses whose mares produce an extraordinarily large quantity of milk. This milk constitutes a very important part of the diet of these inhabitants. This is never used in the fresh condition, but is fermented to make the drink, kumiss, koumiss, kumys, or kumyss.

Kumiss might be called a milk-wine or beer. The name may have been derived from the ancient Asiatic tribes, the Kumanes, Kumyks, or Komans. It is generally made from mare's milk, but other milks may be used. The Mongolian arrag is similar and is made in goatskin bags. Mares, goats, sheep, cows, and sometimes even camels and yaks are milked. Various descriptions have been given to the drinks, some comparing it to a white wine or champagne, others to buttermilk, to beer, and to ginger ale. Obviously, there are several degrees of fermentation of these alcoholic acid milk drinks. There is little doubt that the fermentation is a symbiotic one producing lactic acid by bacteria and alcohol by yeasts. Because of the relatively low acidity, 0.7–1.3%—and rarely 2%—it may be concluded that the lactic fermentation is brought about by strains of *Streptococcus lactis* and *Lactobacillus casei*, rather than the high acid-producing *Lactobacillus bulgaricus* types. However, high acid preparations are described and since conditions vary so much, it is doubtful whether kumiss or arrag can be said to be produced by any specific organism or organisms. Like kefir, kumiss may retain its carbon dioxide and thus have an exhilarating effect.

USE OF FERMENTED MILK AND CREAM

The high food value of milks is too generally recognized to require discussion. Food values of the fermented products are comparable to that of the milk from which they are prepared. Milks from different animal species and breeds of a species differ considerably in nutritive value. Fermented milks differ from the original primarily in that some of the lactose is converted to lactic acid, citric acid is converted in part to diacetyl, and other minor changes occur, none of which decrease materially the original nutritive value. The change in digestibility of fermented milk is ascribed to the precipitation of casein dispersed into finely divided particles. This change occurs with fresh milk as soon as it enters the stomach. The appetizing appeal of good sour milks has been observed even by primitive societies. These societies recognized their refreshing and wholesome qualities.

It has been stated that wherever one finds a society in which wealth is measured in terms of animals one will find a society of relatively hardy and healthy people. The animals may be cows as in Europe, sheep or goats of the dry and mountainous areas of the Middle East, mares of the steppes of Asia, reindeer of the far north, yaks of the Himalayas, or the cattle of the Kenya highlands. Although milk is primarily considered a beverage, the many ways in which the various fermented milks and cream preparations are used is a credit to the ingenuity of the homemaker. There is apparently no limit to the number of ways that sour milks and creams are reported to have been used in various areas of the world. Puddings, porridges, soups, and blends with vegetables and meats are prepared in various countries with sour milk or cream. In a recipe book from the Ukraine, 17 of the 62 recipes call for the addition of sour milk or cream. In the Near East and in India, meats are marinated with sour milk. The American and European housewives use sour milk or cream in baking many kinds of cakes and cookies. The smoothly-curdled sour milk or cream is eaten in many countries with a liberal sprinkling of sugar and cinnamon or ginger. Yoghurt is now sold with 25 different flavors in France. The filbunke and taette of Scandinavia are often eaten with ginger snaps or the hard, dry knäckebröd. A preparation of taette has been described in which herbs are added to the fresh milk before fermentation. The Swedish people smear their serving bowls with filbunke left from the last preparation before adding fresh milk. A viscous, stringy milk is a favorite form. Leben is served with sugar, syrup, cinnamon, pepper, paprika, and various fruits such as bananas, berries, or tomatoes. The English serve curdled milk such as Devonshire cheese with preserved plums, strawberries, or other fruit. The Danish buttermilk soup is flavored with sugar and vanilla. The Bulgarian vegetable soup, tarator, is made with cucumbers, yoghurt, and other foods. The Czech and the Poles also prepare vegetable soup with sour milk or cream. The rømmegrot of Norway is a sour cream porridge. The tarhana of Turkey is a parboiled whole wheat with yoghurt that is allowed to continue fermentation for 4 or 5 days and then served with tomatoes, sweet pepper, onion, garlic, salt, and spice. The Yemenites prepare zhum, a clabbered milk, with flour, garlic, salt, and pepper. In Russia, sour milk or cream is served with blini and many other foods. The black breads of Mongolia and Western Asia are eaten with kumiss. The Ukrainian varenyky is served with fruits and sour cream. Rice is commonly served with sour milk in Iraq, and a lamb-rice dish is served with a sauce of sour milk in Azerbayan. The raitas of India are vegetable-sour milk preparations, one of which is prepared with leavened bread. The curds are served with potatoes, peanuts, coconuts, bananas, lentils, chilies, cucumbers in India, and among those who eat meats, the meats are marinated with curd. The Serbians prepare lamb with yoghurt. The various sour milks or creams are used as salad dressings in many countries; the steamed spinach with leben used in Afghanistan seems to be an odd combination.

In Mongolian tubercular sanitariums the kumiss or arrag is part of the diet and considered curative. In Bhutan, leopard's milk is considered a cure for rabies, and Persian women are reported to use sour milk for cosmetic purposes. Ancient physicians apparently prescribed sour milk preparations for many intestinal disorders long before Metchnikoff carried on his studies.

During the period since 1940, there has been a steady increase in consumption of fermented milk preparations, particularly buttermilk and yoghurt. The wholesome and refreshing characteristics are unquestionable. The therapeutic claims are based largely on the fact that the lactic acid bacteria are helpful in overcoming intestinal disorders and also that these bacteria transform part of the proteins in milk into more digestible forms. Whether or not all claims for yoghurt and similar milks are justified, fermented milks in general have a food value that compares favorably with that of the milk from which it was made. It contains all of the nutrients from the ordinary milk and in a more digestible form. The casein and albumin are, in part, transformed but there is little or no cleavage of casein such as occurs in digestion in the intestinal tract.

All of the fermented milks should be the result of an acid fermentation in which the sugar in the milk is converted to lactic acid and to other products. Following the works of Metchnikoff, the so-called bulgarian milk was produced and sold in large quantities. When it was demonstrated that *Lactobacillus bulgaricus* was not implanted in the intestinal tract, but that greater success was achieved with *Lactobacillus acidophilus,* acidophilus milk was generally adopted. In fact, attempts were made to implant the organisms by various treatments, even in the form of chocolate-coated pills. Lactose feeding was found to be helpful. In recent years, yoghurt, often flavored, has become the common type of fermented milk. In fact, some contend that it is a fad. In France over 200 million gallons are now consumed annually.

BUTTER

Butter is probably older in origin than recorded history. It apparently was made in cool climates by the first people who obtained milk from animals. At an early period it was observed that butter, like sour milk and cheese, could be preserved for a longer time than fresh milks, and each represented an excellent means of storing this valuable food.

On the basis of ancient pictures and writings, it is concluded the Sumerians were probably among the first to make butter. A relief frieze found in the temple in Ur dating back to 3100 B.C. depicts the act of milking and shows two other vessels, one of which may have been a churn and the other a vessel for storing butter. Butter was prepared by the Egyptians as early as 3000 B.C. An alabaster vase containing butter oil was found in a tomb dating back to 2500 B.C. The people of India as long ago as 2000 B.C. were known to use butter as a food and as a holy offering to their gods. Butter made from the milk of the yak is still used in the butter festivals in Tibet. The Hindus churned cream in earthenware vessels by agitation or with a whirling stick. The Arabs churned cream in suspended animal skins either by agitation or swinging back and forth. Centuries later, American immigrants learned to churn butter by utilizing the motion of their wagons on their cross-country treks.

Although the Greeks and Romans did not use butter as such, the neighboring Scythians used it before 400 B.C. Early reports of the use of butter by the Celtic people of Ireland date back to the 5th Century. They packed butter in casks that were buried in peat bogs for ripening. The people of Norway used butter in the 8th Century and exported it in the 13th Century. Holland exported butter shortly thereafter, and this butter became famous all over Europe for its quality. Some of the early butters were flavored with garlic.

All of the early butters were churned on the farm where the milk was produced. Ordinarily, the milk was set in shallow pans in a cool place to permit the cream to rise. Fermentation always occurred during this period, and quality of the butter was influenced greatly by the nature of the fermentation. The fermentation was often hastened by the addition of selected sour buttermilk from a previous churning. The cream was carefully skimmed and transferred to a churn for churning. It is obvious that the quality of the butter varied greatly depending upon the handling of the milk and cream. These methods of manufacture prevailed throughout the world until the beginning of the 20th Century. In the homes of the Pennsylvania Dutch, in which many of the early practices are still used, excellent quality butter is still made essentially by these methods. They take pride in the quality of their butter which is stamped by a mold, and the mold is a family treasure. The pioneer family churns and cheese presses are museum pieces today.

Variations of butters are made by various peoples, and the extremes of these butters are the ghee of India and the beurre noir of France. Ghee is a semifluid butter made by melting the butter and decanting the liquid portion, called ghee. Beurre noir or brown butter is prepared by cooking butter with vinegar until brown. Some references to the early use of butter are actually uses of cheese-like products, instead.

Butter has sometimes been referred to as the balance wheel of the dairy industry. The term is obvious since supplying market milk has always taken precedence, and butter was often a by-product of overproduction of market milk or inability to get milk to the market or to the cheese factory. This has inadvertently led to production of some inferior butters. The necessity of general improvement of quality became obvious years ago. The dairy industry of Denmark and Holland depended to a considerable degree upon export sale of butter, and, it might be added, led the way in advancing techniques and methods of handling milk and cream. At about 1860, creameries in Denmark and Holland began adding buttermilk, selected for its quality, to cream in an attempt to hasten its souring and thus improve the general quality of butter. If the right organisms happened to be present, the buttermilk had the desired effect. On the other hand, if the buttermilk was of poor quality, it was discarded. It was a common practice among these people to obtain buttermilk from a neighbor for a starter when their own supply was poor. These attempts at improvement eventually led to the study of the microbiology of milk and cream, and the general use of starters for butter making.

The flavor of butter is derived not only from the butterfat and small amount of curd, but also from the products of fermentation or ripening that are absorbed by the fat (Fig. 5.5). The ripening of cream refers to the development therein of a certain amount of acid, both volatile and nonvolatile, and of other chemical substances. The chief action is the fermentation by lactic acid bacteria. As a result of this fermentation, the fat globules are more or less liberated from the casein, fat globules agglomerate more readily, and there is less loss of fat in the buttermilk than when cream is churned in an unripened condition.

The bacteriology of butter involves considerations that differ from some of those important in other products of milk. Starters or lactic acid cultures are now widely employed in butter making. They are characterized by rapid production of lactic acid and by flavor development when cultures grow in cream under favorable conditions. As previously stated, the first attempts to control flavor development in cream consisted of the addition of selected clean-flavored buttermilk or sour milk.

The history of the development of starter cultures, while first applied to butter, is important not only in the improvement of butter, cultured buttermilk, and cream, but also in the development of a concept of influence of the right microorganisms in other fermented foods. Conn in 1889 recognized that the process of ripening of cream was a complex one in which something besides souring of the cream took place. In a study of 20 bacterial strains of species, in no case did he find that a single strain produced the typical ripening. In 1896 Conn pointed out that acidity and flavor were not identical in origin and that, although the acid developed by a fermentation of lactose, flavor came from some other source. During this period, Storch in Denmark recognized that among 7 bacterial strains isolated from ripened cream obtained at a creamery known for the high flavor of its butter, only 1 produced an aroma as full as ripened cream possessed. Storch supplied the cultures first used to inoculate the pasteurized cream intended for butter making. The cultures did not yield a butter as fine-flavored as that produced by inoculating cream with buttermilk from a churning of good butter, but his work had an effect in standardizing quality.

Commercial cultures for cream ripening were made available in Europe

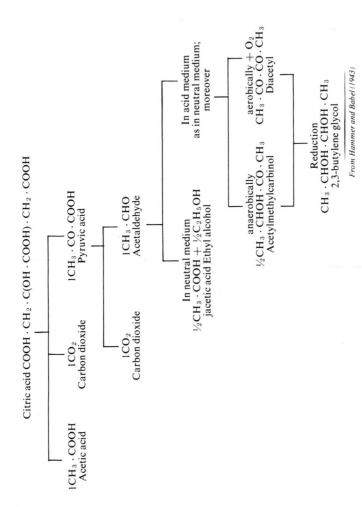

FIG. 5.5. DR. HAMMER AND DR. BABEL ILLUSTRATE THE PRODUCTION OF FLAVOR IN BUTTER FROM CITRIC ACID BY THIS SERIES OF CHEMICAL CHANGES

From Hammer and Babel (1943)

in 1890, and their use increased so rapidly that by 1897, 802 out of 866 butter plants in Denmark were using such cultures. Superiority of Danish butter at that time was attributed to the use of cultures. The influence of pasteurization was recognized also. It was recognized, furthermore, that different pure cultures produced either a clean acid flavor and good keeping quality or a good aroma. Weigmann earlier suggested mixing the two types of bacteria. Conn in the United States pointed out that commercial cultures were not pure cultures, they were not all alike, and some were mixtures of strains with 1 containing no less than 10 strains which could be distinguished from each other.

In 1919 Hammer and Bailey pointed out that lactic acid cultures produced considerable volatile acid in milk, whereas *Streptococcus lactis* did not. They isolated from such lactic cultures a strain which resembled *Streptococcus lactis* morphologically, but it also yielded considerable quantities of volatile acid. In the same year Storch recognized the presence in starter cultures of an organism which he called X bacteria, and Boekhout and Ott de Vries described an aroma-producing bacterium. Since then, various investigators have found that proper development in lactic acid cultures requires the action of two distinct types of bacteria. The following year Hammer reported that citric acid is the primary source of the volatile acid formed in starter cultures, an observation confirmed by several studies that followed. The specific names, *Streptococcus citrovorus* and *Streptococcus paracitrovorus*, were applied to the two species of volatile acid, aroma-producing bacteria. The difference between the two is that *S. citrovorus* does not produce lactic acid in milk whereas *S. paracitrovorus* produces a variable amount of levo lactic acid. Knudsen and Sorensen recognized that these organisms were similar to species of the genus earlier designated by Orla-Jensen as the genus *Betacoccus.* Hucker and Pederson (1930) observed that they produced levorotatory lactic acid, acetic acid, carbon dioxide and alcohol, and were similar to the dextran-producing organism from sugar factories and from fermenting vegetables, and, therefore, should be included in the genus *Leuconostoc* with the specific names *Leuconostoc citrovorum* and *Leuconostoc dextranicum.*

In 1929, Van Niel, Schmalfuss, and Barthmeyer showed that diacetyl is an aroma constituent of butter and demonstrated its formation by oxidation of acetylmethylcarbinol. Acetylmethylcarbinol is an odorless liquid that can be oxidized readily to diacetyl (Fig. 5.5), a volatile yellow liquid with a pungent odor, which, when diluted, suggests the odor of butter. It is generally recognized today that butter containing relatively large amounts of these chemical substances is superior in flavor to those butters in which these substances are absent.

As previously stated, the butter obtained from cream fermented by *Streptococcus lactis* has a uniformly good flavor but lacked the typical aroma of butter produced by species of the genus *Leuconostoc*. The importance of acetylmethylcarbinol and diacetyl in lactic acid cultures was stressed by Michaelian and co-workers at Iowa State. They observed that cultures that developed a satisfactory aroma and flavor always produced considerable quantities of these substances, whereas cultures lacking in flavor produced only small amounts of these chemicals. The bacteria ordinarily associated with souring of milk are typical streptococci. *Streptococcus lactis* converts about 90% of the lactose fermented to dextrorotatory lactic acid. The species of *Leuconostoc* do not grow readily in milk but are favored in growth when some acid is produced by the streptococci. Species of the genus *Leuconostoc* convert slightly less than 50% of the lactose or glucose fermented to levorotatory lactic acid, about 20–25% to carbon dioxide, and 20–25% to acetic acid and ethyl alcohol. Small amounts of other chemical substances are produced. These include the flavor substances mentioned above; however, since they normally convert citric acid to acetylmethylcarbinol and diacetyl, citric acid is sometimes added to milk or cream to insure production of sufficient aroma substance to impart a good butter flavor. These aromatic substances accumulate in significant amounts when the pH is low. Many strains of homofermentative lactic acid bacteria including *Streptococcus lactis* strains will also produce small amounts of these aromatic substances under certain conditions. Hammer has stressed production of volatile acids, primarily acetic acid, although propionic and butyric acids were included. The proportions of sugar converted to ethyl alcohol and acetic acid by the leuconostoc is variable, depending upon the oxidation reduction potential. It is somewhat doubtful whether esters of ethyl alcohol and the acids remain in butter, but they are present in other foods fermented by the leuconostoc. Butter flavor is complex and although acetylmethylcarbinol and diacetyl are the major contributors to flavor, other substances such as lactones are involved. The changes in protein and lipid constituents occurring during fermentation may influence flavor. Some proteolysis is evident in lactic acid cultures. Considerable research has been carried out in regard to handling cultures, the details of which should be obtained from Hammer (1948), from original papers referred to by Hammer, or to the summary presented by Foster *et al.* (1957).

The chemistry and microbiology of butter are most complex. The cream from which it is made is subject to growth of numerous types of organisms, and the flavors these organisms produce are readily absorbed by the butter produced from the cream. The methods of handling cream through the centuries has not been the most conducive to production of the best quality of butter. Much study has been given to methods of isolating organisms responsible for off-flavors and determining the changes that occur. Numerous species of bacteria, yeasts, and molds have been implicated in undesirable flavors. Cheesy, putrid, rancid, nutty, skunk-like, barn-like, yeasty musty, and sour are a few of the descriptions applied to defective butters. Organisms involved in producing these changes include species of the genera *Pseudomonas, Flavobacterium, Streptococcus, Aerobacter, Torula, Monilia,* and several genera of molds. Species of *Pseudomonas* such as *putrefaciens, fluorescens, fragi, mephiticus,* and others are responsible for flavor defects and another, *P. nigrificans,* for a black color defect. These are essentially aerobic low temperature-tolerant species of microorganisms that have been allowed to develop on the surface of cream stored for various periods of time before churning. Many of the defects caused by such organisms have been reduced to a minimum by the newer methods used for making butter.

Methods vary somewhat but generally milk is cooled quickly after milking, delivered to the creamery where the cream is separated, pasteurized, and cooled. The starter, a mixture of streptococci and leuconostoc, is added and fermentation is allowed to continue until the desired acidity is attained. The cream is churned and the resulting butter is immediately washed, salted, packaged, and stored. It may be added that quality is then quite constant.

CHEESE

In general cheese consists of milk curd or coagulum formed by the souring produced by lactic acid bacteria or by souring in conjunction with rennet followed by separation of part of the whey portion from the curd. Sour milk (Fig. 5.6), before separation of whey and sour milk cheeses such as cottage cheese, probably represents the earliest form of preparation of foods from milk. Cheese making began as an art rather than an adaptation of scientific principles and remained so until the advent of the knowledge of microbiology and chemistry during the past century. Long before science had explained some of the complex microbiology and chemistry of milk and the nature of fermentation, methods of making and handling cheese had developed more or less independently in widely separated regions of the world. The practices

FIG. 5.6.   CURD FRESHLY PRECIPITATED FROM CARABAO MILK FILLED INTO
BANANA LEAF CUPS

The fresh milk is inoculated with a rennin preparation prepared by placing pieces of calves' stomachs in a whey obtained from the previous day's cheese making.

adopted were associated with local conditions of climate, agriculture, and habits of people. This resulted in characteristic differences in the cheeses made in different regions, and also gave rise to a great number of names. It became a custom of the early cheese-making people to name their product after the village or country in which it was made. Thus various cheeses were made and sold under more than 700 names. Many of these so-called varieties are unknown outside of the locality in which they are made.

Possibly no other food product prepared from a single type of basic food shows so much variation in kinds of finished products as the varieties of cheese (Fig. 5.7). Environmental factors are responsible to a large extent for these differences. Temperature, moisture, sanitary practices, handling of curds, conditions of storage, and other factors have influenced the nature of the products obtained. Customs and usage have, to a large extent, dictated the type of product which people prefer. The first cheeses may have been made by removing by hand the curd formed when milk fermented (Fig. 5.2). Some of this may have been eaten as the fresh curd, like present-day cottage cheese, and the excess patted and squeezed by hand and placed out in the sun, possibly on a tent flap to dry. The rays of the sun would sterilize the surface, and the cheeses would ripen as they dried to yield a dry cheese of the general cheddar type. Some of the cheeses may have been stored inside where surface fermentations occurred. Others may have been stored as moist cheese in moist areas where mold developed to flavor the product. If others were subjected to cool or to warm conditions, either of these

FIG. 5.7. SOME REPRESENTATIVE TYPES OF CHEESES: (READING CLOCKWISE FROM TOP LEFT) A STRONG WELL-AGED CHEDDAR, A MILD CHEDDAR, A MOLD-CURED ROQUEFORT TYPE, A SWISS TYPE, A HEATED CURD UNCURED MOZZARELLA, A HEATED CURD CURED PROVOLONE, A GOUDA, COTTAGE CHEESE, AND WHEY CHEESE

may have been a factor in microbial differences. Some of the cheese produced in various areas is still made by practices of centuries ago. Present-day cheeses, however, are comparatively recent innovations based upon experience of the past but improved upon by modern scientific knowledge. Today there are more than 700 names applied to different cheeses (Sanders 1953). They almost defy classification because of the great number of variables.

The cheeses will be discussed under 11 headings. The four fermented but uncured table cheeses prepared to consume shortly after making include the soft curd cheeses such as cottage cheese, the soft curd plastic cheese such as Mozzarella, the albuminoid whey cheese such as Ricotta, and the concentrated whey cheese such as Mysost. Methods of preparation of cheeses of the first two types are similar to those used for the cured cheeses, and in many instances the uncured and cured cheeses have the same name.

Cured cheeses to be discussed include the very hard grating cheese such as Asiago, the hard cheese such as Cheddar, the hard plastic cheese such as Provolone, the eyed propionic acid cheese such as Swiss, the softer blue mold-fermented cheese such as Roquefort, the soft mold cheese such as Camembert, and the semihard surface cured cheese such as Limburger. Since some cheeses such as Queso Blanco are not fermented but instead depend upon addition of acid to precipitate the curd, they will not be discussed here.

The large majority of cheeses fall into this general grouping even though there are a number of variables which influence the manufacture and the final character of the cheese. Cheese is made from the milk of many different animals, cow, sheep, goat, mare, llama, buffalo, camel, reindeer, and others. Names of some cheeses are identified with the particular animal species, such as Pecorino, Caprino, and Vacchino, sometimes used by Italians to indicate milk from ewe, goat, or cow.

Rennet or rennin, the milk-coagulating enzyme found in the gastric juice of calves or the young of many other animal species, may or may not be added. Although rennet may come from other sources, such as some microorganisms and some plants, these sources are seldom used. When rennet is added, the milk will curdle at a higher pH and lower acidity and will differ somewhat chemically from the curd formed at the lower pH or higher acidity. Eventually the rennet-coagulated curds may attain the total acidity of the acid-coagulated curds.

The amount of heat and the holding temperature are variables which affect the character of the cheese. The plasticity of some cheeses is brought about by heating. The early type of milk or cheese was not subjected to heating other than that of the atmosphere. Pasteurization of milk is a recent innovation which has improved and standardized the quality of cheese products. In addition, holding of the finished products at controlled low temperatures has had a similar influence.

Originally, milk fermented naturally by the activity of microorganisms which were present in the milk, containers, or surroundings. If these organisms were lactic acid-producing or other desirable types in fresh milk, it was fortunate; but in many cases, organisms were nonlactic types and some even of a pathogenic nature. The use of milk or whey which was souring with a clean and desirable flavor as a starter culture became a general practice in later years. Since the advent of microbiology as a science, the use of pure culture starters (Fig. 5.8) has had a marked effect in improving and standardizing cheese quality.

From Sanders et al. (1950)

FIG. 5.8. ADDING THE STARTER, BACTERIAL CULTURE, TO THE MILK IN
SWISS CHEESE KETTLE

Cheeses are also classified as very hard, hard, semisoft, and soft. Differences in texture are obtained by differences in heating and handling the curd, draining and/or pressing, character of the fermentation and curing, the amount of cream present, and other factors. Texture preferences vary considerably. The very hard cheeses are preferred for grating, softer cheeses are preferred for cooking, others are preferred for cheese spreads, while still others are preferred for the many other ways in which cheese is used.

The amount of cream left in cheese or added to the cheese has a marked effect upon flavor and texture. The length of time and the temperature at which cheese is aged also exerts a marked effect upon flavor and texture. Some cheeses when eaten before curing are very mild but become quite strong when aged. This is a particularly noticeable difference in various cheddar cheeses. The very hard, grating cheeses are usually aged for long periods to impart the strong flavor desired in such cheeses. In contrast, cottage cheeses are very mild in flavor, and their flavors are altered considerably by addition of flavoring substances, spices, flavor extracts, fruits such as pineapple, vegetables and herbs such as onion, chive, pimento, and others, including even beer. A powdered clover leaf preparation is added to the curd to make the sharp, pungent green but pleasing sapsago cheese. The curds for some cheeses are rolled or dipped in herb, spice, or leaf mixtures before curing, an old practice carried on, for example, in making the Syrian shunkleesh.

Cheeses vary considerably in size. Camembert is a very small cheese usually only 4.5 in. in diameter. The small size gives opportunity for the mold mycelia and the enzymes to penetrate readily throughout the cheese. American or cheddar and Swiss cheeses are usually very large, permitting a more or less anaerobic curing and in the case of Swiss, a retention of gases to form the typical eyes.

Finally, there are a number of processed cheeses and cheese foods which are fermented and cured before processing. These will be given no further consideration.

The Fermentation

The lactic acid bacteria are the most important microorganisms in the conversion of milk to cheese. In the ordinary souring of milk, two phenomena are familiar: the formation of acid and the coagulation or curdling of the milk. Different explanations have been offered to explain the changes that take place when the acid curdles the milk (Fig. 5.9). The first one, offered by Scheele in 1790, expressed the view that the coagulum is a compound produced by union of acid and casein. Later, the view was expressed that coagulation is due to reaction of the lactic acid with the calcium previ-

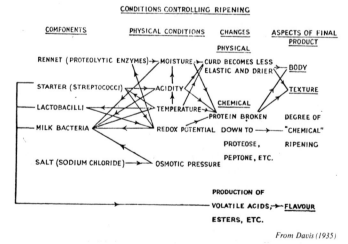

From Davis (1935)

FIG. 5.9. THE CURING OF CHEESE IS A COMPLEX PHENOMENON

Dr. Davis here illustrates the physical, biochemical, and microbiological factors involved by this complex figure.

ously combined with the casein. Van Slyke observed that when milk is first visibly coagulated, part of the casein is in the form of monocalcium caseinate and the rest is uncombined casein. With further increase in acidity the monocalcium caseinate becomes free casein. Lactic acid forms an isoelectric casein curd in milk at pH 4.6. A rennet curd forms a calcium paracasein curd at the higher pH 5.3.

The making of nearly all types of cheese begins with coagulation or curdling of the milk by the acid produced by the lactic acid bacteria with or without the addition of rennet. In the far distant past, before the use of rennet was known, all milk for cheese was curdled by the acid alone. Rennet action was undoubtedly discovered accidentally when someone observed that the milk placed in the bags made from a young animal's stomach curdled more quickly than milk in other containers. It may have been correctly assumed that there was a substance in the young animal's stomach that was responsible for this early coagulation. Rennet has been obtained commonly from the stomach of young animals.

In general, after the curd has been set by the chemical changes produced by the lactic acid bacteria with or without the addition of rennet, it is cut or broken, cooked, and the whey is separated by draining or pressing. The raw curd may be salted, creamed or flavored, and consumed at this stage as uncured cheese, or it may be allowed to cure further by activity of the enzymes of microorganisms present and by natural enzymes of the milk.

The lactic acid bacteria involved in cheese making are species of the genera *Streptococcus, Leuconostoc,* and *Lactobacillus.* Before the days of pasteurization and use of pure culture starters, many other types of microorganisms may have developed in milk. Because of the importance of cheese as a food, attention was given to it by scientists at an earlier date than to many other foods prepared by fermentation processes.

Uncured Cheese

Cottage cheese probably represents the earliest form of cheese making. Cottage cheese is usually a soft, unripened, white cheese made from skim milk. It has always been a popular food in Europe and was introduced into America by the first settlers. It is known as Dutch cheese, country-style cheese, schmierkäse, and pot cheese, and there are any number of variants of this type of cheese.

Manufacture of the modern cottage cheese in industrial plants started in the United States in the early days of this century. Pasteurized skim milk is cooled to 90°F (32.2°C), inoculated with 5% of a starter culture, and rennet is added as the milk is being filled into the fermentation vat. The commercial starter ordinarily used is a blend of *Streptococcus lactis* or *S. cremoris,* the acid formers, and *Leuconostoc citrovorum,* the aroma and acid producers.

In about 5 hr an acidity of 0.5–0.55% acid with a pH of about 4.6 is attained. The curd is cut and then slowly heated to 120°–125°F (48.9°–51.7°C) and cooked for 1–2 hr. The whey is drained, the curd is washed first with a lukewarm water, and then successively with cooler and cold water. Ordinarily, the curds are salted and creamed in a blender. This yields a large curd cheese. If little or no rennet or commercial coagulator is used, a small curd cheese is obtained. The curd then is formed more slowly and a higher acidity must be developed before the curd sets to a desirable consistency. The cheese must be cooled immediately and stored at refrigerator temperatures until consumed. The cheese is uniform in consistency and has a pleasing aroma.

Considerable variation exists among uncured cheeses of this type. It may be marketed as a creamy homogenized product or an extruded product. It is often flavored with chives, onion, pimento, pineapple, or other fruit, and even flavored with other cheeses such as the Roquefort type.

Cottage cheese, that is a fermented uncured cheese, has been made for centuries, long before pasteurization and starters were devised, and even before rennet was first used. Cheeses of this type have been prepared in farm homes for centuries, both from regular milk and from buttermilk. Some of these were fully as pleasing as the commercially prepared product available today; however, many were inferior. Rennet was not used in many of such cheeses. Although the housewife relied upon the natural souring of milk, previously soured milk or good quality buttermilk was sometimes used as a starter.

The effect of rennet from the calf's stomach was observed centuries ago. One such cheese made from buffalo milk was described in which the unpasteurized fresh milk was inoculated with a rennet from the calf's stomach (Fig. 5.6) that was soaked in whey saved from the previous day's cheese making. Understandably, the whey was very acid. Microscopic examination showed numerous cocci and rods which were presumably *Streptococcus lactis* and *Lactobacillus casei.* Shortly after adding the whey, the milk began to set, the soft curd was dipped into banana leaf cups, drainage of whey occurred, and the curd became increasingly sour and firm. During the first 24 hr, the fermentation was obviously the action of lactic acid bacteria, but the odor and flavor of yeasts and alcohol became increasingly pronounced. This is cited because it illustrates the more primitive method of preparing the curd and the rapidity with which products of this nature change. Furthermore, it undoubtedly represents more nearly the conditions under which curds are produced in many areas of the world where facilities for pasteurization and pure culture inoculation are unavailable. Regardless of these facts, such milk products have supplied basic protein foods for millions of people throughout countless centuries. Whether such cheeses are considered spoiled is strictly opinion based upon the consuming experience of the past. Certainly when organisms that cause illness predominate, the product must be considered spoiled.

Soft, unripened cheeses such as cream cheese, carré, Fromage a la Créme, and the Neufchâtel when eaten fresh, are similar to cottage cheese in the method of fermentation but always contain more cream. They may be made with or without rennet. The smooth texture may be obtained by homogenizing.

The soft cheeses produced in Italy such as Mozzarella and Scamorze and some types of Ricotta are often consumed before curing. These cheeses are most suitable for cooking purposes. Mozzarella, once made from buffalo milk, is produced with rennet and a carefully controlled acidity development. After the whey is drained the curd is heated with hot water to produce the plastic type of curd so common to Italian cheeses. These curds may be molded, stretched, or pulled into various shapes. Although cheeses of this type are often cured, they generally retain the same name.

About 30 of the cheeses briefly described by Sanders (1953), including those already mentioned, are similar to these two groups of cheeses.

Ricotta, Ziger, Schottenziger, Recuit, Broccio, Hudelziger, Brocotte, Sérac, Ceracee, and Mejette are albumin cheeses made from the whey drained from the curds. Before the wheys have developed high acidity, they are heated in a steam jacketed kettle to 200°F. After the albumins coagulate and are separated from the liquid, starters may be added when the coagulate cools to about 100°F. They may be marketed as somewhat moist cheeses or dried.

The fourth type of uncured cheese is also prepared from whey. Mysost, made from cow's milk whey, and gjetost, made from goat's milk whey in Scandinavia, consist primarily of caramelized lactose and the fat, protein, and minerals left in the whey. The whey is concentrated to about ¼ of its original volume or until it becomes quite creamy. During this heating the albumin coagulum is skimmed and later added after heating is completed. Upon cooling it sets as a semi-hard to firm, brown and sweet cheese entirely unlike the ordinary cheeses. Cream is often added during later stages of cooling to produce a smoother, richer product known as flφtost or primost.

Cured Cheese

The cured cheeses may be divided into two categories, the lactic acid bacteria-fermented cheese and the lactic acid bacteria-fermented cheese cured further by activity of microorganisms other than the lactic acid bacteria. Possibly 150 of the names applied to cheeses, as recorded by Sanders, belong in the first group. About 25 of these are very hard dry cheeses suitable for grating. The others are divided between the cheeses which include the cheddar types and the plastic curd type.

Cheese Cured by Lactic Acid Bacteria.—This discussion begins with some of the oldest types of cured cheeses. Since there are no written records, it can only be surmised what may have occurred before the advent of rennet. It can be surmised that starters were obtained from the previous fermentations. At a much later date, pasteurization and pure culture starters were employed.

Great numbers of species of microorganisms find milk an ideal growth medium. In fresh milk there is always competition among the many microbial species. Many species cause undesirable flavors, aromas, gassy conditions, digestion, and even illnesses to the consumer. The lactic acid bacteria do not always gain ascendancy immediately, but when and if they do, they will acidify the milk and lower the oxidation-reduction potential so that great numbers of other species either cannot survive or cannot grow. The first physical change is a curdling of the casein and, depending upon conditions, possibly an extrusion of whey. Strains of the species *Streptococcus lactis* produce sufficient acid to accomplish this curdling. The smooth curd milk at this stage is consumed and relished by many peoples as a sour milk drink. If the milk curd is broken up by stirring, the whey may be separated readily and a cottage cheese type curd is obtained. During the initial fermentation by species of *S. lactis,* other acid-producing bacteria, particularly strains of the species *Lactobacillus casei,* grow and may gain ascendancy even before the maximum potential acidity produced by *S. lactis* is attained. When the curd is separated in one way or another, the species *L. casei* may become the predominant organism in the curd continuing the fermentation and curing the cheese. The importance of this organism was observed by von Freudenreich over 70 yr ago and stressed by Orla-Jensen and others. Naturally, other species of lactic acid bacteria may be present, depending upon environmental conditions. For example, in the very warm climates, *Streptococcus thermophilus,* closely related to *S. lactis,* and *Lactobacillus bulgaricus* or other thermoduric species may exert the major action. Many of the strains of organisms called *L. bulgaricus* in early microbiological literature were undoubtedly strains of *L. casei.*

Possibly the first advancement in primitive cheese making was the use of rennet. Also, the first starter may have been the whey or milk from the previous day's cheese making. Heating to kill microorganisms was an early advance. The grana-type cheeses were fermented at temperatures of 90°–100°F, and the plasticity of provolone-type cheeses was an early advancement. Pasteurization procedures as practiced today began within the past century, and the use and development of starter cultures came later. These innovations are not practiced in many societies; some cheeses are still made by methods practiced centuries ago. In the Middle East the curds are separated, usually by hand, flattened by patting, and then placed on tent flaps to dry and cure. When dry, they are stored until consumed. While exposed on the tent flaps, aerobic organisms, molds, yeast, and bacteria are inhibited by the effects of the sunlight. Sometimes such cheeses are spiced, rolled in herbs, or treated in other ways to protect or impart flavor. Practices undoubtedly varied not only from society to society, but even among the peoples of a particular group.

Cheddar cheese is one of a great number of fermented and cured cheeses. It is of comparatively recent origin and as prepared today, represents the results of the present day application of scientific knowledge. Possibly more than 75% of the cheese made in the United States is cheddar, commonly called American cheese. Cheddar is the name of a village in England where this cheese was first made in the latter part of the 16th Century. Cheddaring, the unique step in processing, is the cutting of the curd followed by turning and matting the curd by piling and repiling for a period of about 2 hr. During this period the acidity is increasing and undesirable microorganisms are killed. Cheddar cheese is a hard white to yellow cheese which may vary from a mild to a markedly strong flavor, depending to a great extent upon the length of time it has been aged.

Pasteurized milk is cooled to a setting temperature of 86°–88°F and run into the cheese vat. An active starter, 0.5% *Streptococcus lactis*, is added. The milk is allowed to ferment for about ½ hr when rennet and color, if used, are added. In about 1½–2 hr the curd is sufficiently firm so that it can be cut into cubes about ¼ in. wide. About 15 min later the curd is gradually heated with stirring to 100°F (37.8°C) and held at this temperature for a maximum of 45 min. The curd is trenched or pushed back so that the whey can drain more readily. The curd is recut and trenched at about 15 min intervals for the next few hours so that whey may drain more readily. Fermentation by lactics continues during this period while the whey is draining and extraneous microorganisms are removed in the whey. The curds are finally milled, salted, hooped, and pressed for further curing.

Curing or ripening of cheddar cheese occurs in a moist room at about 36°–60°F. Enzymes active during this curing include proteinases, peptidases, lipase, decarboxylase, deaminases, and others so that a series of new chemical substances are produced which characterize the cheese. Amino acids, amines, fatty acids and alcohol, diacetyl, acetylmethylcarbinol, acetic acid, and ethyl alcohol contribute to the flavor that appeals to a great number of cheese consumers. The extent of development of these flavors depends upon the temperature and time of aging. Some consumers prefer the mild-flavored cheese while others prefer the sharp well-aged product.

American-type cheese, Monterey or jack, Chester or Cheshire, Colby, Cornhusker, Coon, Derby, English Dairy, granular or stirred-curd, Herkimer, Lancashire, Leicester, Warwickshire, Wiltshire, and others are cheeses of similar type. In addition, there are a number of spiced or flavored cheeses such as Sage that may be included in this grouping.

Plastic Curd Cheeses.—These cheeses, usually retailed as Provolone cheese, include Caciocavallo, Provole, Provatura, panedda, moliterno, cured scamorze and mozzarella, and several others. The cheeses are fermented with rennet at the fairly high temperature of about 96°F, the curd is cut into small pieces, in about ½ hr the whey is removed, and the curd is pressed. The compressed curd is transferred to a kettle in a warm room and covered with hot whey. An active fermentation occurs. As soon as the curd becomes plastic enough to stretch, the curd is immersed in hot water. The curd remains plastic in nature and may be molded. Provolone is made into various shapes. One typical one is pear-shaped and weighs from 6–14 lb. Provolone is often smoked, and its aromatic flavor is suitable as a dessert or sandwich cheese.

The high temperatures used in preparing these cheeses are particularly favorable for growth of *Streptococcus thermophilus* and *Lactobacillus bulgaricus* and for the rapid suppression of other microorganisms. These species are used in starters. A rapid fermentation occurs with high acidity development that is later reduced in washing procedures. Some of these cheeses are consumed as fresh cheese, but others are cured for various periods of time.

Grana-type Cheese.—Grana-type cheeses may have originated in the Po Valley before 1200 A.D. They are exported primarily under the name Parmesan and are characterized by their distinct granular structure, hard rind, and pleasing, biting, piquant flavor. These are dry cheeses particularly adapted to grating and for flavoring many food products. The grana-type cheeses, primarily of Italian origin, include the dry, very hard type cheeses usually retailed as Parmesan or Romano in the United States. Asiago, Lodigiano, Lombardo, Reggiano, Parmigiano, Emiliano, Bagozza, Veneto, and Venezza are similar, differing somewhat in size, shape, amount of fat, and slightly by method of manufacture.

Parmesan is generally made from the skim milk from cows. The milk is warmed to 90°–98°F, and a starter containing the heat tolerant *Streptococcus thermophilus* and *Lactobacillus bulgaricus* is added with rennet to curdle the milk in 20–30 min. The curd is cut into particles $\frac{1}{8}$–$\frac{3}{16}$ in. in diamater and heated slowly for 35–50 min to approximately 115°–125°F or higher if necessary to firm the curd. The whey is drained, the curd is pressed for 18–20 hr, and salted first in the hoop and later in a brine for 12–15 days. It is then dried for 8–10 days and then stored and dried for 4 months or more. The cheeses may be coated with lamp black, burnt umber, and dextrin dispersed in wine. The cheeses keep very well since they contain only about 30% moisture.

Romano and Asiago differ very little in method of manufacture; however, Romano has a slightly higher moisture content of 34% and Asiago still higher, 40%. They also may be coated with olive oil or paraffin containing lamp black or burnt umber. Asiago is somewhat like Swiss in certain respects.

Swiss Cheese Types.—Swiss or Emmenthaler, known by the Swiss as the King of Cheeses, are characterized by round to oval shiny eyes, sweet, fragrant, nutty flavor and somewhat plastic texture. They are made in large wheels weighing from 160–230 lb. They constitute one of the few types of cheese in which the gas formation is considered desirable. The gas, however, is caused by the propionic acid bacteria. Gassiness in other cheeses is often caused by strains of the coli-aerogenes bacteria. As far as is known, Swiss cheeses were made in the 15th Century, and by the 17th Century they were being exported.

Swiss cheese is one of the most difficult cheeses to make. It is made in large steam jacketed kettles that hold from 2000–3000 lb of milk. Good quality milk is essential. The milk is usually clarified. As the milk flows from the clarifier into the kettle, steam is turned on, and the milk is warmed to 88°–90°F. Stirring is begun, the starter is added, and then sufficient rennet is added to firm the curd enough to cut in about 30 min. Care is taken to mix thoroughly the cream with the curd. The curd is cut into long rectangular strips, then into cubes and then into small pieces of about $\frac{1}{8}$ in. The curd is then stirred for 30–60 min or until it acquires the desired firmness. Next the curd is slowly heated again to a temperature between 120° and 127.5°F, stirring continuously for $\frac{1}{2}$–1 hr. When the particles of curd can be broken apart easily, the curd is dipped. Stirring is stopped to allow the curd to settle and some of the whey is drawn off. The curd is enclosed in a large, coarsely woven, dipping cloth which is hoisted over the kettle so that excess whey may drain into the kettle. It is then transferred to the hoop, a press board is placed on top, and the curd is pressed. After about 5 min the hoop is removed, the heavy burlap dipping cloth is replaced by a clean light cloth, the cheese is turned over, another press board is placed on it, and it is pressed again. This process is repeated for a period of 24 hr. The cheese is then salted by placing it in a salt brine tank for 2 or 3 days. It is then placed on a circular board on a shelf in a room at about 65° to 72°F at a relative humidty of 80–85%. The final ripening occurs on this board. Every few days the cheese is washed and salted. Eye formation, an indication of proper curing, begins to form in about 3 weeks, but the curing is continued at the above temperature for 4–6 weeks after which it is placed in a room at about 40°F to continue curing for another 6–10 months.

Three species of bacteria are used as starters, *Streptococcus thermophilus, Lactobacillus lactis,* or *Lactobacillus bulgaricus,* and *Propionibacterium shermanii.* The first two are lactic acid bacteria, the third, the propionic acid, carbon dioxide-producing eye-former.

Swiss cheese is made in many countries other than Switzerland, including the United States, France, Germany, Italy, Sweden, Finland, Austria, and others. Many names have been applied to cheeses of this type, e.g., Allgäuer-Rundkäse, Appenzeller, Battelmatt, Piora, Gruyére, Fontina, Montasio,

Bitto, Spalen, Sbrinz, Saanen, Lapland, Lüneberg, Walliser, Comté, Herrgårdsost, Runesten, and Västgotaöst.

Edam and Gouda, particularly, have been generally accepted in the United States for their pleasingly sweet, mild flavor. These are sweet curd, renneted cheeses made from partly skim milk of excellent quality. The high pH of these cheeses, of pH 5.3–5.4, does not offer the repressive action of lactic acid as do the cheeses of the cheddar type. The curds are heated to 90°–95°F and when firm, are placed in pressing molds. When removed, they are dipped in warm whey at 125°–130°F before salting, effecting a heating similar to that of Swiss cheese. Commission, Bergquara, Danish export, Geheimrath, Patagras, Piaro, and Sveciaost are said to resemble Edam or Gouda.

Surface-ripened Cheeses.—In the manufacture of the majority of these cheeses, the characteristics are imparted by growth of lactic acid bacteria under reduced oxygen tension followed by surface-ripening microorganisms that are responsible for the characteristic flavor and aroma. Yeasts which reduce the acidity are followed by the bacterum, *Brevibacterium linens*.

Limburger-type cheeses are semisoft, surface-ripened cheeses which develop a characteristic strong flavor and aroma that many people learn to relish. Limburger cheese was first made in Limburg, Belgium, but it is now made in various countries especially Germany, Austria, and the United States. Surface-ripened cheeses of the Limburger type include Bachsteiner, Brioler, Frühstuck, Harzkäse, Hervé, Romadur, Leiderkranz, Marienhofer, Poona, Prattigau, Royal Brabant, Schloss, Schützenkäse, Spitzkäse, Stangenkäse, Steinbuscher, Tanzenberger, Void, Weisslacker, and others.

The method of manufacture varies somewhat in different localities, but the initial fermentation and preparation of curd result from a typical lactic acid fermentation. The curd is dipped into small metal or wooden forms differing in size depending upon the type. When the cheeses are sufficiently firm, they are removed from the form and salted on salting tables. While in contact with the tables or shelves where they are turned frequently and salted, surface-ripening microorganisms, primarily *Brevibacterium linens* and yeasts, grow on the surface. The enzymes penetrate the cheeses to produce the typical flavor and aroma. The attention given to manufacturing procedure varies.

Brick cheese is a sweet curd, semisoft, surface-ripened cheese with a mild but pungent flavor resembling Limburger. It is made in somewhat the same way as Limburger. Box, Chantelle, Güssing, Gautrias, Lüneberg, Mainauer, Münster, Oka. Port du Salut, Piora, Providence, Prattigau, Rangiport, Tilsiter, Trappist, Twdr Sir, and Wilstermarsch are apparently other mild but pungent-flavored surface-ripened cheese somewhat similar to brick cheese.

Bel Paese is a soft, sweet, surface-ripened cheese known also as butter cheese in Canada. Bella Alpina, Bella Milano, Bel Piano Lombardo, Bel

Piemonte Fior d'alpe, Savoia, Vittoria, Schönland, Königskäse, Cacio Fiore, Caciotta, Stracchino, Crescenza, Chantelle, and St. Stephano are similar.

Mold-ripened Cheese.—Roquefort and Camembert represent the two important types of mold-ripened cheese. Initially they are fermented with or without rennet and with or without lactic starter. After the curd is formed, mold is permitted to grow on the surface and inoculated into the cheese to impart the typical flavors of these cheeses.

Penicillium roqueforti is the blue mold of Roquefort cheese while *Penicillium camemberti* is the white mold of Camembert. Other types of mold or bacteria may develop on the surface of these types of cheese. For example, *Penicillium candidum* is the mold of Gournay cheese, a Camembert type, and *Penicillium candidum, Penicillium camemberti* and *Mycoderma casei* have been isolated from aged Neufchâtel, also a Camembert type.

Roquefort type cheeses include the better known Fromage Bleu, Blue, Stilton, Gorgonzola, Gammelost, and the lesser known Ambert, Bellelay, Tête de Moine, Blue d'Auvergne, Laguiole, Gex, Mont Cenis, Sassenage, Septomoncel, St. Flour, Castelmagno, Cotherstone, Dorset, Moncenisio, Olivet, Paglia, Pannarone, Sarrazin, Tignard, Wensleydale, and others.

Camembert type cheeses include the Gournay and aged Neufchâtel previously mentioned, Brie or Fromage de Melum, Coulommiers, Carré, Carré Affine, Double Crème Carré, Carré d l'Est, Chaource, Ervy, Barberry, Macqueline, Monthéry, Rollot, Thenay, Vendôme, Werder, and others.

French regulations limit the use of the name, Roquefort, to cheese made from ewe's milk in the Roquefort area. Roquefort cheese is characterized by the mottled, blue-green veins of the mold *Penicillium roqueforti* and by the sharp, peppery, salty, piquant flavor.

BIBLIOGRAPHY

BREED, R. S. 1928. Bacteria in milk. *In* The Newer Knowledge of Bacteriology and Immunology, E. D. Jordan, and I. S. Falk (Editors). University of Chicago Press, Chicago.
BREED, R. S. 1939. Foreign-type cheese should be more generally made and consumed in America. New York State Agr. Expt. Sta. Circ. *987.*
BREED, R. S., MURRAY, E. G. D., and SMITH, N. R. (Editors) 1948. Bergey's Manual of Determinative Bacteriology, 6th Edition. Williams & Wilkins Co., Baltimore.
BREED, R. S., MURRAY, E. G. D., and SMITH, N. R. (Editors) 1957. Bergey's Manual of Determinative Bacteriology, 7th Edition. Williams and Wilkins Co., Baltimore.
BUCHANAN, R. E., HOLT, J. G., and LESSEL, E. F. (Editors) 1966. Index Bergeyana. Williams & Wilkins Co., Baltimore.
CURRAN, H. R., ROGERS, L. A., and WHITTIER, E. C. 1933. The distinguishing characteristics of *Lactobacillus acidophilus.* J. Bacteriol. *25,* 595–621.
DAVIS, J. G. 1935. Some biochemical aspects of cheese-ripening. J. Soc. Chem. Ind. *54,* No. 27, 631–635.
ELLIKER, P. R. 1949. Practical Dairy Bacteriology. McGraw-Hill Book Co., New York.
FOSTER, E. M. *et al.* 1957. Dairy Microbiology. Prentice-Hall, Englewood Cliffs, N.J.
FRAZIER, W. C. 1967. Food Microbiology, 2nd Edition. McGraw-Hill Book Co., New York.

HAMMER, B. W. 1948. Dairy Bacteriology, 3rd Edition. John Wiley & Sons, New York.
HAMMER, B. W., and BABEL, F. G. 1943. Bacteriology of butter cultures: A review. J. Dairy Sci. 26, No. 2, 83–168.
HUCKER, G. J., and PEDERSON, C. S. 1930. Studies on the coccaceae. XVI. The genus *Leuconostoc*. New York State Agr. Expt. Sta. Tech. Bull. 167.
KOSIKOWSKY, F. V. 1966. Cheese and Fermented Milk Foods. Published by the author. Ithaca, N.Y.
ORLA-JENSEN, S. 1919. The lactic acid bacteria. Mem. Acad. Roy. Sci. Lettres, Danemark. Sect. Sci. 8, No. 5, 81–196.
PEDERSON, C. S. 1947. The relationship between *Lactobacillus acidophilus* (Moro) (Holland) and *Lactobacillus casei* (Orla-Jensen) (Holland). J. Bacteriol. 53, No. 4, 407–415.
SANDERS, G. P. 1953. Cheese varieties and descriptions. U.S. Dept. Agr., Agr. Handbook 54.
SANDERS, G. P., BURKEY, L. A., and LOCHRY, H. R. 1950. General procedures for manufacturing Swiss cheese. U.S. Dept. Agr. Circ. 851.
WEBB, B. H., and JOHNSON, A. H. 1964. Fundamentals of Dairy Chemistry. Avi Publishing Co., Westport, Conn.

CHAPTER 6

Fermented Vegetable Products

Although their exact origins are unknown, the methods of preparation and preservation of vegetable substances by fermentation presumably originated in the Orient. The preservative and organoleptic properties of salt were recognized long before history was recorded. The use of salt for preservation of vegetables and meats is an ancient practice, whereas methods of preparation and purification of salt are relatively recent innovations. Since salt formerly was expensive and impure, often containing sand as well as salts of elements other than sodium and chloride, it is surprising that methods of fermentation were developed at a very early period. Certainly when low quantities of salt were added to vegetables, it must have been observed that the brines became cloudy, and the product acquired an acid flavor. This acid flavor partially balanced the excessively salty flavor and, therefore, undoubtedly appealed to the consumer.

Probably the vegetables were eventually packed in brine solutions, not only because dry salt withdrew insufficient water from the vegetable pieces or chunks, but also because the use of brine would permit settling out of sand and other insolubles introduced with the salt. The housewife, without doubt, recognized that when salt withdrew some water from the vegetable, it obviated the need to add large quantities of brine. Practices were individualized and although some interchange of ideas occurred, the practices were far from standardized. From this beginning, it may be stated three practices developed: high salt brine salting, low salt brine salting, and dry salting. Dry salting is probably comparatively recent in origin.

Practices continued for centuries with little standardization; household methods were passed on from mother to daughter. The scums formed on brine surfaces were considered by many, even in recent times, to be responsible for fermentation, since, when agitated, they produced cloudiness similar to that produced by true lactic acid bacteria. Actually, there was little standardization of practices until the early period of this century, and there still is a considerable quantity of foods packed by ancient methods. The application of microbiology to pickling and the use of names for the bacteria and yeasts present in fermenting vegetable substances started in the early 1900's.

Improvements in techniques were developed during a period of centuries. Advancement in production methods and development of superior and disease-resistant strains of vegetables, particularly during the past 60 yr, have been largely responsible for providing adequate and satisfactory vegetable substances for preservation by fermentation. The major improvements in vegetable fermentation during the past 40–50 yr began with developments in microbiological science beginning about 100 yr ago. These culminated with the conclusion that more than one species of lactic acid bacteria is responsible for vegetable fermentation (Pederson 1930; Pederson and Albury 1969). There are similarities and also important differences among the species and genera of lactic acid bacteria that develop in a natural sequence. The growth and fermentation by these species are influenced by environmental factors, especially salt concentration and temperature. Furthermore, the nature of vegetable fermentation and the microbiological, physical, and chemical changes that occur have been elucidated in large part.

Formerly the fermentation of vegetable products lacked the background of research and observation that until recent years was generally associated with various other fermented foods such as dairy products; it is, therefore, hoped that in this discussion the similarities and differences in the course of fermentation of various vegetable substances may be correlated. The studies conducted by scientists at the University of Wisconsin, Michigan State University, University of California, Cornell University at Geneva, and the U.S. Dept. of Agr., particularly at Raleigh, N.C., are basic to these comparisons, but there also have been many other excellent studies conducted in laboratories in this country as well as abroad.

The history of fermented vegetables extends so far into antiquity that no precise time can be established for its origin. Presumably, it was observed that when vegetables flavored with salt or salt brine were packed tightly in a vessel, they changed in character but remained appetizing and nutritious (Fig. 6.1). The fact that fermentation occurred was entirely unrecognized by these ancient people.

FIG. 6.1. SOME COMMON VEGETABLE SIDE DISHES INCLUDING SWEET AND DILL PICKLES, OLIVES, PICKLE RELISH, SAUERKRAUT, AND A SAUERKRAUT RELISH CALLED "KILARNEY KRESS"

From the meager information available it seems that the Chinese may have been the first to preserve vegetables by a fermentation process. When the Emperor Ch'in Shih Huang Ti was constructing the Great Wall of China in the 3rd Century B.C., a portion of the coolies' rations consisted of a fermented mixture of vegetables, possibly cabbages, radishes, turnips, cucumbers, beets, and whatever other vegetable materials were available. From the fact that vegetables of the Orient are now nearly all fermented in salt brines, it is assumed that dry salting was a later development of the art. Pickling of cucumbers supposedly originated in Southeast Asia.

The Tarters under Genghis Khan are sometimes credited with introducing fermented vegetables to Europe, but there is little doubt that these methods of preserving vegetables were introduced much earlier. A fondness for pickled products always has been characteristic of people throughout the world. Pliny's writings mentioned spiced and preserved cucumbers. They were a delicacy enjoyed by Caesar's legions; Emperor Tiberius ate pickles regularly; and Cleopatra is said to have delighted in eating pickles. They are a significant part of the diet of the Far Eastern peoples, the Europeans, and Americans today; in fact, it is estimated that at least 75% of American families eat fermented vegetable products at least once a week.

Drawings and carvings from Ancient Egypt depict cabbages placed on temple altars as offerings worthy of the gods. Greek doctors used cabbage as a general cure for illnesses, and it was a common vegetable in both Greek and Roman gardens. The fact that early descriptions indicate that they had solid, white-headed varieties would indicate that cabbage must have been domesticated for a long time in order to have undergone the extensive modification from the wild types to the cultivated types. Although cabbage is native to many parts of Europe, its use as a cultivated vegetable in Northern Europe is attributed to the Romans. Cucumbers and other vegetable materials have an equally ancient origin.

The Romans used olive oil in cooking, and they also valued the olive pickled in brine. Pickled olives were found among the buried stores of Pompeii. Olives were used to enhance the flavor of wines, but it may be presumed that initially when wines soured, olives were added to the sour wines to partially balance or mask the sour flavor.

There are no figures for consumption of fermented vegetables throughout the world. In many areas of the world fermentation is the only method used for preserving vegetables, and they are prepared almost entirely in the home. In Korea, for example, a blend of pickled mixed vegetables, called kimchi, is second only to rice in feeding the population. It is prepared and stored in crockery jars of varying sizes. The Chinese developed a special jar with a moat around the top so that the cover could be placed over it with the flange immersed in water in the moat to exclude air. Vegetables are of equal importance in other areas of the Orient. In Europe and America, pickles, sauerkraut, and olives have become the favorites, although pickled onions, cauliflower, peppers, and others are enjoyed, especially in mixed pickle products.

The preparation and preservation of these foods have presented problems of spoilage and abnormal off-flavors with subsequent losses of the product. Since vegetables are relatively inexpensive and the fermented products have not presented health problems like those of milk, research to improve quality and prevent spoilage was delayed. Until recently the study of vegetable products has not kept pace in research with that of cheese, butter, bread, wines, and beers.

Although these products have an ancient heritage, the microbiological and chemical changes that occurred during curing were unknown until recent years. No doubt the clouding of brines, the evolution of gases, the changes in flavor, and the development of surface scums were observed by the ancients and possibly were associated with environmental conditions. Undoubtedly, the relationship between the presence of surface growths and the flavor, odor, and texture characteristics of the foods was observed. The importance of properly covering the fermenting products (Fig. 6.2) eventually became apparent, but some characteristics, now considered undesirable, were then undoubtedly acceptable.

FIG. 6.2. THE EXCLUSION OF AIR FROM THE SURFACE OF FERMENTING VEGETABLE MATERIALS IS ACCOMPLISHED BY USE OF PLASTIC COVERS FILLED WITH WATER

Although the science of microbiology may be traced to von Leeuwenhoek, the actual relationships of microorganisms to natural phenomena became evident to Pasteur, Koch, and others about 100 yr ago. For years that science was applied primarily to illnesses and to food spoilage. The correlation of microorganisms with the fermentation of vegetables, as practiced for centuries, was ushered in with the studies of Aderhold, Kossowicz, Henneberg, Wehmer, Rahn, and others in the early years of this century. Approximately another 20 yr elapsed before much progress was made that lead to our present-day knowledge of vegetable fermentations.

PRESERVATION PRINCIPLES OF VEGETABLES

Preservation of vegetable materials by fermentation depends upon reduction of the activity of the enzymes native to the vegetable, i.e., those that are responsible for the normal changes involved in maturation, the inhibition of certain oxidative chemical changes, and the inhibition also of growth of microorganisms that may produce deteriorative changes in the food.

Environmental factors important in the fermentation of vegetables consist essentially of the establishment of anaerobic conditions, a suitable salt concentration, proper temperature, cleanliness, and the presence of suitable' lactic acid bacteria. The initiation of fermentation by the species *Leuconostoc mesenteroides* (Fig. 2.5, Chap. 2) has been found recently to have a marked effect upon the fermentation and quality of the product. This will be discussed in more detail in the section on sauerkraut.

Preservation was attributed by many individuals to the inhibitory effect of salt; by others it was assumed that the acid was entirely responsible. The failure to appreciate fully the role of a variety of factors, particularly the combined effect of acid and salt, has led to considerable losses in the past.

It has been realized too slowly that the preservation of fermented vegetables depends upon the combined effect of acid, salt, carbon dioxide, the low oxidation-reduction potential, and other minor factors. In order to attain this combined effect, the vegetable must be properly fermented. Trimming and cleaning are essential for removing the numbers of extraneous microorganisms; sufficent salt or salt brine must be added to withdraw liquid from the vegetable and to deter softening, but the content of salt must be kept low enough (Fig. 6.3) to allow a rapid initiation of the fermentation in order to produce acid rapidly and to reduce oxygen tension. The product must be properly packed and covered to exclude air and retain the anaerobic condition produced by fermentation. Temperature is an important factor, but since

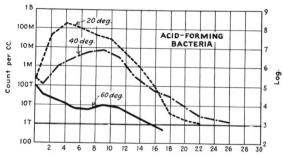

From Etchells and Jones (1943)

FIG. 6.3. GROWTH OF LACTIC ACID-FORMING BACTERIA IN CUCUMBER
FERMENTATIONS AT INITIAL BRINE CONCENTRATIONS OF 20°, 40°, AND 60°
SALOMETER

the lactic acid bacteria have a wide temperature range for growth, the effects of temperature differences are often minimal.

In order to attain optimum conditions, two general methods of packing are practiced. These are referred to as dry salting and brine salting. In either case, the amount of salt used must be consistent with the other requirements. In general, shredded or chopped cabbage is dry-salted in making sauerkraut, while cucumbers, olives, and other vegetable materials in large pieces are brined.

Microbiology

Relatively few species of bacteria are responsible for the fermentation of the majority of our vegetable products. They develop in a natural sequence of species; the relative role of each species is governed primarily by environmental conditions. The species responsible for fermentation are originally present in comparatively low numbers on the surfaces of the vegetables. In fact, the proportions of these in relation to undesirable aerobic species are so low that in order to estimate the numbers of fermentative organisms, special techniques must be used to inhibit the growth of aerobic species while allowing growth of the fermentative species. Since growing vegetables present an aerobic environment, in intimate contact with air, soil, and moisture the majority of the bacteria present on their surfaces are strains of aerobic soil and water species of genera such as *Pseudomonas, Flavobacterium, Achromobacter, Aerobacter, Escherichia,* and *Bacillus.* When vegetables are cut for harvesting, small amounts of protoplasm exude on the cut surface. Fermentative species, particularly the leuconostoc, find this a satisfactory growth medium and in some instances growth of *Leuconostoc mesenteroides* is sufficient to produce a dextranous, slimy condition in these exudates. In packing vegetables for fermentation, it is essential to establish environmental conditions unfavorable to the aerobic flora while at the same time, favorable to the lactic acid bacterial flora. By centuries of experience and comparatively recent research, these conditions are fairly well defined today. The absence of air and the proper concentration of salt are of utmost importance. Nearly all vegetable material, including fruits handled like vegetables such as cucumbers, tomatoes, and olives, are fermentable by lactic acid bacteria. They contain sugars and are adequate nutritionally to provide an adequate substrate for growth of the lactic acid bacteria and other microorganisms. Early microbiologists attributed the fermentation to two species of bacteria, the homofermentative lactic acid-producing species called *Bacillus cucumeris fermentati* and the heterofermentative species named *Bacillus brassicae fermentatae* by Henneberg. A number of closely related strains have been given specific names and may be found under the accepted names used today, e.g., *Lactobacillus plantarum* and *Lactobacillus brevis* (Breed *et al.* 1948).

As the names cucumeris and brassicae imply, the first species was considered the cucumber fermenter and the second, the cabbage fermenter. Later studies have shown that both species are involved in nearly all vegetable fermentations. In fact, Wehmer, a contemporary of Henneberg, applied the name *Bacterium brassicae* to the strain he isolated from sauerkraut, although it is actually similar to *Bacillus cucumeris fermentati.* Wehmer's name, *Bacterium brassicae,* was used by LeFevre in his reports some 20 yr later.

The two species differ in that the first is a homofermenter while the second is a heterofermenter; the primary end products from sugar are lactic acid by the homofermenter and lactic acid, acetic acid, ethanol, and carbon dioxide by the heterofermenter. Both may be isolated at some stage in the fermentation of nearly all fermenting vegetable materials; however, *Lactobacillus plantarum* will usually be present in greatest numbers following the initial fermentation and will produce the maximum acidity. For this reason, so many fermentations were attributed entirely to this species. *Lactobacillus brevis,* appearing in intermediate and later stages of fermentation, is active in utilizing pentose sugars. Fermentative yeasts often appear in limited numbers during later stages of fermentation, and aerobic yeasts and molds appear on the surface of improperly covered vegetable materials.

The profuse evolution of carbon dioxide in most vegetable fermentations was first ascribed to yeast growth and in part to the bacterial species *Lactobacillus brevis.* In the early stages of fermentation neither sufficient numbers of yeasts nor of long rod lactic acid bacteria resembling *L. brevis* can be observed ordinarily by microscopic examination or by isolation procedures to account for the copious evolution of carbon dioxide. Microscopic examination during early stages of fermentation will generally show, instead, great numbers of coccoid bacteria often appearing in pairs. It was assumed at first that these were species of the genus *Streptococcus;* therefore, in some of the early studies of the use of starters, strains of the species *Streptococcus lactis* were used.

This apparent inconsistency finally led to an observation which has changed the concept of vegetable fermentations.

Pederson (1930) demonstrated that the sauerkraut fermentation was initiated by the bacterial species *Leuconostoc mesenteroides.* This coccoid organism produces carbon dioxide, lactic acid, acetic acid, and ethyl alcohol (Fig. 3.1, Chap. 3). The significance of the role of this species was not fully appreciated until further study revealed the complete role of the species and the importance of environmental conditions in its growth (Pederson and Albury 1954, 1969). Known previously only as a spoilage organism in sugar processing operations, its value as an important and useful microorganism in the fermentation of foods was unsuspected. This species initiates growth in vegetable fermentations more rapidly than other lactic acid bacteria and over a wide range of temperatures and salt concentrations. It produces carbon dioxide and acids that rapidly lower the pH, thereby inhibiting the development of undesirable microorganisms and the activity of their enzymes that may soften vegetables. The carbon dioxide replaces air and provides an anaerobic condition favorable to stabilizing ascorbic acid and the natural color of the vegetable. The growth of this species apparently changes the environment, making it more favorable for the growth of other lactic acid bacteria in a bacterial sequence. The combination of acids, alcohol, esters, and other growth products imparts a unique and desirable flavor. The species converts excess sugars to mannitol and dextran that are generally nonfermentable by organisms other than the lactic acid bacteria. Mannitol and dextran lack free aldehyde or ketone groups that combine with amino acids to initiate darkening of the food. The species produces a higher pH level for any given level of acid than the homofermentative bacteria. Elucidation of the role of the species *Leuconostoc mesenteroides* has changed and standardized certain practices in the vegetable fermentation industry. Its role in other fermentations is not appreciated fully.

Since 1930, this species has been found important in initiating the fermentation of many other vegetable products, both dry- and brine-salted. This includes beets, cucumbers, turnips, chard, cauliflower, green beans, sliced green tomatoes, whole head cabbage, Brussels sprouts, and mixed vegetables such as kimchi and paw tsay and others that may include even soybeans. Although some of these vegetables may be relatively unimportant fermented foods in the West, some are extremely important in the Orient.

The complex changes that occur in vegetable fermentation are produced by the growth of a sequence of lactic acid bacteria. The growth of each species depends upon its initial presence on the vegetable, the sugar and salt concentrations, and the temperature. While the importance of *Leuconostoc mesenteroides* in the sequence is stressed, this should not be construed to imply that the roles of other lactic acid bacterial species are unimportant. *Lactobacillus plantarum* is the high acid-producing species and, together with the lesser known *Pediococcus cerevisiae,* plays a major role in fermentations, particularly those in brine. *Lactobacillus brevis* is important in imparting character to fermented vegetables and is often characterized by its ability to ferment pentose sugars. The distinctive value of *Leuconostoc mesenteroides* in many salt brine fermentations may be its contribution in establishing proper environmental conditions for continuation of lactic acid bacterial fermentations rather than producing flavorful growth products.

These species vary in their characteristics, particularly in regard to their tolerance to salt and acid and to their temperature ranges for growth. These characteristic differences must be considered in the fermentation of any of the vegetable products and particularly when fermented by dry salting, in contrast to brine salting. The primary problems have centered around softening of vegetables on one hand and abnormal fermentations, bloating, and discolorations on the other. Many of these problems have centered around misconceptions of the role of salt. The studies and numerous publications of Fabian and of Etchells and their associates concerning cucumber fermentation, of Vaughn and associates with olive fermentation, and of Fred and Peterson and of Pederson and their associates with sauerkraut have done much to elucidate some of the problems.

In many of the early studies of fermenting vegetables, the species *Leuconostoc mesenteroides* was not observed either because plating to isolate the organism was not started early in the fermentation or unsatisfactory media were used. The species prefers fructose to glucose; therefore, in the fermentation of sucrose, the fructose moiety is fermented leaving the glucose molecules to intercombine to form the slimy, water-insoluble dextrans. These vary from an almost rubber-like solid gelatinous mass to a watery slime surrounding the bacterial cells. The fermenting food in some cases is very slimy during intermediate stages. In further fermentation the dextran is utilized. In sugar factories these growths not only reduce the sugar yield, but they are so severe at times that they clog pipelines and other equipment and reduce the efficiency of operation, which at times necessitates interrupting operations to remove the slime. Their growth may be observed on cut or bruised surfaces of sugar canes or beets and frequently on other vegetables.

As previously stated, the gas evolution observed at the surface of vats of fermenting vegetable material initially is produced by *Leuconostoc mesenteroides* as well as by the release of gases from within the vegetables themselves. Later, *Lactobacillus brevis* may produce some of the gas evolved. An acidity of 1.0–1.2% acid, calculated as lactic acid, may be attained in a drysalted fermenting vegetable by the activity of *Leuconostoc mesenteroides.* Lesser amounts of acid are produced in brined vegetables since the buffer content is lower. *Lactobacillus brevis,* sometimes *Pediococcus cerevisiae,* and at all times *Lactobacillus plantarum* will develop simultaneously while the initial fermentation is proceeding, provided that the temperature is suitable. An acidity of 2.0–2.5% will be produced in a dry-salted vegetable if sufficient sugar is present. In brined vegetables the acid and carbon dioxide produced inhibits the growth of undesirable aerobic microorganisms. The replacement of air by carbon dioxide produces an anaerobic condition favorable to stabilizing ascorbic acid and inhibiting oxidation and darkening of the color.

To summarize, the species *Leuconostoc mesenteroides* imparts desirable flavors derived from the acids, alcohol, and other products of fermentation. It apparently produces conditions favorable to growth of other lactic acid bacteria in the growth sequence. The bacterial growth sequence includes *Leuconostoc mesenteroides, Lactobacillus brevis,* often *Pediococcus cerevisiae,* and *Lactobacillus plantarum.* Occasionally strains of streptococci, related to *Streptococcus faecalis,* are present in the early stages but when present, they play only a very minor role. The peak populations of the various species will occur in the order of their acid tolerance, but there will always be considerable overlapping of species. In fermentation in brine, an acidity beyond 1% is rarely attained. However, as is quite common, if the salt content is raised too rapidly in brined vegetables, the combined inhibitory action of salt and acid may stop acid development at a lower acidity.

THE SAUERKRAUT FERMENTATION

The fermentation of cabbage to sauerkraut will be discussed in greater detail than other fermentations because the principle of initiation of fermentation by a single bacteria species, *Leuconostoc mesenteroides,* and the sequential growth of bacterial species applies to other vegetable fermentations. Establishment of proper environmental conditions (Fig. 4.2, 4.4, Chap. 4) is essential in ensuring a desirable fermentation. The basic facts discussed herein apply to many other fermentations.

Cabbage has always had a peculiar place in the diet as an adjunct to make other foods more agreeable and digestible rather than for its own nutritional value. Cato, in his manuscript *De re rustica* written about 200 B.C., lauds cabbage as the most important vegetable the Romans had under cultivation. During a period beginning about 200 B.C. and continuing until about 450 A.D., it was the principal plant used in the Roman Empire for treatment of disease. The term sauerkraut is originally German, but the assertion so often made that cabbage in the form of kraut is of Germanic origin is not substantiated. These peoples were nomadic when they first came in contact with the Romans about the beginning of the Christian era. Cabbage is a crop native to many areas and a temperate climate crop; the large solid white heads are common in the north Temperate Zone.

Sauerkraut is literally acid cabbage. A discussion of the factors involved in the preparation of sauerkraut is important because the principles apply not only to sauerkraut but also to the preparation of many other food products by fermentation processes. Although easy to prepare, its preparation is subject to many errors that may result in poor quality. These are discussed in more detail in a recent paper by Pederson and Albury (1969).

In the beginning, sauerkraut differed markedly from that prepared at present. Among the earliest forerunners were cabbage leaves dressed with sour wine or vinegar. In another preparation the cabbage was broken or cut into pieces, packed in vessels, and covered with verjuice, sour wine, or vinegar. The early French choucroute was prepared in this way. Then salt was added, sometimes to such an extent that the pieces had to be soaked before they could be consumed. Gradually the sour liquids were entirely replaced by salt, and a spontaneous fermentation occurred. *Le Thresor de Santi* published in 1607 gives what seems to be a description of the immediate precursor of present-day sauerkraut. The Germans, in order to prepare a food for winter, cut the cabbages, loosened them, placed shreds in layers with salt, juniper berries, spice, barberry, and pepper roots, using a layer of cabbage and the next layer of these ingredients. Each layer was pressed firmly for better contact. To these ingredients, apparently, a salt brine was added.

The first description of sauerkraut manufacture, comparable to the methods used commercially today, is that of the Dutch zoorkool given in 1772 by James Lind in *A Treatise on Scurvy, Third Edition.*

It is not surprising that cabbage became the only ingredient used in preparing sauerkraut, in view of the health characteristics ascribed to it by the Greeks and Romans and the plentiful supply of that vegetable in so many areas of Europe. The other related vegetables apparently originated at later dates; cauliflower at about 1600; broccoli at about 1700 and Brussels sprouts at a somewhat earlier date. The selection of mild-flavored, sweet, solid, white-headed cabbage is chosen for its superiority in making sauerkraut. Total yield with freedom from disease has been the controlling factor in selection in the past 50 yr.

Sauerkraut can be made by adding salt to shredded cabbage then packing in containers and allowing it to ferment. Probably an edible fermented product would be available within a few weeks, but unless it is prepared carefully with the correct amount of salt, packed tightly, and properly covered, probably the quality will be inferior.

The present-day practices used in sauerkraut manufacture have evolved from an increasing knowledge of the underlying facts. Very little is known about the ancient methods used in fermenting vegetables in the Orient. Without doubt, they were developed and improved by trial and error. The stone crock with a moat encircling the top and used by the Chinese was one advancement in the art of preparing fermented foods, others were the advent of crude pottery-making and the subsequent invention of the potter's wheel and later, fire-polishing. Until recently, sauerkraut making has remained a rule-of-thumb procedure laden with much inferior quality and spoilage. Studies beginning about 1900, designed to learn more about the process and the influence of environmental factors, have resulted in methods that have somewhat standardized the product. Types of cabbage have been developed primarily for making sauerkraut. Solid, white-headed, sweet varieties of many strains that mature at different times yield uniform products.

The cabbage is harvested when properly matured and brought to the factory. The modern factory is a clean, mechanized plant in which the products can be handled efficiently. The solid white heads are trimmed to remove green, broken, or dirty leaves. The cores are cut mechanically by a reversing corer that leaves the core in the cabbage. The cabbage is sliced to about $\frac{1}{32}$ of an inch, 2.25% salt is added, and the shredded cabbage is packed into tanks or vats. Brine begins to form shortly after shreds are salted, and when the tank is full, there is sufficient brine present. When the tank is full, the cabbage is covered with a plastic sheet large enough to extend over the edge of the tank. Water or brine is placed in this sheet or cover to serve as a weight as well as an effective seal (Fig. 6.2). The salt in solution in the first brine formed increases the specific gravity of the brine so that the cabbage shreds tend to float on the brine. Sufficient water must be placed in the plastic cover to cause the shredded cabbage to be submerged in the brine.

A fermentation will proceed, initiated by the bacterial species *Leuconostoc mesenteroides,* followed by *Lactobacillus brevis, Pediococcus cerevisiae,* and *Lactobacillus plantarum* (Fig. 4.2, 4.4, Chap. 4). The environmental conditions, numbers and kinds of organisms present, cleanliness of cabbage and vat, salt concentration and distribution, temperature of the shredded cabbage, and covering will influence the course of fermentation (Pederson and Albury 1954, 1969). These factors will be discussed more fully because they are also important in other vegetable fermentations.

Microorganisms on Cabbage

The microorganisms on the surfaces of vegetables may be extremely variable both in numbers and types. A count of 13×10^6 per gm has been recorded from the outer leaves of cabbage. The surfaces of root vegetables may contain more. Species of the genera *Aerobacter, Pseudomonas, Achromobacter,* and *Flavobacterium* generally predominate on the outer leaves, but other species, particularly aerobic spore-formers, are also usually present. The numbers of microorganisms on leaves decrease toward the center of the head. Cut surfaces such as the cut stem will ordinarily show a small amount of sap or protoplasm exuding from the vascular system. The bacterial counts in this exudate may be extremely high and will ordinarily contain high numbers of lactic acid bacteria: leuconostoc, lactobacilli, and pediococci.

Certain chemical substances present in the protoplasm of cells of cabbage and other vegetables are inhibitory of growth of Gram-negative bacteria (Fig. 6.4) and may even destroy some of the aerobic species present on the vegetable. The amounts of unknown chemical identities are variable, but it is established that their effect is beneficial.

Manufacturing procedures have been and continue to be improved to standardize the product. Much attention has been given to the exclusion of air, control of the salt content, and to observing the relationship between temperature and rate of fermentation.

Probably the most critical factor in producing a sauerkraut acceptable to the majority of consumers relates to the effect of air. Kraut fermentation depends on lactic acid bacterial fermentation and is essentially an anaerobic one. Contact with air permits growth of yeasts and molds on the vegetable surfaces and generally results in softening, darkening, and development of un-

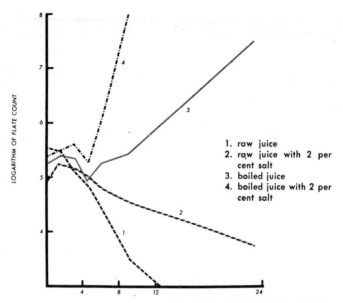

FIG. 6.4. BACTERICIDAL ACTION OF CABBAGE JUICE TOWARDS A GRAM-
NEGATIVE BACTERIUM OBTAINED FROM CABBAGE SURFACE

Bottom scale represents time in hours.

desirable flavors that may diffuse throughout the entire mass of sauerkraut. Only a few individuals are conditioned to desire this flavor development.

Wooden tanks are at present the most common containers used for fermentation. Recently several reinforced concrete vats with plastic coatings have been installed in some factories. Glazed tile vats are used in some factories in Europe. In many factories there may be several tank rooms so arranged that the temperatures may be controlled. Wooden vats are very often paraffined or treated in a similar way to close the pores and cracks in the wood. Such crevices may harbor millions of microorganisms. If they are lactic acid bacteria, they may be homofermentative strains that should not initiate the fermentation and are, therefore, undesirable. Also, undesirable yeasts and molds are sometimes harbored in these crevices.

Salt withdraws water and nutrients from the vegetable tissue. The nutrients furnish the substrate for growth of the lactic acid bacteria. Salt in conjunction with the acids produced by fermentation inhibits the growth of undesirable microorganisms and delays enzymatic softening of tissue. A satisfactory salt concentration favors the growth of the various lactic acid bacteria in their natural sequence and yields a kraut with the proper salt-acid balance.

Insufficient salt not only results in softening of the tissue but also yields a product lacking in flavor. Too much salt delays the natural fermentation and, depending upon the degree of oversalting, may result in an acrid flavor, darkening of color, or growth of pink pigment-producing yeasts.

In most factories, salt is weighed and applied at a 2.25% level to the shredded cabbage. Considerable attention has been given to efficient methods of distribution of salt. The uniform distribution of salt is fully as important as the amount of salt used. Salt concentration in different areas of vats may vary from as high as 5% to below 1%. Soft kraut, a condition associated with insufficient salt, has been observed in areas only in a few feet from other areas of pink kraut, a condition associated with excess salt. Such conditions are due to the uneven distribution of salt. In some factories, the shredded cabbage is weighed on belt lines, and by means of a suitable proportioner, salt is sprinkled on the shredded cabbage in the desired amount as it is conveyed along the belt to the vat. In this way it is distributed more uniformly. Even by this method the first brine formed, always containing a high concentration of salt, tends to seep to the bottom. In other factories, the cabbage is transported to the vat in carts which are weighed occassionally to check their capacity. The shredded cabbage is dumped into the vat, distributed by forks, and then salted with a specific weight of salt. This method tends to give a lower concentration of salt in the area where the cabbage is dumped and a high concentration around the periphery of this site.

The importance of proper salt concentration and distribution cannot be overemphasized. Excess salt, i.e., 3.5% or more, is detrimental to growth of all the lactics but more so to the one important species that initiates fermentation, *Leuconostoc mesenteroides*. Excessive salt may result in proportionately excessive growth of homofermentative types that produce little carbon dioxide, so important in establishing anaerobiosis. Yeast growth, including the pink pigment-producing strains, becomes more prevalent in high salt concentration brine. A low concentration of salt permits rapid growth of *Leuconostoc mesenteroides* but withdraws less water from the cabbage shreds before equilibrium is attained; therefore, acid is required in conjunction with the salt to inhibit the softening due to enzymatic activity.

In preparing cabbage for sauerkraut, the temperature of the shredded cabbage in the vat will be governed primarily by the temperature of the cabbage beforehand and the air temperature at the time of vat-filling. In commercial operation, cabbage is handled so rapidly that the shredded cabbage has little opportunity to warm during slicing and packing. Cabbage, a temperate climate crop, is usually harvested in the fall months when temperature conditions vary considerably. It is not unusual for the temperature of shredded cabbage in vats to be below 50°F. The rate of fermentation will depend to a great extent upon the temperature of the shredded cabbage. Fortunately, the species that initiates fermentation, *Leuconostoc mesenteroides,* can grow at relatively low temperatures. At 45.5°F fermentation is brought about by this species and an acidity of 0.4% calculated as lactic acid may be attained in about 10 days, and 0.8–0.9% in less than a month. Attainment of these acidities coupled with saturation with carbon dioxide is sufficient to provide the environmental conditions necessary for preservation and later continued fermentation. It is not clear why this species initiates growth so rapidly. Population studies of raw vegetables show a preponderance of Gram-negative aerobic species, but also all three species of lactic acid species are found when cabbage is packed for kraut making. Possibly the leuconostoc are in the logarithmic phase of growth brought about by their growth in the liquid exudates on the freshly cut stem of freshly harvested cabbage. Species of the genera *Lactobacillus* and *Pediococcus* grow slowly, if at all, at this low temperature; therefore, it may require 6 months to 1 yr before fermentation is completed. Such kraut, when fully fermented, is usually excellent in quality because it remains cold. In fact, in good commercial operation this variation is used to advantage, in that the processor is thereby supplied with new, completely fermented kraut through the year.

With increasingly higher temperatures, rates of fermentation will be greater. At an average temperature of about 65°F with salt concentration of 2.25% an active fermentation will occur, initiated by the species *Leuconostoc mesenteroides* and continued by the other three species. An acidity of about 1.2% will be attained in a few weeks and about 2.0% in about 2–3 months. At a still higher temperature of 73.4°F, the rate of fermentation will be more rapid and a brine acidity of 1.0–1.5% may be attained in 8–10 days. Active growth of both *Lactobacillus brevis* and *Lactobacillus plantarum* may be initiated in less than 3–5 days, and the kraut may be completely fermented in approximately 1 month. At a still higher temperature of 89.6°F, fermentation will be rapid, and a brine acidity of 1.8–2.0% acid calculated as lactic acid may be attained in 8–10 days. At this high temperature the major proportion of the acid is produced by the homofermentative species *Pediococcus cerevisiae* and *Lactobacillus plantarum*. The flavor of the kraut will be inferior, resembling acidified cabbage. Because of its high temperature, the kraut will darken readily and unless canned immediately will have a poorer shelf-life than the kraut fermented at lower temperatures. Kraut fermented at the higher temperatures will often have low percentages of acetic acid, alcohol, and esters and will not attain the maximum kraut acidity even though the pH remains lower. It is more subject to yeast spoilage and may be low in ascorbic acid.

The quality of sauerkraut fermented at 65°F or lower is generally superior in flavor, color, and ascorbic acid content because the heterofermentative lactic acid bacteria exert a greater effect, particularly at a salt concentration of 2.25% or less.

Many methods have been tried over a period of years to exclude air and thereby reduce surface fermentation; they include use of tight wooden covers, cloth, papers, oil coatings, and others. The general use today of plastic bag-like covers containing water to provide weights has proven to be the most satisfactory for exclusion of air and the prevention of surface mold and yeast growth. The use of plastic covers to cover the shredded cabbage and seal off contact with air during fermentation is a major development in sanitation and in improvement of the quality of fermenting vegetable matter. The cover is much larger in diameter than the vat itself. It is placed firmly against the shredded cabbage with the edges draped over the sides of the vat to form an open bag (Fig. 6.2). Sufficient water or salt brine, about 700 lb in a 10 × 10 ft vat, is placed in this bag so that the shredded cabbage is forced down into the brine until the brine is level with the surface of the shreds. This provides an essentially anaerobic condition, particularly after carbon dioxide is produced by fermentation. This method of covering is a great improvement over the older methods of using boards and weights or plate and stone. Aerobic yeasts and molds cannot grow to produce undesirable flavors and odors and destroy the lactic acid; occasionally, however, the water may be insufficient for one reason or another, and the kraut shreds are permitted to rise above the sauerkraut brine level. Certain undesirable changes may then occur in sauerkraut improperly covered.

The fermentation of many vegetables and vegetable blends are very similar to that of sauerkraut. This will be noted in subsequent sections.

FERMENTATION IN SALT BRINE SOLUTIONS

There is a great variety of pickled products prepared from vegetable materials including cucumbers. For the purpose of this discussion they will be classified in four categories. The first includes those pickled products prepared directly from vegetables without undergoing fermentation. The second includes those products fermented in a relatively weak brine solution. These include dill pickles and many vegetable blends consumed without further alteration. The third category includes those vegetable products fermented in a relatively high salt brine, such as salt stock pickles that are at a later date refreshed and converted to some type of finished pickle. The fourth previously discussed group includes those products prepared by dry salting with a relatively low salt content such as sauerkraut.

Bulky and whole vegetables and those with a low water content are usually placed in brine solutions for fermentation. Traces of acid in the form of vinegar are sometimes added, especially with leafy products that are low in sugars. This is not recommended since the acid will interfere with normal fermentation. In the United States, pickles prepared from cucumbers and olives are by far the most important of the products fermented, cured, and preserved in brine solutions. In the Orient, blends of vegetables are more important (Fig. 6.5).

Pickles have been prepared in homes throughout the world for centuries

FIG. 6.5 EXPERIMENTAL FERMENTATION OF A BLEND OF
VEGETABLES

and they remain the universal favorites among the fermented vegetable products. There are so many possibilities of types of products that they almost defy classification. The combinations of sugar, salt, acid, spice, dill, and other flavoring substances may yield an endless variety of products. Because of their general acceptance, they were the first fermented vegetable product prepared on a large commercial scale. Furthermore, research on cucumber pickling started at an earlier period, and more work has been conducted to improve the products and avoid losses. Cucumbers, particularly of the more desirable, smaller size, are comparatively expensive and unfortunately, when spoilage occurs, it may often affect the entire content of a barrel or vat. Losses are enormous at times.

Among the many pickled products prepared in the home and now to an increasingly larger extent commercially are those prepared from fresh vegetables. The bread-and-butter type of pickle is the most common one in this group. Since such pickles are not fermented, they will not be included in the discussions to follow. Some of this type of pickle are prepared from cucumbers and other vegetables after soaking in a salt brine for a day or more. Since a slight fermentation and cure often occurs during this short period, they should be included in the second category. Most housewives do not realize that cucumbers in brine ferment and cure rapidly after 24 hr fermentation will have occurred. Some of the desirable characteristics of such pickles are the result of these short fermentations. At room temperatures, fermentation is initiated within 24 hr (see Fig. 4.1, Chap. 4) and 0.1–0.2% acid may be produced. In a week, sugar fermentation may be complete, although the cucumbers are seldom well-cured.

Curing involves changes in physical and chemical properties recognizable in several ways. Cucumbers undergo a typical lactic acid bacterial fermentation. Most obvious is a change in the color of the cut cucumber from a variable white, opaque translucency to a uniform darker, olive-green trans-

FIG. 6.6. CUCUMBERS UNDERGO A TYPICAL LACTIC
AND BACTERIAL FERMENTATION AND CHANGE
INTO A UNIFORM DARK OLIVE GREEN TRANS-
PARENCY

parency (Fig. 6.6). Texture changes also occur. When cucumbers are placed in brine, water is withdrawn and the cucumbers lose weight and become somewhat flexible, rubbery, and tough. They will remain this way until during curing they absorb salt brine that restores the lost weight, and they become firm and crisp. Well-cured pickles will break upon bending, and they may show an actual gain in weight because of absorbed water and salt. Such fermented cucumbers or pickles are more permeable than uncured cucumbers and will absorb sugars, vinegar, and spices rapidly. Permeability changes are the result of changes in the pectins.

Since these cucumbers are ordinarily started in a low salt-content brine, they do not have sufficient salt even in conjunction with the acid produced by fermentation to assure inhibition of enzymes that may soften the cucumber tissue. Dill pickles, sour pickles, and some spiced pickles that are fermented in low salt brines and consumed without further processing will soften unless they are stored at cold temperatures or pasteurized to destroy enzymes after fermentation and curing are complete. Salt stock pickles that are to be refreshed at some later date must have additional salt to inhibit enzyme action.

Effects of Weak Salt Brine Solution

Henneberg (Breed *et al.* 1948) applied the trinomial *Bacillus cucumeris fermentati* to the lactic acid-producing bacterium and recognized that it produced primarily lactic acid by fermentation of sugar. The profuse foaming and evolution of carbon dioxide (Fig. 6.7) were considered products of yeast fermentation by the early scientist. In 1913 Rahn stated, "The secret of pickle curing lies in bringing about quickly the normal fermentation, and after this is over in keeping the acidity as high as possible. Ordinarily the fermentation will come by itself." Rahn recognized the need for sugar in fermentation and also suggested covering the vat with paraffin to exclude air.

FIG. 6.7. PROFUSE EVOLUTION OF CARBON DIOXIDE AT THE START OF CU-
CUMBER FERMENTATIONS

Furthermore, he suggested inoculating a fresh tank of cucumbers with brine
from a fermenting tank. The concepts of Rahn were restated later by Le-
Fevre, Fabian, Etchells, and their co-workers and others in numerous publications. These scientists as well as many others recognized that growth of
the lactic acid bacteria was inhibited by high salt concentrations. In spite of
these observations, variations in salting schedule still occur and spoilages
continue. Cucumbers are sometimes packed in 40°–50° brines, 10.6–13.2%
salt and even higher, and plant operators wonder why fermentation is so slow
or may not even occur.

Following the publication of the book by Henneberg (1926) in which he
recognized species of the genera *Lactobacillus* and *Pediococcus,* Wüstenfeld
and Kreipe (1933) divided the lactic acid bacteria into the genera *Bacillus*
(Lactobacillus), *Streptococcus,* and *Pediococcus.* Vahlteich *et al.* (1935) attributed the high acid production to species of the genus *Lactobacillus.* They
considered that one or more of their isolates were similar to the species of the
genus *Leuconostoc* to which a few years earlier Pederson (1930) had attributed the initiation of the sauerkraut fermentation. Etchells and Jones (1946)
identified all of their isolates from cucumber fermentation as strains of *Lactobacillus plantarum.* Pederson and Ward (1949) and Pederson and Albury
(1950) demonstrated that strains of the species *Leuconostoc mesenteroides*
were active in the early stages of the cucumber fermentation except at times
in high salt concentration brines or in fermentations carried at high temperatures. Faville and Fabian (1949) theorized that a reduction in oxidation-
reduction potential was essential for rapid growth and fermentation by the
high acid-producing lactobacilli. Possibly another important role for the
species *Leuconostoc mesenteroides,* besides producing carbon dioxide and
acid that lower the pH, may be its ability to reduce the oxidation-reduction
potential rapidly. Branen and Keinan (1969) demonstrated that growth of
streptococci stimulated growth of lactobacilli. The leuconostoc may perform

a similar role. The species *Leuconostoc mesenteroides, Pediococcus cerevisiae,* and *Lactobacillus plantarum* are now recognized as important in the cucumber fermentation. Although the presence of *Lactobacillus brevis* is recognized, it is not considered significant.

Several types of pickles such as dill pickles are fermented in low salt brines and may be consumed after fermentation without further processing. They undergo a typical lactic acid bacterial fermentation during which changes occur in acidity, texture, and flavor. The curing involves changes in physical and chemical properties recognizable in several ways. The most obvious is a change in the appearance of the cut surface from a variable white, opaque translucency to a uniform darker, olive-green transparency (Fig. 6.6). Texture changes occur concomitantly.

To quote a statement by Fabian and Wickerham (1935), "Genuine dill pickles are one of the most delightful relishes. When properly cured, they have a characteristic flavor which is lacking in other relishes. The reason for this flavor in genuine dill pickles is not entirely due to the dill and other spices . . . is directly dependent upon the delicate flavor developed during the curing process. The flavor of genuine dill pickles, like that of good beer or fine wines, is the result of biochemical changes produced by microorganisms during the fermentation and subsequent aging process." These statements could also apply to other types of pickles fermented in low salt brines.

Genuine dill pickles have been referred to as the "champagne" of pickles. It is difficult to buy a genuine dill pickle in markets today. Genuine dill pickles are produced by the natural fermentation of fresh cucumbers in a salt brine to which has been added cured dill and a mild blend of mixed spices. At room temperature it requires from 3 to 6 weeks to ferment, cure, and absorb the dill flavor. In fermentation, about 0.7–1.0% acid calculated as lactic acid is formed. Traces of other fermentation products are produced that, when blended with the dill, spice, and salt, yield a pickle that has not been artificially duplicated. It is unfortunate that there are so many hazards in making genuine dill pickles, and that relatively few people value a good dill pickle in its own cloudy brine.

There are many variations used in the preparation of dill pickles, but basically they are fermented in a low salt-content brine of 5% or even less salt. In the past most dill pickles have been made in 50-gal barrels by placing alternate layers of dill weed and spices between layers of clean cucumbers and then filling the barrel with a 5% salt brine. The selection and amount of dill and spice vary with individual preference. In some cases a small amount of vinegar is added to the brine but in general this is not recommended. The fermentation is rapid in the brine containing only 5% salt. The cucumbers will absorb salt until an equilibrium of acid and salt between brine and cucumbers is attained. The salt concentration in the brine will be reduced to about 3%; an acidity of 0.6–1.0% calculated as lactic acid, and a pH 3.4–3.6 will be attained within 1 or 2 weeks depending upon the temperature. The

TABLE 6.1

TWO SELECTED EXAMPLES SHOWING THE EFFECT OF TEMPERATURE AND OF SALT UPON DEVELOPMENT OF ACID AND CHANGES IN BACTERIAL FLORA IN CUCUMBER FERMENTATIONS

Time, Days	Total Acid, %	pH	Total Plate Count 100,000 per Ml	Aerobic Species Gas−	Aerobic Species Gas+	Leuconostoc mesenteroides	Streptococcus faecalis	Pediococcus cerevisiae	Lactobacillus brevis	Lactobacillus plantarum Typical	Lactobacillus plantarum Low Acid
Ser. I: 76°F, 5% Salt, 223 Strains Identified											
⅙	0.02	5.69	10.9	10.9							
1	0.07	5.41	6.1	5.8							
1¼			8.3	5.5	0.3	1.4	1.4				
2	0.07	5.20	1570	15.7	157	314		730	392	195	550
3	0.20	4.80	1460			292	73	940			365
4	0.33	4.18	1800			98	97	1326			470
5	0.38	4.02	1560					1410			234
6	0.44	3.92	2230							820	
8	0.62	3.63	640					840		640	
10	0.71	3.53	4000							3160	
13	0.91	3.47	1970						21	1970	
21	0.74	3.52	21								
150	0.75	3.42									
Ser. H: 65°F, 5% Salt, 232 Strains Identified											
⅙	0.01	5.90	5.1	4.8			0.3				
2	0.05	5.29	100		45	747	55				
3	0.12	5.10	2100			560	1107				
4	0.16	4.92	2100			1467	1400				
5	0.21	4.50	2610			1846	815				
6	0.27	4.46	2950			3380	552	240	368		246
8	0.35	4.27	4100			400		820	240	200	140
10	0.35	4.30	3450					320	1015	730	326
13	0.64	3.90	1610			45	45	180	320	550	184
16	0.57	3.82	910			137			45	137	240
21	0.60	3.72	1230						546	120	1015
28	0.74	3.67	530						350	247	240
35	0.78	3.62	493						246	404	90
42	0.77	3.60	520						116	35	410
54	0.62	3.52	102						67		60
150	0.60	3.58									

Source: Pederson and Albury (1950).

pickles will change in appearance, color, and translucency and eventually regain the lost weight. The cured pickle will have a distinct, dill-spicy, mildly acid, and salty flavor.

During fermentation, the brine becomes increasingly cloudy for the first few days due to growth of bacteria, and an effusive liberation of gases occurs with foaming. Later, if the brine is not covered to exclude air, a filmy yeast growth will often occur on the surface. This growth is undesirable since the yeasts utilize the lactic acid, neutralize the brine, and make possible the growth of other microorganisms. Softening may result from such growth. Without yeast growth, the lactic acid bacteria normally settle onto the pickles and to the bottom of the container.

Fabian and Wickerham (1935) in their microscopic study of the changes that occur during fermentation noted a definite sequence of bacterial populations in genuine dill pickle fermentation. At the beginning, Gram-positive cocci predominated. These, in turn, were replaced by short rods and these, by long rods. From the relatively high volatile acid production reported it may be assumed that the cocci were strains of *Leuconostoc mesenteroides*. Later Pederson and Ward (1949) and Pederson and Albury (1950) (Table 6.1) observed initiation of fermentation in many instances by *Leuconostoc mesenteroides*, particularly at lower temperatures or lower salt concentrations. At higher temperatures or salt concentrations fermentation was often initiated by *Streptococcus faecalis, Pediococcus cerevisiae, Lactobacillus plantarum* and, to a lesser extent, *Lactobacillus brevis*. The latter two species are responsible for the high acidity attained.

Genuine dill pickles are very susceptible to softening on the one hand and bloating on the other. The larger-sized cucumbers and those packed in high concentration salt brines are prone to bloat. When cucumbers are packed in lower concentration brines, they may soften because of enzymatic action produced by yeasts during natural fermentation or as a result of the native enzymes of the cucumber. During fermentation, yeasts may develop readily on the surface of brines and cause softening. For many years dill pickles have been prepared in barrels which were headed immediately, leaving only a small opening for gas to escape. The barrels were rolled at intervals to cause the yeasts to settle, the brine was added so that the barrel was always full, decreasing the opportunity for yeast growth. An anaerobic condition is now attained by the use of plastic covers as practiced in the sauerkraut industry. The total acid produced by fermentation in conjunction with the salt is not great enough to inhibit completely enzymic activity. Softening is usually considered an enzymatic change, and the source of the enzymes is generally attributed to the cucumber itself and to those secreted by microorganisms, either bacteria, yeasts, or molds. This will be further discussed in the section on salt stock pickles. Alum is sometimes added to firm the fermented pickles. Softening can be delayed by pasteurization to inactive enzymes or by cool storage to inhibit their activity. Genuine dill pickles pasteurized at 165°F for

15 min followed by rapid cooling will remain firm and crisp for long periods (Fabian and Wickerham 1935; Jones *et al.* 1941).

Considerably more attention could be given to the effect of temperature, not only upon the fermentation but also upon storage of the finished product. It is generally recognized that dill pickles soften more readily in warm weather or in dry, warm years than in cool weather. Some packers prepare dill pickles during the cooler season of late fall. The fermentations at the lower temperature of 65°F or even lower are apt to be initiated by *Leuconostoc mesenteroides* while those at higher temperatures may be initiated by *Streptococcus faecalis,* followed by the homofermenters (Pederson and Albury 1950). While the rate of fermentation is slower at the lower temperature, the quality of the product will be superior. These colder dill pickles may be kept in the brine for longer periods of time before softening occurs. Furthermore, if dill pickles are stored at the low temperatures of even 35–40°F, they will remain firm and crisp for an extended period. Because of the possibility of softening, some recommendations advise covering the fresh cucumbers with a higher salt concentration brine with a 36°–40° salometer (9.54–10.6% salt). When fermented and cured, this will yield a finished pickle containing between 0.7 and 1.0% of acid and between 4.25 and 5.5% salt. Many consumers object to the saltiness of such a pickle because the salt tends to mask the delicate flavor of dill, spice, and acid.

There are numerous recipes for making dill pickles, and in many less than even a 5% brine is used. A lower salt concentration brine of 20°–25° (5.3–6.6%) will yield a finished pickle containing 3.6–3.3% salt, a balance of salt, acid, and dill-spice that is acceptable to the consumer. To retain the natural flavor, pickles should be marketed in their own brine. It is unfortunate that consumers are so accustomed to clear solutions on foods that they fail to recognize that finest quality is often dependent upon keeping a product in its natural condition.

Dill-flavored pickles are frequently prepared from fresh cucumbers, sometimes after soaking the cucumbers for several hours. Dill-flavored pickles are also prepared from salt stock after refreshing. The latter pickles are fermented. They are discussed more fully under salt stock.

The various sour pickles and spiced pickles differ from dill pickles only in the flavoring substances added. Although at one time it was believed that spices and herbs affected the growth of microorganisms, this has not been substantiated.

Green tomatoes as well as other vegetables have been fermented in weak brines using dill and spices. No other vegetable, however, has been found as satisfactory as the cucumber for this type of fermented food.

FERMENTATION OF BLENDS OF VEGETABLES

Nearly all vegetable substances whether leafy, tuberous, or those containing seeds provide sufficient nutrients for the growth of lactic acid bacteria. The leafy vegetables sometimes may be low in total sugars; the root vegetables high in starch and sugar sometimes may be deficient in other nutrients; and some seed vegetables such as beans and peas may be so highly buffered that during fermentation a considerable amount of acid must be produced before a marked change in pH is attained.

In the Orient, considerable quantities of vegetables are raised and preserved for future use by fermentation. Various blends of vegetables are prepared. Apparently these blends utilize the vegetables currently available. There obviously are planned variations also. For example, in Korea the mild-flavored blends of kimchi are strikingly different from the hot peppery blends of kimchi. The large white radishes, turnips, Chinese cabbage, cabbage, peppers, and other vegetables are used in the various blends. In some preparations, seaweed, fish, and even nuts are reportedly added. Paw tsay, yentsai, nukamiso, sajur asin, karanyshe dong, so chican, dûa chua, and kimchi are some of the names applied to the various blends of the Orient apparently prepared by the majority of the people. Rice liquid is used in some blends. In the fermentation of sajur asin in Indonesia, the criterion for completion of fermentation is the clearing of the rice liquid. Sajur asin means salty vegetables. Green leaves of a variety of brassica are used for this preparation. Other grain infusions as well as milk are added to these blends in some areas. Apparently, the tarhana prepared in Turkey is one variation of this type. Rice bran is sometimes used in the nukamiso vegetable blends prepared in Japan.

In general, the preparation of these fermented vegetable blends is conducted by methods that have been in practice for centuries. There is little doubt that considerable variation occurs and that the various housewives take pride in the quality of their product. In general, these are home-prepared products; therefore, they are not subjected to the scrutiny of research that follows commercial production. Hsio Hui Chao (1949) studied the fermentation of the vegetable blend prepared in China; however, it is doubtful that his methods of preparation resembled those used in the homes. The more recent studies by Kim and others in Korea (Kim and Whang 1959) indicate that the fermentation of the vegetable blend, kimchi, is very similar to that occurring in sauerkraut, initiated by *Leuconostoc mesenteroides* and continued by the other lactic acid bacterial species.

Kim and Whang have provided considerable information in regard to the vegetable content of blends. Tongkimchi is made from Chinese cabbage, sliced radishes, red pepper, fish, and other ingredients. Dongchimi kimchi is prepared with long white radishes as a major vegetable. A winter kimchi is made with white radishes, Chinese cabbage, salted fish, and various flavorings including leek, onion, red pepper, ginger, water cress, and certain seaweeds. Chestnuts, mushrooms, oysters, starch solutions, beef extracts, and octopus extracts are used. The vegetables, spices, and flavorings are sandwiched between the leaves of the Chinese cabbage, and then the leaves are pulled together and tied. These are packed tightly in fresh salt brine for fermentation.

Kim and Whang (1959) ascribed the fermentation to species of *Leuconostic mesenteroides, Lactobacillus brevis,* and *Lactobacillus plantarum* but also isolated strains of species of *Streptococcus* and *Pediococcus.* They reported a low acidity attainment of 0.4–0.6% acid. This is not surprising since the blends are packed in brine and the ingredients are not high sugar-containing substances. In addition to lactic acid, they reported the presence of citric, malic, succinic, gluconic, and fumaric acids.

Hsio Hui Chao (1949) reported a lactic acid bacterial fermentation of paw tsay. Cabbages, radishes, turnips, dandelion greens, persimmons, cucumbers, bean sprouts, water chestnuts, pepper, garlic, onion, eggplant, and lotus ginger are all used in preparing paw tsay.

In a more recent study of the fermentation of vegetable blends by Orillo *et al.* (1969), typical lactic acid bacterial fermentations similar to that occurring in sauerkraut were demonstrated (Table 6.2).

TABLE 6.2

MICROBIOLOGICAL AND CHEMICAL CHANGES DURING THE FERMENTATION
OF VEGETABLE BLEND WITH SOYBEAN (BLEND NO. V)[1]

							Estimated Number of Each Species of Bacteria $\times$ 10^6			
Time Hr	Temp C	Salt %	Total Acid %	pH	Total Count $\times 10^6$	Aerobic Species	Leuconostoc mesenteroides	Lacto-bacillus brevis	Pedio-coccus cerevisiae	Lacto-bacillus plantarum
3	30	.	0	5.90	7	6.3	0.7			
10		2.77	0.17	5.10	470		470			
21	30	2.23	0.49	4.64	3700		3700			
27	32	2.34	0.5	4.50	2700		2600			100
34	33	2.25	0.77	4.30	1200		864	46	144	144
45	31	2.49	0.99	4.15	1630		68	204	340	1020
51	31	2.23	1.14	4.09	2100		84	336	336	1343
60		2.40	1.24	3.94	1540			124	62	1354
68		2.55	1.32	3.94	1660			330		1330
76	31	2.23	1.41	3.90	2500			900		1600
94	31	2.27	1.50	3.86	1320			50		1270
119	30	2.30	1.51	3.86	1340			60		1280
143	30	2.34	1.63	3.75	570			250		320
165	30	2.23	1.83	3.78	64			6		58
213		2.30	1.85	3.80	171			64		107
285	29	2.34	1.86	3.78	90			35		55

Source: Orillo *et al.* (1969).
[1]Approximate percentages of vegetables used in blend No. V: cabbage, 32; singkamas, 21; carrot, 10; pepper, 4; cooked soybean, 31.

In the Philippines, shredded vegetable blends are frequently prepared for special occasions. These are acidified with vinegar and sweetened. Among the seven vegetable blends studied by Orillo *et al.* (1969), various proportions of white-headed cabbage, the turnip-like vegetable singkamas, large white radishes, carrots, pepper, asparagus, beans, and mustard greens were used. Cooked soybeans were used in two blends and mango bean sprouts in another to increase the protein content of the blends. Since soybeans are so high in protein and have a low water content, the beans when added to a relatively high water-content vegetable blend, will markedly increase the protein content based on dry weight. The soybeans acquire a somewhat chewy peanut-like flavor. It seems unfortunate that more attention is not given to the possibilities of utilizing these various vegetables by fermentation processes. In the Orient it seems possible that high protein-content foods, particularly fish, could be used more frequently in these blends than they are at present.

All of these blends were fermented in a manner similar to that of sauerkraut, and for two preparations acidities of 1.8–2.0% were attained in 2 weeks. The 2 blends containing soybeans were particularly interesting in that the total protein was increased yielding 34 and 23% on a dry basis. The pH was 3.78 and 3.75 respectively, with acidities of 1.86 and 1.80% acid as lactic acid.

OTHER FERMENTED VEGETABLES

Beets, turnips, radishes, chard, cauliflower, Brussels sprouts, mustard leaves, lettuce, sliced green tomatoes, green beans, fresh peas, okra, and other vegetables have been fermented separately in brines or by dry salting methods. In all of these studies the fermentations were initiated by the species *Leuconostoc mesenteroides* and continued by the other lactic acid bacterial species.

Beets are fermented in certain areas in Europe and in the borscht prepared in Russia. When cleaned, sliced, salted, and packed in a container, sometimes it is essential to add some water. Because of their high sucrose content, a considerable amount of dextran is formed causing the slices to become very slimy. There is no objection to this condition other than the physical character. The same type of sliminess sometimes occurs in the early stages of the sauerkraut fermentation and disappears as fermentation continues; it is, however, undesirable when some of the fresh low acid krauts are used in homes or retailed in bulk but only because of its abnormal appearance. Sliced cauliflower and tomatoes when fermented also become slimy. Okra because of its nature is in the same category. Chard, celery, and Chinese cabbage are more or less tasteless, and because of their usual low sugar content, little acid is produced by fermentation.

Experimental studies of two fermented vegetables, lettuce and mustard, have been reported in greater detail. In the fermentation of lettuce reported by Hohl and Cruess (1942), *Leuconostoc mesenteroides* was isolated in 1 of 2 natural fermentations and *Lactobacillus brevis* and *Lactobacillus plantarum* predominated. A maximum count of 17.8×10^7 was reported for the 5th day of fermentation with an acidity of 1.13% acid reported as lactic acid of which 0.152% was volatile acid. Sugar was reduced from 1.85 to 0.075%. This may be an instance of a vegetable fermentation that was not initiated by *Leuconostoc mesenteroides.*

Mostasa, commonly prepared from mustard leaves in the Philippines, was studied by Palo and Lapuz (1955). The organism they described as a species of the genus *Streptococcus* is actually a typical strain of *Leuconostoc mesenteroides.* This product was later studied by Orillo and Pederson (1966) and found to ferment in a manner similar to other vegetables; however, because of its low sugar content, a high acidity cannot be attained. Pickled mustard greens, pak-gard dong is prepared in Thailand and tang-chai is a blend of mustard greens and cabbage.

A number of vegetable fermentations by lactic acid bacteria have been observed over a period of years. They are not used for preparation of foods but are interesting in relation to the forementioned fermentation. Some are considered spoilage conditions such as the lactic acid fermentation of peas, beans, corn, and other vegetables being prepared for freezing. In order to fill containers to a proper weight, some of the blanched vegetables are placed on the side of the belt line, from which small amounts may be added or removed from each package. Spontaneous fermentation, particularly by *Leuconostoc mesenteroides,* often occurs in such vegetables and, although usually not serious, may become so if the packages or cans of vegetables are not frozen quickly. In one such case involving diced carrots frozen in 30 lb tins, acid fermentation by lactics did occur causing a considerable loss. Coincident with this, a patent has recently been obtained for carotene concentration from carrot juice in which a fermentation by lactic acid bacteria is involved. Following fermentation the solids are readily separated from the liquid portion, and the latter may then be concentrated and added to the puree to concentrate the carotene.

The seeds of many vegetables have been separated in the past by a lactic acid fermentation in which *Leuconostoc mesenteroides* plays a major role. After fermentation of the pulpy materials seeds are readily removed.

In the preparation of tomato juice in the early days of that industry, lactic acid fermentation occurred frequently and at times caused considerable loss. With the introduction of modern, streamlined equipment such losses have been reduced to a minimum. Flat-sour spoilage by the spore former *Bacillus thermoacidurans* is now a more serious cause of losses. A fermented tomato juice was prepared for a short time but discontinued. This involved fermentation by yeasts and lactic acid bacteria to produce a pleasing effervescent acidic low-alcohol beverage. Sliced green semiripe and ripe tomatoes are fermented readily by the lactics. Sliced green and red tomatoes with seeds removed are used in some areas. Many other similar fermentations are utilized in some areas.

A variety of wild plants have been reported to have been used in the past for pickling. Included are roots of wild onion, cattail, false spikenard, solomon's seal, Indian cucumber, live-forever, bugleweed, Jerusalem artichoke, succulent leafy or young plants or fleshy branches of sapphire, pokeweed, sea-pursline, sea-milkwort, and flower buds of marsh marigold, barberries, redbud, elder buds, and ash buds. These have usually been placed in salt solutions and later packed with weak vinegar. There is little scientific evidence to corroborate such reports.

VARIATIONS OF FERMENTED CABBAGE

Fermented cabbage has been used for many years and in many forms. A knowledge of the fermentation and preparation has not always been available in order to produce the lactic acid-fermented product available today either in cans or as a fresh product or even as the product preserved by chemicals in a plastic container. In some societies, the people have become so accustomed to either the well-aged product or to one in which yeast fermentation has yielded a marked, even cheesy, flavor that it is these types of products that are acceptable to them. In various areas of Europe, spices or fruit, particularly sliced apples or pears, are added to the shredded product. The choucroute, originally prepared by the French, consisted of chunks of cabbage immersed in wine. Later, salt brines became more common. In other areas, the whole heads of cabbage, often red varieties, were fermented in brine solutions and when fermented, were either cut or leaves were removed to prepare the particular dish such as the sarma and the podvarak of Serbia. The brine from kraut, particularly that made from red cabbage, has a pink wine color and is relished as an appetizer. Many people drink canned sauerkraut juice from canned kraut, but such juice is not equal in character with the spritely effervescent product obtained directly from a vat of freshly fermented sauerkraut.

Fermentation of whole heads of cabbage in salt brines is common in some areas of Europe. Studies by Pederson *et al.* (1962) on the fermentation of whole heads of cabbage in different concentrations of salt and whole heads mixed with shredded cabbage have demonstrated a fermentation similar to that of shredded cabbage. The fermentation was initiated by *Leuconostoc mesenteroides* and continued by *Pediococcus cerevisiae, Lactobacillus brevis,* and *Lactobacillus plantarum.* The Gram-negative aerobic species of bacteria persisted for a longer period, e.g., three days, in brines containing only the whole heads. This is further evidence that the bactericidal substance of cabbage active for these organisms resides in the protoplasm. Yeasts were more numerous in these fermentations than in shredded sauerkraut. Softening of the core areas and of leaves adjacent to the cores was common, indicating that salt did not penetrate well and that rapid acid fermentation did not occur within the head.

CAULIFLOWER

Cauliflower has been one of the most difficult of the vegetables to handle in brining. Numerous methods of brining have been and continue to be used. One common method is covering the cauliflower with a 60° salometer brine so that no fermentation occurs and continuing the addition of salt as the salt is absorbed by the cauliflower holding the concentration at 60°. Since no fermentation can occur in such brine, no acid is produced, and a low oxidation-reduction potential is not attained. The cauliflower usually darkens; therefore, sodium sulfite is frequently used in refreshing the product. Other processors use lower salt concentration brines to allow a limited fermentation to occur, and then they raise the salt level rapidly to 60° brine concentration.

Cauliflower packed in a brine with sufficient salt to give 2.5% concentration will initiate fermentation fairly rapidly with profuse growth of *Leuconostoc mesenteroides* followed by growth of the other lactics in order. Since considerable amounts of dextran are frequently produced by the leuconostoc, the brine and cauliflower may become very slimy. Softening may occur also. In higher concentration brines the growth of leuconostoc is suppressed and the rate of fermentation is retarded, but dextran production is retarded. In one such fermentation started in a 5% brine (20°), the highest count of leuconostoc was 42 $\times$ 10^6 per ml. This is only about $\frac{1}{20}$th of the number of leuconostoc frequently attained in a sauerkraut fermentation. A maximum brine acidity of only 0.39% was attained and the minimum was pH 4.10 after 22 days when it was deemed advisable to begin raising the brine concentration to 60° or about 15%. After four months the cauliflower was firm and crisp with a clean, salty, acid flavor and a slightly gray-pink color. During refreshing, partly in 0.2% sodium sulfite solution, this color was reduced to a satisfactory white color.

Cauliflower is frequently used in mixed pickles. The amounts produced are relatively small. However, this fermentation warrants further study.

OLIVES

The olive is one of the oldest fruit crops grown in the Mediterranean area. Most of the crop has been used for oil, but the pickled olive industry is growing rapidly and now is a sizeable industry in the United States. Olives have, however, been consumed for centuries but not always prepared in the manner used today. Pickled olives were found in the ruins of Pompeii.

The continuous increase in olive production in California in the latter part of the past century resulted in a demand for information on the treatment of the crop in pickling. Cruess (1924) classified six processes for pickling olives in Mediterranean countries as follows: Spanish green olive, French brine, dry salt, water, and Italian dried. In the first two processes, lye is used to destroy the bitter glucoside, oleuropein. Vaughn (1954) mentions six types of processed olives: California green-ripe, Greek-type, California ripe, brined Greek-type, Siciliano-type green, and Spanish-type green. The first two are not fermented and will not be discussed further. The Spanish-type green olive as packed in California is an example of practices as a result of application of scientific knowledge. Vaughn et al. (1943) reviewed the literature and presented the results of their investigations. They reasoned since so much sugar was removed during lye treatment, sugar must be added to the brine before a satisfactory fermentation could be attained. There was little need to add a pure culture of brine from a fermenting barrel of olives as an inoculum unless sufficient sugar was present.

Like cucumbers, olives are fruits, but they are classed as vegetable substances and are fermented in a manner similar to other vegetable materials. The fruits are harvested when fully developed but still green. They are graded, usually into eight size grades. The fermented green olives are more common than the fermented ripe olives that are treated in a different manner.

Spanish-type green olives are subjected to a lye treatment; Siciliano-type green olives are fermented without lye treatment. The green olives are covered with a 1.25–2.0% lye solution at a temperature of 60°–70°F (15.6–21.1°C). When the lye has penetrated to ½–¾ of the way toward the pit, they are leached with several changes of water to remove the lye. The bitterness caused by the glucoside, oleuropein, is reduced by the lye treatment and leaching. Some bitterness is desirable. The olives are then covered with a salt brine and allowed to ferment. Since the Sevillano and Ascolano varieties are subject to shrinking, they are covered with the weaker brine of 20°–25° salometer (5.2–6.25% salt), and after fermentation has been initiated the salt concentration is raised to 6–8%. Manzinello and Mission varieties are started in 40°–50° brine (10–15% salt). Since the olives will absorb salt from the brine, a 5–6.25% brine will be reduced to about 2.5–4.0% and a 10–15% brine will be reduced to about 6–9% after equilibrium is attained between brine and olives. Vaughn et al. (1943) pointed out that Cruess had stated as much as 65% of the fermentable sugar is leached out, and fermentation will not occur rapidly in cold weather or in high salt content brines. By addition of sugar, fermentation can be completed in about 3 weeks during which time an acidity of 0.7–1.0% acid calculated as lactic acid and a pH of 3.8 or lower may be attained.

Needless to say, the rate and extent of fermentation will depend upon the salt concentration, the temperature, and the amount of sugar present. Vaughn et al. (1943) and Vaughn (1954) stated that olives are fermented by the same group of bacteria that ferment sauerkraut and pickles, that is, Leuconostoc mesenteroides, Lactobacillus brevis, and Lactobacillus plantarum. However, other species of the genera Aerobacter, Escherichia, and Bacillus and yeasts persist for longer periods than in sauerkraut and pickles, and spoilage is apparently more apt to occur. Vaughn et al. (1943) divided the fermentation into 3 stages, the first stage lasting for 7–14 days, during which little change in pH and acidity occurs and when organisms other than lactics are present in large numbers. In the second stage the species *Leuconostoc mesenteroides*, *Lactobacillus brevis*, and *Lactobacillus plantarum* predominate; counts in the hundreds of millions as well as an increase in total acids are attained. During the third stage *Lactobacillus plantarum* is the predominant species; but even then the desirable high acidity and low pH may not be attained. The addition of pure cultures of *Lactobacillus plantarum* has been suggested. It is further brought out that Manzinello, Mission, and Barouni olives that usually ferment so slowly, would ferment rapidly if placed in low concentration brines. Since Sevillano and Ascolano olives shrivel more readily in high concentration salt brines, possibly this fact has made them more acceptable since low concentration salt brines must be used.

Fermented green olives after sorting, grading, and washing are packed in glass jars under vacuum, covered with a 28° salometer brine, and pasteurized at 140°F or covered with a brine at 175–180°F. Ripe olives are ordinarily low in acid and are, therefore, processed at 240°F for 1 hr.

Gassiness and bloating were noted by Cruess (1930) and partly ascribed to use of strong lye solution. West *et al.* (1941) found that strains of the genus *Aerobacter* were the more common coliforms responsible for gassy deterioration. Later, Vaughn *et al.* (1943) associated yeasts with such conditions.

The cheesy, malodorous spoilage referred to by Cruess (1924) as "Zapatera" was associated with growth of species of the genus *Clostridium* by Kawatomari and Vaughn (1956). Besides lactic and acetic acids, formic, propionic, butyric, and succinic acids were present in such brines. Later, the cheesy odor was associated with growth of species of the genus *Propionibacterium* by Plastourgus and Vaughn (1957). These malodorous spoilages are difficult to characterize. A similar cheesy malodorous fermentation was described in sauerkraut by Vorbeck *et al.* (1961), who found about 1000 times as much n-butyric acid in spoiled samples as in good samples. Other members of the lower fatty acid series were also present in abnormal amounts.

Vaughn *et al.* (1953) identified the so-called white yeast spots so often seen under the skin of olives as colonies of *Lactobacillus plantarum*. These also occur at times under the skin of cucumber pickles and green tomato pickles. Sometimes yeasts cause similar spots.

Cruess and Guthier (1923) stated that Sevillano, Ascolano, and Manzanillo olives are more subject to softening than Mission olives. They related the condition to a bacterial decomposition and recommended an adequate pasteurization. Nortje and Vaughn (1953) related softening to species of the genus *Bacillus* and studied their pectolytic activities. Later, Balatsouras and Vaughn (1958) related softening to various species of molds. It may be concluded that enzymes produced by microorganisms are the primary cause of softening. It must also be realized that any food product stored for extended periods, particularly at raised temperatures, will soften.

These various spoilage conditions may be associated with the relatively slow fermentation. Vaughn *et al.* (1943), among other recommendations, advised the careful control of salting, addition of sugar to low-sugar olives, and careful control of lye solutions.

The Sicilian type olives are not lye-treated to remove oleuropein, are covered with a relatively weak brine, and salt is added until a concentration of 7–8% is attained. Barrels are kept full of brine to suppress yeast growth. They ferment rapidly since the sugar has not been removed by a lye treatment. The high final salt concentration and the acidity of 0.4–0.6% suppress yeast growth. Higher salt concentrations are used with Manzinello and Mission olives than with Ascolano and Sevillano olives; therefore, the fermentations are slower. The same sequence of bacterial growth occurs as that which occurs in Spanish-green olive fermentation.

Little information is available in regard to the fermentation of Greek-type olives that are started in 7–10% salt brines and increased to 15%. The olives cure slowly, and the yeasts possibly destroy the little acid that is produced. The highly-colored to jet-black color is obtained by lye treatment followed by aeration. The natural bitterness of this type of olive does not appeal to the American consumer.

Olives to be converted into ripe olives are also picked when green to straw-colored and immediately barrelled and covered with a 5–10% brine. Lactic acid fermentation takes place in the brine in a manner similar to the Spanish-green olives. Later, the olives are given a lye and aeration treatment followed by leaching to remove the oleuropein. The first lye barely penetrates the skin. This is followed by an aeration treatment to darken the skin. Further lye and aeration treatments are used to penetrate to the pit, removing nearly all of the glucoside and producing the final dark-colored olive. After each treatment the lye is leached out with water and often with weak acid.

Vaughn *et al.* (1943) discussed the various spoilage conditions that occur during fermentation. Gassy deterioration, malodorous fermentation, Zapatera spoilage, yeast spots, and softening are among the more important. In consideration of the rate of fermentation and the lye treatment with accompanying removal of sugars and other soluble components, it is not surprising that olives may be subject to spoilage in one form or another. It has been demonstrated that vegetables contain cellular bactericidal substances. If olives contain such substances, what may happen to them during lye treatment and subsequent leaching is questionable. Certainly many of the solubles necessary for growth of bacteria of the lactic acid group are leached away. Vaughn has even suggested addition of nutrients other than sugar.

SALT STOCK

The preparation of salt stock is a commercial development designed to handle the large quantities of cucumbers within the short harvest season. At a later date the pickles are refreshed and converted to any one of a number of types of pickles. The cucumbers are immersed in a 20°–40° salt brine, and then additional salt is added at intervals in order to raise the salt concentration to about 60° or about 15% concentration. The final concentration of salt should be high enough so that, in combination with the acid produced by fermentation, they will inhibit enzyme action and the brine will not freeze during the winter. Pickle stock may be stored for as much as 1 yr or more before conversion to finished pickles. When needed, the salt stock pickles are refreshed in several changes of water. At least one change of water is hot enough to inactivate enzymes. After refreshing they are made into any type of pickle desired.

One commercial salting procedure used in the past is briefly as follows. From 1 to 3 ft of a 40° brine, e.g., a 40% saturated salt solution, is placed in the bottom of a large wooden vat. The vat is filled with fresh cucumbers, and during filling either dry salt or more 40° brine is added to keep the brine concentration at 40°. When the vat is full, it is covered with a wooden cover, weighted with stones, or keyed down to hold the cucumbers in the brine. Enough additional brine is then added to cover the wooden cover with brine. After filling, fermentation occurs and the brines become cloudy due to growth of the bacteria responsible for the fermentation. In commercial operation the brine strength is increased, usually daily at first and then at weekly intervals to raise the salt content to 60° or higher in about 6 weeks. Under these conditions the salt stock may be held for considerable periods of time. The high concentration of salt prevents freezing of stock during winter months, and the combination of salt with the acid prevents further enzyme action and microbial growth if properly covered to exclude air. There are many variations in the method of filling and strength of brine used. In the opinion of the writer, some of these variations result in a better fermentation and cure while others inhibit a natural fermentation and result in inferior products.

Considerable losses may occur due to softening of the stock, to bloating (Fig. 6.8), blackening, and other forms of spoilage. This has led to many variations in practices in the industry to avoid such losses. This applies to the concentration of salt used and the methods of adding salt. Unfortunately, there has been much misunderstanding in the industry in regard to the role of salt. It has been assumed by many that salt alone was responsible for preservation and that acid development was only incidental. A critical analysis of various practices should be considered in deciding what concentration of salt is to be used in the brine for fermenting cucumbers. The factors which should be evaluated are: the effect of salt in brine on the growth of micro-

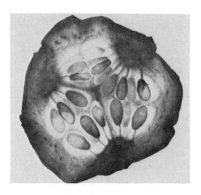

FIG. 6.8. SLICE OF LARGE CUCUMBER SHOWING TYPICAL SEPARATION OF
LOCULES, THE FIRST STAGE IN BLOATING

organisms; the inhibition of salt and acid on the activity of enzymes that cause
softening of cucumbers; the relationship between brine concentration and
extent of bloating; the physical effect of brines on the cucumbers, and the
weight required to submerge or hold cucumbers below the surface of the
brine.

Lactic acid bacteria require the presence of small amounts of salt for optimal growth. Salt concentrations below 2% have little effect upon the extent
of growth, but above this concentration inhibition varies with the specific
strains and species; therefore, no absolute tolerance can be specified beyond
which no growth will occur. A little growth of lactic acid bacteria should
occur in the first 24 hr, but subsequently the rate of growth and fermentation
will increase rapidly. Brine acidities of 0.7–1.0% may be attained in a week
in low salt concentration brines and suitable temperature. Actual curing of
the stock will require a month or more. Since many strains of lactic acid
bacteria are unable to survive and grow in 10.6% or 40° salt brine (Fig. 6.9),
initiation of growth and fermentation is curtailed. In high concentration salt
brines, above 40°, growth and fermentation by lactic acid bacteria may be
retarded sufficiently to allow undesirable changes to occur. Restriction of
fermentation is a function of the combined effects of salt and acid concentration. In an 8% salt solution there is less acid produced than in a 2% salt solution. It should be realized that the addition of dry salt either during filling
the vat or in subsequent salting will dehydrate all bacteria in immediate contact with the salt and drastically affect growth of the lactic acid bacteria.
The concentration of salt in the brine, 20°, 30°, or 40°, is of critical importance. It is a well-known fact that the higher the salt concentration, the
greater will be its inhibiting effect upon growth of lactic acid bacteria and the
lower the amount of acid that will be produced. Although this has been noted
in the reports of several researchers, it was summarized adequately by
Etchells *et al.* (1943) in their statement that there is a direct correlation be-

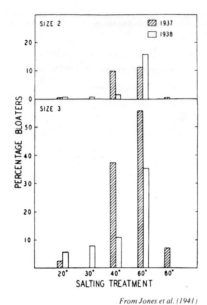

From Jones et al. (1941)

FIG. 6.9. RELATION BETWEEN THE SALTING TREATMENT THAT IS FOL-
LOWED AND THE PROPORTION OF BLOATERS FORMED IN SALT STOCK
CURED IN SMALL VATS

tween salt concentration and bacterial population. This is adequately substantiated by the later work of Pederson and Ward (1949).

Just as the growth of microorganisms is a function of the concentrations
of salt and acid so also is the activity of enzymes that cause softening of cucumber tissue. Enzymes indigenous to the cucumber will continue their activity until the combined inhibitory action of acid and salt deters further
activity. In low concentration brines, enzyme activity may be so rapid at
higher temperatures that some softening may occur before acid is produced.
Similarly, in high concentration brines, growth of lactic acid bacteria may be
so delayed that softening may occur.

The effect of salt concentration on the extent of bloating was demonstrated
by Jones *et al.* (1941) and confirmed by many other studies and observations.
Jones *et al.* observed less bloating in 20° brines than in 30° brines and less in
30° than in 40° brines. When cucumbers are placed in salt brines, water is
withdrawn from the cucumber until an equilibrium is attained between brine
and cucumber. The higher the salt concentration the greater the amount of
water that must be withdrawn; therefore, in high concentration salt brines
or in dry salting procedures, salt may practically dehydrate the cucumber.
As a result, it will close the vascular system and dry and toughen the epidermis so that salt brine cannot be reabsorbed by the cucumber (Pederson and
Albury 1962). The smaller-sized fresh cucumbers seldom present a problem

of bloating. They are ordinarily brined as soon as they are obtained, and they absorb salt rapidly and equilibrium is attained quickly. With large-sized cucumbers, separation of locules as a triangular split at the center may occur before the cucumbers are even placed in the brine. When such cucumbers are placed in high concentration brines, this locule area increases in size, microorganisms may enter and grow, and a true bloater is produced.

The salt in brine increases the specific gravity of the brine; the higher the salt content is the higher the specific gravity will be and therefore, the greater the weight required to hold cucumbers in brine. Tremendous pressure is exerted on cucumbers when placed in high salt concentration brine to hold them in the brine. This may be great enough so that the cucumbers may be permanently distorted.

The factor seldom considered in brining is this buoyant effect of the brine on cucumbers and the relative weights required to submerge the cucumbers in the brine. In a 10 × 10 ft vat, approximately 1½ tons of weight are required to hold cucumbers down in a 10% brine. Only ½ of this weight is required in a 5% brine. Absorption of brine should begin immediately and as cucumbers absorb salt, the weight required decreases. Absorption of salt is most rapid with small fresh cucumbers. Large and overly mature cucumbers absorb salt slowly. The physical effect of great weights is obvious and, when considered in relation to the dehydrating effect of salt, accounts for the relatively flattened state of cucumbers.

In the opinion of the writer, based upon these factors and upon the observations of Fabian, Etchells, and their colleagues and of many others who have conducted research in cucumber fermentation, salting is the most critical factor in preparing salt stock. The starting of cucumbers in a low brine concentration, 30°, and waiting until fermentation has started and the salt is partially equilibrated before adding salt gradually day by day seems a more logical method of salting. During filling of the vat, 30° brine should be added instead of dry salt. After filling, daily additions of small amounts of salt to raise the salt concentration only 1° should be made. This will require more frequent attention and a longer period of time to restore the salt concentration to 30°, but such additions will have little adverse effect upon growth and fermentation. By this time, fermentation should be advanced and equilibrium will be established. Salt can then be added in larger quantities to eventually raise the concentration to 60° brine strength.

This procedure would allow a rapid fermentation so that acid and salt could inhibit activity of enzymes inherent in the cucumber, there would be less dehydration of cucumbers, and there would be less pressure exerted on the cucumbers.

SOFTENING OF PICKLES

Every year the pickling industries experience considerable losses due to softening. The microbiological and cytological changes that occur in softening have been a subject of investigation for almost a century. Softening may occur within a few days after packing cucumbers in brine or may not be apparent until months after they are packed. It may appear in bloated stock and in heat-processed pickles. It is generally assumed that softening of pickled products is due to the decomposition of pectic substance of the cucumber. In extreme cases, entire vats of pickles must be discarded since the product becomes so mushy that it cannot be used in any by-product. At other times some of the cucumbers may be utilized for relish stock. Softening occurs in several forms. It may be apparent as a partial softening at the blossom end; it may occur in the seed-containing locular area; it may occur as a general breakdown of the seed area (specifically, bloated stock); it may occur in the area adjacent to the epidermis, causing a sloughing of the epidermis; it may occur as a slippery epidermis; or it may occur as a general softening throughout the pickle. Softening was attributed to bacteria by Van Tieghem 90 yr ago. This was followed by the observations of Vignal, Aderhold, Kossowicz, Jones, Rahn, and LeFevre concerning the influence of various organisms on softening. Rahn and later LeFevre emphasized the role of the lactobacilli in producing the acid essential for inhibiting the activity of these pectolytic bacteria. In contrast, later Lesley and Cruess (1928) associated softening with high acidity. Obviously, these types of deterioration are not to be considered similar. It is doubtful that the various conditions can be attributed to any single specific agent, although in general, softening is often ascribed to enzymatic activity. The enzymes may be native to the cucumber, including the blossom residue or enzymes secreted by bacteria, yeasts, or molds. Other types of softening may be related to a general change in the food product due to the slow chemical changes that occur in any processed food.

Fabian et al. (1932) observed certain structural differences among various lots of pickles. The most obvious histological change was a conspicuous lack of the pectin cementing material in soft pickles in contrast to normal stock. This condition progressively increased until there was no material left in the middle lamella. After the disappearance of the presumed pectic material, the cellulose of the cell wall was attacked until finally all of the parenchymatous tissue was destroyed. The most resistant parts of the cucumber were the seeds and vascular bundles. Fabian and Johnson (1938) observed that the softening of pickles was not necessarily an invasion of the tissue by a bacterium, in this case Bacillus mesentericus fuscus as was previously theorized, but was more strictly an enzymatic decomposition of pectic materials in the tissue. This condition occurred within 6 to 24 hr after cucumbers were placed in brine. They further observed that the organisms were more resistant to salt than their enzymes were but that the enzymes were more acid-tolerant than the organisms. The optimum temperature for activity of these implicated enzymes was 37°C. Above this temperature there was a gradual diminution of activity. Apparently, this condition was quite commonly observed previous to their study.

Fabian and Johnsons' histological study and calcium pectate analysis demonstrated a rapid change of insoluble pectic material to soluble pectic materials. They further observed that softening occurred more rapidly at the blossom end of the cucumber. Histological study indicated that the epidermal cells in the blossom end differ from those in the stem end and are more subject to softening.

Wenzel and Fabian (1945) described a softening of dill pickles that involved enzymatic breakdown of the pectic materials and associated this with the presence of enzymes secreted by molds and bacteria introduced by the garlic.

In further studies by Bell et al. (1950), Bell (1951), and later Etchells et al. (1955), Vaughn (1954), and Bell et al. (1958), softening was associated with the activity of pectolytic enzymes native to the fruit, particularly in their seeds, flowers, and ripe fruit. Later, this enzyme was associated particularly with the blossoms adhering to the cucumbers at brining.

Softening of pickle stock lying at or near the surface of vats of pickles has generally been associated with growth of aerobic yeasts in the upper layers. In the review of Bell and Etchells (1956) several studies are cited that indicate pectin hydrolysis by yeasts; however, in their studies there was little evidence to indicate that surface and subsurface yeasts were a potential source of the softening enzyme, polygalacturonase. Softening of salt stock may be attributed almost entirely to the enzymes native to the cucumber plant itself. Etchells, Bell, and Jones (1954) implicate the enzymes intrinsic to male flowers, fertilized female flowers, seeds, and ripe cucumbers.

The most serious condition ever observed by the writer occurred in 1967. The cucumbers were warm, about 90°F, before they were brined, and the temperature of the 40°–45° brine was about 90°F. Although the brine remained clear, extreme softening occurred in 24 hr and the stock was a complete loss. It was recommended that a lower concentration brine of 30° be used in order to speed fermentation and that cooler brine should be used. The recommended modification resulted in rapid fermentation and complete elimination of the problem. In these observations the rapidity of change and the fact that the brine remained clear indicated that enzyme activity was due to the enzymes indigenous to the cucumber, even though the cucumbers closely packed in containers in the damp climate were obviously subject to surface growth of microorganisms before they were brined.

BLOATING OF PICKLES

At times, losses due to bloating of pickles in brine may be equally as serious as the losses due to softening. Bloated pickles are referred to as bloaters, floaters, and hollow pickles. The degree of bloating may vary from a slight triangular separation of the locule sections to serious bloating with a collapse of the entire seed area and sometimes may involve a softening or breakdown of the parenchymal tissue (Fig. 6.8 and 6.10). Bloaters that have not collapsed will usually float on the surface of the brine; when a slight pressure is applied, however, they will collapse and usually sink into the brine. The pressure of vat weights above the brines will cause their collapse. In milder cases the pickles may remain sufficiently firm for use in mixed pickle stock. More seriously bloated stock may be used in relishes, but the most serious may represent a total loss.

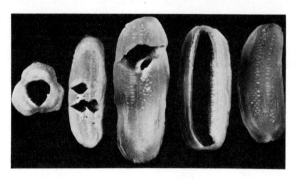

From Etchells et el. (1951)

FIG. 6.10.   THE FOUR CUCUMBERS ON THE LEFT ARE SPECIMENS OF "BLOAT-ER" DAMAGE CAUSED BY YEASTS WITH AN UNDAMAGED PICKLE ON THE RIGHT FOR COMPARISON

Several suggestions have been offered to explain the cause of bloater formation. These include the physical structure of the cucumber, gas formation by microorganisms, the effect of the pressure of the brine upon their surfaces, and lately, the belief that varietal and maturity differences are responsible.

Jones *et al.* (1941) demonstrated that factors that favored the development of a gaseous fermentation induced the production of appreciable quantities of bloaters. Among the significant observations, they observed the least bloating in 20° brines and increasing extent of bloating in brines of 30°, 40°, and 50° concentration. This would indicate a direct correlation with rate of acid production. The most rapid lactic acid fermentation occurs in 20° brines, while in 50° brines lactic acid fermentation is very slow; however, yeast fermentation and aerobacter fermentation may occur in high salt concentration brines that inhibit a lactic acid fermentation. The addition of vinegar or lactic acid to brines increased bloating. These acids would inhibit aerobacter fermentation but not yeast fermentation. Analysis of the gases sometimes demonstrated production of carbon dioxide alone but usually carbon dioxide and hydrogen. The former would indicate yeast fermentation and possibly fermentation by heterofermentative lactics, the latter, aerobacter fermentation.

The results obtained by their research further implicate a physical relationship, the toughening effect on the skin as a result of the action of the brine. This was further substantiated by the studies of Pederson and Albury (1962) that demonstrated a reduction of bloating when the vascular system remains open to permit absorption of brine by pickles through this system. High concentration salt brines will tend to close the system as well as to harden the skin. Drying of cucumbers in storage before brining will have a similar effect.

Jones et al. (1954) demonstrated a varietal relationship to bloating. Samish et al. (1957) observed the presence of gas and acid-producing organisms within bloated cucumbers and observed less bloating when benzoate was added to a brine. Benzoate is particularly inhibitory toward yeasts. Pederson and Albury (1961) stated that the extent of bloating can be correlated with the relative influence of heterofermentative and homofermentative species. The summary paper by Etchells et al. (1954) indicates the relationship between yeast growth and bloating, and furthermore, that an active lactic acid fermentation will reduce the extent of bloating.

OTHER SPOILAGE ORGANISMS

Various other forms of spoilage of less serious nature have occurred in pickle brines. These have usually occurred in brines that have been neutralized by destruction of the lactic acid by yeasts. Fabian and Nienhuis (1934) have described a Gram-negative organism that may cause ropiness in brine and an aerobic spore-former that may cause blackening of pickles in the presence of iron.

YEASTS

In Cucumber Fermentation

Part of the microbiological activity involved in the fermentation of pickle brines has been attributed to yeasts. Their importance and their degree of activity are discussed quite fully in the papers of Etchells and Bell (1950), and Etchells et al. (1952, 1953). The growth of yeasts is of considerable importance to the vegetable fermentation industry. Surface yeasts growing with luxuriant scum formation on pickles, sauerkraut, and other vegetables result in destruction of the lactic acid, hydrolysis of pectic materials, and decomposition of proteins and lipid materials with resulting undesirable flavor and texture changes. Bloating, softening, color change, and formation of white colony spots beneath the skin of cucumbers and olives are usually the result of yeast growth.

Salt stock pickles are frequently fermented in vats exposed to direct sunlight in order to suppress the growth of yeasts.

In Vegetable Fermentation

In the first microbiological study of vegetable fermentation, Henneberg (1926) described and named strains of yeasts that were believed to produce the gases evolved in the early stages of fermentation. Since then, yeasts have been considered essential to fermentation by many industrial operators and researchers. There is little doubt that yeasts can be isolated from any vegetable fermentation, but their numbers in good fermentations are so low in comparison to the numbers of lactic acid bacteria (Pederson and Albury 1969) that the importance of their role as a desirable microorganism is questionable.

There is no doubt that various species of yeasts are responsible for many undesirable conditions in vegetable fermentation. The aerobic yeasts forming the scums on the surface of fermenting vegetable materials obtain their energy by fermentation of lactic acid, thus neutralizing the brines and permitting other organisms to grow. The changes in protein, lipids, and other constituents are a major cause of undesirable flavors and odors. Peterson and Fred (1923) and later Pederson and Kelly (1938) attributed the development of pink color in kraut to yeast growth. Many species of yeasts may develop on the surface of the brines (Etchells and Bell 1950; Etchells et al. 1952, 1953).

In order to preclude faulty fermentations, many individuals have attempted to control the fermentation by inoculation, using brine from a previous fermentation or adding a pure culture of a lactic bacterium, usually *Lactobacillus plantarum*. Henneberg actually produced and sold a pure culture in the early part of the century even though he stated that it did not improve the quality of sauerkraut. LeFevre, Fabian, Fred, and Peterson and associates, Pederson, Vaughn and associates, Etchells and coworkers, and others have used inocula with varying degrees of success. A number of patents have been granted on various ideas using pure culture.

REDUCTION OF SPOILAGE

Sufficient research and observation have been conducted with the various vegetable fermentations so that losses could be reduced by application of a few basic principles. The practices used in other food fermentation industries such as cheese, butter, wine, beer, and other industrial processes may serve as guiding principles in the vegetable fermentation industry. The conditions resulting in excessive growth of surface yeasts, softening of vegetables, bloating of pickles, and other spoilage conditions are known and may be corrected. Spoilages that formerly were common in the sauerkraut industry have been reduced to a minimum in many factories by the careful application of a few basic principles such as proper concentration and distribution of salt and adequate covering of the kraut in vats. Soft kraut, pink kraut, dark kraut, and most important of all, the rancid krauts with cheese-like odors and flavors have markedly affected the acceptability of this food, but these spoilages are seldom encountered today in well-operated sauerkraut factories. With the realization of the course of fermentation, the organisms involved, and the chemical changes effected by the individual species of microorganisms, certain practices have been instituted that favor the growth of the desirable species in their proper sequences. Similar adaptation of practices could eliminate problems in other industries.

Microbiologically, the organisms responsible for fermentation must be distinguished from each other and from those species responsible for spoilage conditions. The lactic acid bacteria are the acid producers; they are anaerobic, and they vary in their characteristics. They require a complex medium for growth. This is usually supplied from the nutrients withdrawn from the vegetable by the plasmotic action of salt. The lactic acid bacteria require small amounts of salt in the medium for optimal growth but excess salt is inhibitory to growth. An exact limit of salt concentration cannot be specified since the various strains of the species differ in their tolerance to salt. Restriction of fermentation is a function of the combined effect of salt and acid. Regardless of the strain, more rapid initiation of fermentation will occur in 2.5% salt (10° brine) than in 5% salt. In 10% salt many strains are unable to initiate fermentation; in fact, fermentation may be retarded sufficiently to permit undesirable changes to occur.

In general, *Leuconostoc mesenteroides* will initiate growth more rapidly than species of the genera *Lactobacillus* and *Pediococcus;* however, this species is not equally tolerant to salt. Since it is desirable to produce acid as rapidly as possible to lower the pH and to produce a low oxidation reduction potential, it is logical to use salt concentrations low enough to favor the growth of leuconostoc. This species not only produces lactic acid but also carbon dioxide which tends to flush out entrapped oxygen. There is also evidence to indicate that the leuconostoc tend to produce conditions favorable for rapid growth of the pediococci and lactobacilli. The rapid growth of lactic acid-producers and a lowered oxidation reduction potential are unfavorable to growth of yeasts as well as the nonspore and spore-forming bacteria.

Just as the growth of microorganisms is a function of the concentrations of salt and acid, so also is the activity of enzymes that cause softening of vegetables. The comparative effects of acid and salt in the inhibition of enzyme activity are unknown; however, softening of tissue in sauerkraut fermentations in which high acidity is produced is much less pronounced than in cucumber stock in which comparatively lower acid is produced, even though much less salt is used in kraut fermentations. This may be a function of types and concentration of enzymes present. Enzymes native to the cucumber will continue their activity until the concentration of salt is at a much higher level than that attained in a kraut fermentation. This activity will be greater at higher temperatures; in fact, softening may occur within the first day after packing at temperatures of 90°F.

In fermentations of cut or sliced vegetables in which all of the brine is obtained by dry salting, a level of 2.25% salt by weight has been found by repeated tests to yield the best fermentation, considering flavor, texture, and general acceptability. At favorable temperatures an acidity of 0.7–1.0% expressed as lactic acid will be attained in about a week and 2.0–2.3% in about 4–6 weeks. In brine salting such high acidities are never attained. In very high concentration salt brines little or no fermentation occurs, and oxidative changes with darkening and sometimes spoilage due to nonfermentative microorganisms are serious problems.

The cucumber fermentation is more complex than the sliced vegetable fermentation, complexed by the fact that it involves fruits in various stages of maturity, salted in warm weather; their physical structure is variable; and the enzyme systems obviously are more complex.

BIBLIOGRAPHY

BALATSOURAS, G. D., and VAUGHN, R. H. 1958. Some fungi that might cause softening of stored olives. Food Res. *23*, 235–243.

BELL, T. A. 1951. Pectolytic enzyme activity in various parts of the cucumber plant and fruit. Botan. Gaz. *113*, No. 2, 216–221.

BELL, T. A., and ETCHELLS, J. L. 1956. Pectin hydrolysis by certain salt-tolerant yeasts. Appl. Microbiol. *4*, No. 4, 196–201.

BELL, T. A., ETCHELLS, J. L., and COSTILOW, R. N. 1958. Softening enzymes activity of cucumber flowers from northern production areas. Food Res. *23*, 198–204.

BELL, T. A., ETCHELLS, J. L., and JONES, I. D. 1950. Softening of commercial cucumber salt stock in relation to polygalacturonase activity. Food Technol. *4*, 157–163.

BRANEN, A. L., and KEINAN, T. W. 1969. Growth stimulation of *Lactobacillus* species by lactic streptococci. Appl. Microbiol. *17*, 280–285.

BREED, R. S., MURRAY, E. G. D., and SMITH, N. R. (Editors) 1948. Bergey's Manual of Determinative Bacteriology, 6th Edition. Williams & Wilkins Co., Baltimore.

CRUESS, W. V. 1924. Olive pickling in Mediterranean countries. Univ. California Coll. Agr. Circ. *278*.

CRUESS, W. V. 1930. Pickling green olives. Univ. California Coll. Agr. Bull. *498*.

CRUESS, W. V., and GUTHIER, E. H. 1923. Bacterial decomposition of olives during pickling. Univ. California Coll. Agr. Bull. *368*.

ETCHELLS, J. L., and BELL, T. A. 1950. Classification of yeasts from the fermentation of commercially brined cucumbers. Farlowia *4*, No. 1, 87–112.

ETCHELLS, J. L., and JONES, I. D. 1943. Bacteriological changes in cucumber fermentation. Food Ind. *15*, 54–56.

ETCHELLS, J. L., and JONES, I. D. 1946. Characteristics of lactic acid bacteria from commercial cucumber fermentations. J. Bacteriol. *52*, 539–599.

ETCHELLS, J. L., BELL, T. A., and JONES, I. D. 1953. Morphology and pigmentation of certain yeasts from brines and the cucumber plant. Farlowia *4*, No. 3, 265–304.

ETCHELLS, J. L., BELL, T. A., and JONES, I. D. 1954. Studies on the origin of pectinolytic and cellulolytic enzymes in commercial cucumber fermentation. Food Technol. *9*, No. 3, 14–16.

ETCHELLS, J. L., BELL, T. A., and JONES, I. D. 1955. Cucumber blossoms in salt stock mean soft pickles. N. Carolina Agr. Expt. Sta. Progr. Rept.

ETCHELLS, J. L., COSTILOW, R. N., and BELL, T. A. 1952. Identification of yeasts from commercial cucumber fermentations in northern brining areas. Farlowia *4*, No. 2, 249–264.

ETCHELLS, J. L., JONES, I. D., and BELL, T. A. 1951. Brigands in brine. N. Carolina Agr. Expt. Sta. Res. Farming *10*.

ETCHELLS, J. L., JONES, I. D., and HOFFMAN, M. A. 1943. Brine preservation of vegetables. Proc. Inst. Food Technol. 176–182.

FABIAN, F. W., and JOHNSON, E. A. 1938. Experimental work on cucumber fermentation. IX. A bacteriological study of the cause of soft pickles. X. Zymological studies of the cause of soft cucumbers. XI. Histological changes produced in pickles by bacteria, acids, and heat. XII. Chemical changes produced in the pectic substances of pickles by bacterial enzymes. Michigan State Agr. Expt. Sta. Bull. *157*.

FABIAN, F. W., and NIENHUIS, A. L. 1934. Experimental work on cucumber fermentation. VI. Factors influencing the formation of ropy brine in cucumber fermentation. Michigan State Agr. Expt. Sta. Bull. *140*.

FABIAN, F. W., and WICKERHAM, L. J. 1935. Experimental work on cucumber fermentation. VIII. Genuine dill pickles—a biochemical and bacteriological study of the curing process. Michigan State Agr. Expt. Sta. Bull. *146*.

FABIAN, F. W., BRYAN, C. S., and ETCHELLS, J. L. 1932. Experimental work on cucumber fermentation. IV. Morphological studies on spoiled cucumber pickles. Michigan State Agr. Expt. Sta. Bull. *126*.

FAVILLE, L. W., and FABIAN, F. W. 1949. The influence of bacteriophage, antibiotics, and Eh on the lactic fermentation of cucumbers. Michigan State Agr. Expt. Sta. Bull. *217*.

HENNEBERG, W. 1926. Handbook of Fermentation Bacteriology, Vol. 2. Paul Parey, Berlin. (German)

HOHL, L. A., and CRUESS, W. V. 1942. A study of organisms found in lactic acid fermentation of lettuce. Food Res. *7*, 309.

HSIO, H. C. 1949. Microbiology of paw tsay. I. Lactobacilli and lactic acid fermentation. Food Res. *14*, 405–412.

JONES, I.D., ETCHELLS, J. L., and MONROE, R. J. 1954. Varietal difference in cucumbers for pickling. Food Technol. *8*, No. 9, 415–418.

JONES, I. D., ETCHELLS, J. L., VELDHUIS, M. K., and VEERHOFF, O. 1941. Pasteurization of genuine dill pickles. Fruit Prod. J. *20*, No. 10, 304.

KAWATOMARI, T., and VAUGHN, R. H. 1956. Species of *Clostridium* associated with zapatera spoilage of olives. Food Res. *21*, 481–490.

KIM, H., and WHANG, K. 1959. Microbiological studies on kimchis. I. Isolation and identification of anaerobic bacteria. Sci. Res. Inst. Bull. *4*, 56–63.

LESLEY, B. E., and CRUESS, W. V. 1928. The effect of acidity on the softening of dill pickles. Fruit Prod. J. *7*, No. 10, 12.

NORTJE, B. K., and VAUGHN, R. H. 1953. The pectolytic activity of species of the genus *Bacillus:* Qualitative studies with *Bacillus subtilis* and *Bacillus pumilus* in relation to softening of olives and pickles. Food Res. *18*, 57–69.

ORILLO, C. A., and PEDERSON, C. S. 1966. Unpublished data.

ORILLO, C. A., SISON, E. C., LUIS, M., and PEDERSON, C. S. 1969. The fermentation of vegetable blends. Appl. Microbiol. *17*, No. 1, 10–13.

PALO, M. A., and LAPUZ, M. M. 1955. On a new gum-forming streptococcus, with studies on the optimal conditions for the synthesis of the gum and its production in coconut water. Philippine J. Sci. *83*, 327–357.

PEDERSON, C. S. 1930. Floral changes in the fermentation of sauerkraut. New York State Agr. Expt. Sta. Bull. *168*.

PEDERSON, C. S., and ALBURY, M. N. 1950. The effect of temperature upon bacteriological and chemical changes in fermenting cucumbers. New York State Agr. Expt. Sta. Bull. *744*.

PEDERSON, C. S., and ALBURY, M. N. 1954. The influence of salt and temperature on the microflora of sauerkraut fermentation. Food Technol. *8*, 1–5.

PEDERSON, C. S., and ALBURY, M. N. 1961. The effect of pure-culture inoculation on the fermentation of cucumbers. Food Technol. *15*, 351.

PEDERSON, C. S., and ALBURY, M. N. 1962. The absorption of salt by cucumbers during fermentation. Food Technol. *16*, No. 5, 126–130.

PEDERSON, C. S., and ALBURY, M. N. 1969. The sauerkraut fermentation. New York State Agr. Expt. Sta. Bull. *824*.

PEDERSON, C. S., and KELLY, C. D. 1938. Development of pink color in sauerkraut. Food Res. *3*, 583–588.

PEDERSON, C. S., and WARD, L. 1949. The effect of salt upon the bacteriological and chemical changes in fermenting cucumbers. New York State Agr. Expt. Sta. Bull. *273*.

PEDERSON, C. S., NIKETIĆ, G., and ALBURY, M. N. 1962. Fermentation of Yugoslavian pickled cabbage. Appl. Microbiol. *10*, 86–89.

PETERSON, W. H., and FRED, E. B. 1923. An abnormal fermentation of sauerkraut. Centr. Bakteriol. Parasitenk. Abt. II *58*, 199–204.

PLASTOURGAS, S., and VAUGHN, R. H. 1957. Species of *Propionibacterium* associated with zapatera spoilage of olives. Appl. Microbiol. *5*, 267–271.

SAMISH, Z., DIMANT, D., and MARANI, T. 1957. Hollowness in cucumber pickles. Food Manuf. *32*, 501.

VAHLTEICH, H. W., HAURAND, C. H., and PERRY, G. A. 1935. Modern science applies itself to cucumber salting. Food Ind. *7*, 334–336.

VAUGHN, R. H. 1954. Lactic acid fermentation of cucumbers, sauerkraut, and olives. *In* Industrial Fermentations, L. A. Underkofler, and R. J. Hickey (Editors). Chemical Publishing Co., New York.

VAUGHN, R. H., DOUGLAS, H. C., and GILILLAND, J. R. 1943. Production of Spanish-type green olives. Univ. California Coll. Agr. Bull. *678*.

VAUGHN, R. H. *et al.* 1953. *Lactobacillus plantarum*, the cause of "yeast spots" on olives. Appl. Microbiol. *1*, No. 2, 82–85.

VORBECK, M. L., MATTICK, L. R., LEE, F. A., and PEDERSON, C. S. 1961. Volatile flavor of sauerkraut. Gas chromatographic identification of a volatile acidic off-odor. J. Food Sci. *26*, 569–572.

WENZEL, F. W., JR., and FABIAN, F. W. 1945. Experimental work on cucumber fermentation. XIII. Influence of garlic on the softening of genuine kosher dill pickles. Michigan State Agr. Expt. Sta. Bull. *199*.

WEST, N. S., GILILLAND, J. R., and VAUGHN, R. H. 1941. Characteristics of coliform bacteria from olives. J. Bacteriol. *41*, 341–351.

WÜSTENFELD, H., and KREIPE, H. 1933. Experiments on the softening of sour pickles. Deutsche Essigindustrie *37*, 77–81. (German)

CHAPTER 7

Fermented Sausage

Primitive man seldom had reason to be concerned about an adequate supply of protein food, because his diet consisted primarily of meat, fish, eggs, and shellfish that furnished protein of good quality. Although he subsisted on foods of animal origin at times, he also ate cereals, fruits, nuts, leafy and root vegetables, and berries. He invented snares, traps, and lines to catch the animals he could not procure directly. Although possibly little more inventive and skillful than the higher apes, his knowledge was accumulative and progressive since he could communicate with his fellow man.

Modern man, *Homo sapiens,* appeared sometime about 30 to 60 thousand years ago. With an increase in population and seasonal changes in food supply, the changes in food supply affected his nutrition. The succession of cultures saw many changes in the eating habits of people. They became food gatherers and developed skills in basketry, pottery, and the use of implements including weapons and snares. Man became more sedentary, learned to domesticate the more docile animals, and learned to cultivate the wild grasses to feed himself and his herds. As fields or areas became exhausted, he moved to new areas with his livestock and cereal grasses; thereby, animal husbandry and cereal cropping became essential features of civilization. When he domesticated his herds and cultivated his plants, he insured against the uncertainties of varying supplies of food. Primal man enjoyed the pleasure of eating without knowledge of the function of food in nutrition. When later he learned the art of cookery, he relished cooked foods.

ORIGIN OF SAUSAGE

The people of the pre-Christian era knew that unless the meats were consumed shortly after slaughter of the animal, they would spoil and be unfit for later consumption. So they learned to cut or grind the meat, to season it with salt and spices, and to allow it to dry in rolls. This is the forerunner of our present-day sausages such as the salamis and other summer sausages (Fig. 7.1). Sausage is one of the oldest forms of processed foods. Basically, all sausage is comminuted meat. The preserved sausages consumed by the ancient Babylonians, Greeks, and Romans without doubt were fermented and dried so that they became safe for consumption. It has been recorded that King Nebuchadnezzar enjoyed munching on a sausage similar to salami. The recorded history of sausage manufacture begins in the 9th Century before Christ, as established by statements in Homer's Odyssy; however, it has been stated that sausage was prepared and consumed by the ancient Baby-

FIG. 7.1 CURED OR FERMENTED SAUSAGES ARE PREPARED IN A VARIETY
OF SIZES

Included here are variations from a moist thuringer to a dry pepperoni
variously called thuringer, summer sausage, beerwurst, beer summer
sausage, Lebanon bologna, Genoa sausage, and many others.

Ionians as far back as 1500 B.C. and by the peoples of ancient China. These
statements would be difficult to refute. Grecian literature after Homer's
time makes frequent mention of sausage, or oryae. The play Orya written by
Epicharmus about 500 B.C. mentions oryae, and Aristophanes mentions it
in his play written 423 B.C. Since the time of Christ, sausage is frequently
mentioned in literature.

Drying of meat was a commonplace practice along the shores of the
Mediterranean, and the preparation of sausage may have developed from
this practice. Incidentally, the American Indians chopped up dried meat,
blended it with dried berries and herbs, and used this when other food was
scarce. Without doubt, sausage manufacture developed gradually by a series
of steps.

Salamis, a city on the east coast of Cypress, was destroyed in 449 B.C.
It is reasonable to assume that the term "salami" originated there. Sausage
is derived from the Latin, salsus, meaning salt. The Romans of the period of
the Caesars knew that fresh meat tainted very quickly and became unsafe
after a few days. The Romans were fond of sausage made with fresh pork and
white pine nuts chopped fine and seasoned. This was often used as fresh sausage and was associated with certain ritualistic pre-Christian practices which
were later banned by the Christian Church.

Breasted (1938) stated that the success of Caesar's legions in the conquest
of Gaul can be attributed to their use of dry sausage for their meat supply and
that this aided in the retention of their vigor and health. The Roman butchers
cut their beef and pork into small pieces, added salt and spices, packed the
blend into skins, and placed these in special rooms to dry. They knew that
they obtained better and more uniform sausages when the meat was prepared
and dried in these rooms. Today, it can be surmised that the equipment used,
the shelves, tables, and even the air itself were thoroughly impregnated with

lactic acid bacteria as well as micrococci. These bacteria fermented the meat, imparted the characteristic tang, and prevented growth of undesirable types of microorganisms. The presence of microorganisms, the science of microbiology, and the observations that certain bacteria and yeasts caused desirable changes in food products are knowledge that was revealed hundreds of years later.

The Roman butchers had guilds for their protection and undoubtedly their methods were secret. Their shops were sanitary and they were controlled in their practices by the government. Their shops may have been comparable to the shops of today. Preparation and spicing of sausage and sausage making became a culinary art in Mediterranean countries and later in upper Europe. Relatively large operations appeared in various cities. The practices led to the production of our present-day dry and semidry fermented sausages.

VARIETIES OF SAUSAGE

During the Middle Ages, sausage making was practiced in many areas of Europe. Many manufacturers became so adept in blending that their sausage became distinctive, and their fame spread; thereby, hundreds of varieties of dry and semidry sausage were developed, as there are hundreds of varieties of cheese. The names of many of these sausages were derived from the cities or areas where they were developed. Although the principles of fermentation and curing are similar, the proportions of different meats, the salt, spicing, smoking, and curing vary considerably. Sausage types like so many other products were developed in certain areas because of the predominance of certain animals, the local climatic conditions, and the preferences of the people. Artificial refrigeration was relatively unknown, canning processes had not yet been developed, and the sausage makers were forced to develop products that would keep under the conditions prevailing in the area. Every area developed its own particular style of sausage. The people of the warm Mediterranean shores desired sausage heavily seasoned and objected to a highly smoked product, while the people further north desired a product slightly spiced, heavily smoked, moist, and well salted. Because in the colder areas sausage was made in the cold winter months and stored and aged until summer months, it was called summer sausage. These types of sausage so commonly consumed in the United States are of Northern European origin. The preponderance of this type of product in distinct areas is in conformity with climatic conditions.

Although there are several distinct types of sausage, they may be classified into five groups. Fresh sausage, as the name implies, is ground, chopped, or minced meat with or without spices and is prepared and cooked immediately before consumption. Cooked sausage is heat-processed immediately after preparation. Cooked specialties are prepared from uncured meats and processed. Smoked sausage is sometimes sold as fresh smoked sausage. Since these three classes of sausage, ordinarily referred to as wet sausages, are not fermented, they will not be considered further.

The other two types of sausage are cured by fermentation and represent the types that have been prepared for centuries. These sausages will keep for long periods, particularly if held in a cool and fairly dry atmosphere. New condition sausage or, more precisely, the semidry sausages differ from dry sausages principally in that curing includes smoking at a relatively low temperature during which time active fermentation occurs. This is followed by cooking in the smokehouse and subsequent air-drying for a comparatively short time. The present-day thuringer cervelat is a common example of this type. It should be emphasized that there is no clear line of distinction between the classes of cured sausage.

The principal dry and semidry sausages are salamis or cervelats. The former are generally more highly seasoned. There are hard and soft styles in either group. They also vary considerably in taste; in fact, one may be found to suit practically every taste. Such sausage is prepared from selected meats, comminuted, blended with curing ingredients and spices, then held for a day or more at a relatively low temperature, e.g., 36°–38°F, filled into casings, and dried by air-drying or smoke-drying at higher and carefully controlled temperature and humidity conditions. It is during the latter period when active fermentation by lactic acid-producing bacteria occurs. This imparts the distinctive and tangy flavor. Some of these sausages are often referred to as summer sausage.

Early sausage makers in Europe and neighboring Mediterranean countries were limited in their sales to the area near their homes. Each used his ingenuity or imagination to prepare sausage to suit the taste of his customers; therefore, each sausage maker developed a brand unique in its character. Some of these were duplicated by neighboring butchers. Over a period of years, many distinctive types of sausage were developed, and their general acceptance led to the use of specific names for these types, the name usually derived from the city or area where they were developed. These names were generally retained as large commercial companies developed. The demand for a particular name type, such as thuringer, apparently led to several formulations and methodologies. Salami and cervelat have been used as names for a variety of types.

The Italian salamis, Genoa, Milano, Siciliano, and cappicola, are prepared primarily with lean pork, sometimes moistened with wine, but usually heavily flavored with garlic and other spices, and then dried. Alessandro and alpino are Italian types of American origin. Mortadello originally was made from flesh of pigs which fed on chestnuts and acorns, but now it ordinarily contains beef as well as pork, and it is smoked. Pepperoni is highly seasoned, ordinarily with ground red pepper; Arles and Lyons are of French origin; while the chorizos are of Spanish or Portuguese origin. The German and

TABLE 7.1

MICROBIOLOGICAL CHANGES DURING FERMENTATION OF CHORIZO-TYPE SAUSAGE[1]

Time in Hours	Stage of Processing	pH[2]	Plate Count × 10^5/Gm	Total Aerobic Species	Micrococcus Species	Total Microaerophilic Species	Leuconostoc mesenteroides	Streptococcus Species	Pediococcus cerevisiae	Lactobacillus plantarum
						Estimated Number of Each Type of Bacteria × 10^5/Gm				
0	Initial	5.98	342	192	100	150		150		
8	8 hr at 41°C	5.70	1090	220	110	870		870		
16	16 hr at 41°C	4.96	2540	690	148	1850	265	1320	265	
24	24 hr at 41°C	4.80	1200	200[3]		1000		210	790	
36	12 hr in smokehouse at 50°–55°C	4.58	844	166		678	36		606	36
48	24 hr in smokehouse at 50°–55°C	4.49	890	440		450		45	360	45
72	48 hr in smokehouse at 50°–55°C	4.22	860	360		500		25	475	
96	1 day at 15°C	4.18	582	182		400		20	380	

Source: Sison (1967).
[1] Original moisture 38.46%; final 24.63%.
[2] Original acidity as lactic acid 0.34%; final acidity 1.32%.
[3] Presumably micrococci.

Hungarian sausages are less highly flavored but contain traces of garlic and are heavily smoked. Thuringer, Holsteiner, farmer, and other semidry mild summer sausages are more delicately seasoned and subjected to mild smoking. Göteborg and medwurst are of Swedish origin and are usually heavily smoked. Landjaeger is of Swiss origin. These are often referred to as cervelats.

MICROBIOLOGICAL ACTIVITY

Obviously, the Babylonians, Greeks, Romans, and later peoples from many other areas knew nothing about the microscopic plants that grew in their meat products, producing the changes in their sausages that they relished. It is difficult to conceive of small plants growing so rapidly that numbers in the hundred millions may be present per gram of meat after a 24 hr period to produce sufficient acid to impart a marked effect on flavor (Table 7.1). Even during the advances made in the science of microbiology by many scientists during the past 100 yr, little attention or thought was given for a number of years to the fermentative changes that occurred in sausage making. The application of microbiological research to the improvement in meat science had its beginning in the 1890's and was established for control purposes. Sausage making had become a winter industry largely because of lack of refrigeration. Chemical laboratories in the larger packing houses were concerned with analytical studies.

The contribution of bacteria to sausage processing was considered in the 1930's comparable to the changes brought about in the cheese industry. In

TABLE 7.2

THE TOTAL VIABLE BACTERIAL COUNTS OF SOME FINISHED SAUSAGES

No.	Bacteria $\times 10^6$/ Gm	Predominant Flora	pH	NaCl (%)	H_2O	Na-NO_2 Ppm	Na-NO_3 (%)	Sugar (%)
P—Summer sausage	48	Lactobacillus	5.3	2.5	32.3	tr	0.04	0.37
Q—Summer sausage	190	Lactobacillus	5.3	2.6	24.9			0.26
T—Summer sausage	234	Lactobacillus	4.9	2.0	48.9	tr	0.03	0.66
F—Summer sausage	362	Lactobacillus and Pediococcus	4.8	2.6	38.9			0.61
Th—Thuringer	50	Pediococcus	5.0	1.7	58.7	tr	0.02	0.12
Th2—Thuringer	18	Pediococcus	4.9	1.6	54.9			0.12
A—Salami	28	Lactobacillus	5.2	3.1	40.5			0.55
L—Salami	14	Lactobacillus	5.4	4.4	27.9	tr	0.04	0.66
A—Cervelat	25	Lactobacillus and Pediococcus	5.3	3.0	23.0	14	0.07	0.51
M—Cervelat	14	Lactobacillus	4.8	3.3	30.9			0.92
A—Genoa	38	Lactobacillus	4.9	3.8	32.5			0.08
I—Goteborg	109	Lactobacillus	4.8	2.0	45.0			0.41
J—Lebanon	320	Lactobacillus	4.7	2.1	51.0	tr	0.03	2.10
L—Lebanon	0.1	Streptococcus	4.7	1.5	63.5	tr	0.03	0.27

Adapted from Deibel *et al.* (1961A).

the 1920's, because a few scientists believed that the fermentation was due to yeast growth, the inoculation of sausage meat with yeasts was suggested. Too little attention was given to the important contributions of Jensen and Paddock (1940), who used various species of lactic acid bacteria to standardize and improve the character of sausage. These scientists suggested pure culture starters of bacteria as used in the cheese industry. This was not generally practiced industrially until after the studies of Niven (1951), Deibel and Niven (1957), Niven *et al.* (1958), and Niinivaara (1955) (Table 7.2).

The tangy character, taste, and aroma of the dry and semidry sausages result in large part from the changes that occur during fermentation and curing. These changes are now generally considered the effect of lactic acid bacterial fermentation although the Finnish workers stress the importance of micrococci in establishing suitable oxidation-reduction conditions. The products of fermentation and changes that occur are so complex, qualitatively and quantitatively, that relatively too little progress has been made in determining the chemistry of the flavor, aroma, and other organoleptic characteristics of the cured product.

Control of Microorganisms

An inherent characteristic of successful sausage production is the strict control of microorganisms. Successful manufacture of sausage depends upon fermentation and curing for the production of the typical, cured meat color and suitable proportions of lactic acid and other fermentation products responsible for the characteristic flavor.

The preparation of fermented sausage is complex and dependent upon a combination of interdependent processes, one of which is the metabolic activity of microorganisms. Centuries before mankind realized the contributions of bacteria, yeasts, and molds to fermented foods, it was observed that sometimes foods became unfit for consumption. For centuries every attempt was made to create conditions that would either delay or prevent such food spoilages. The problem was more difficult because the people were unaware of the causes for such changes. With recently acquired knowledge, the problem of preventing growth of undesirable microorganisms while encouraging growth of desirable forms has been simplified.

Comminuted meats are more subject to growth of microorganisms than uncut meat because comminution distributes the organisms throughout the meat, provides a considerably larger surface area for microbial growth, and releases meat juices containing the nutritive cell protoplasmic constituents. Meat juices are ideal media for the growth of numerous types of organisms, not only because they are moist and contain nutrient cell constituents, the proteins, lipids, vitamins, and sugar, but also because they have a nearly neutral reaction.

Since comminuted meats are such an excellent medium for the growth of microorganisms, any of a great variety of species can be expected to grow. These organisms may come from equipment, atmosphere, hands of workers, or the meat itself. Since there are so many possibilities, the microbiology can be discussed only in terms of generalities and the practices that have been followed throughout the years. These practices have been developed through the experiences of the people. It is interesting to note how well they fit into present-day scientific knowledge.

Certain environmental conditions are important in determining the species of organisms that may flourish. Among these, the populations of the species, the optimal temperatures, their relationship to air and acidity, and the presence of inhibitory agents are primary. Meats handled in a sanitary manner contain relatively few organisms on their surfaces. If the surfaces of meats to be ground are relatively dry and the meats are kept cool, the comminuted meats may contain low microbial populations, and the microorganisms will be relatively uniformly distributed. The populations remain low for several hours during the lag phase of growth. After passing through the lag phase, growth may be rapid if conditions are favorable. Various analysts have reported populations of 10^4 to 20^6 per gm of ground meat. In contrast, microorganisms grow rapidly at high relative humidities, and surface slimes may develop on the meat surface before grinding. Maximum populations of 50×10^6 per sq cm have been reported. Comminuting such meats introduces and distributes great numbers of organisms in the ground product; in addition, since such organisms may be in their rapid growth phase, they may continue growth at a very rapid rate.

The ancient sausage makers did not have these facts at their disposal. Despite this, it is surprising how practical their methods were.

USE OF BACTERIAL STARTERS

Many sausage makers are no longer satisfied with reliance upon chance inoculation by lactic acid bacteria to produce the acids so essential for flavor and preservation or upon chance inoculation with nitrate-reducing bacteria to adjust the oxidation-reduction potential for necessary color stability. The addition of nitrite as well as nitrate is now permitted in some countries. The assurance of inoculation with lactic acid bacteria was not so readily attained. It is true that equipment, working tables, and storage shelves or rooms used over and over again for sausage making are undoubtedly heavily seeded with the bacteria necessary for fermentation. Furthermore, some sausage makers apparently added some of their fermenting sausage to the freshly prepared blends. Still, producers were no longer satisfied to rely upon chance inoculation to achieve the desired results. Tanner (1944) cites the early work of Cesari and Guillermond in 1919 and 1920. They believed that yeasts contributed to the flavor of sausage and, therefore, inoculated fresh meat with yeasts in an attempt to improve flavor and prevent putrefaction. Tanner stated that Moser in 1935 also related flavor of sausage to the presence of bacteria and yeasts.

Jensen and Paddock (1940) in their patent observed that several species of the genus *Lactobacillus* could be utilized as starters. Jensen (1942) stated that the pleasant tangy character of thuringer-type sausage is formed by several species of the genera *Lactobacillus* and *Leuconostoc*. Jensen stated that many of the steps in processing tangy sausages could be eliminated or shortened and that then the finished sausage would be more uniform. He also stated that chance inoculation is never economical.

Although Niinivaara (1955) used a culture of *Micrococcus aurantiacus* to reduce nitrate to nitrite, he depended on chance inoculation of the lactic acid bacteria; however, since the industry is now using nitrite in their cures, the use of a nitrate-reducing organism is not so essential.

Deibel and Niven (1957) reported success with the use of the bacterial species, *Pediococcus cerevisiae* as a starter culture for summer sausage. This species, a coccoid bacterium, is a true lactic acid bacterium similar in many respects to the species *Lactobacillus plantarum*, but it produces less acid. Further studies by Deibel *et al.* (1961A) were reported with the ultimate aim of employing a starter culture to effect greater control of the fermentation. Later, Deibel *et al.* (1961B) developed lyophilized cultures of *Pediococcus cerevisiae* that performed satisfactorily under current manufacturing conditions. Since then, this has proved its usefulness in commercial operation. *Pediococcus cerevisiae* grows readily in meat products and produces less acid than some of the *Lactobacillus* species that may produce excessive acid.

The use of the starter culture *Pediococcus cerevisiae* was approved by the Meat Inspection Division of the U.S. Dept. of Agr. Merck and Co., of Rahway, New Jersey, has produced and distributed the culture commercially under the trade name "ACCEL."

Merck and Co. (Anon. 1938) has suggested several basic meat formulae for thuringer, summer sausage, cervelat, Lebanon bologna, and pork roll. Except for the addition of the starter culture their suggestions do not differ materially from the other new methods used for manufacturing sausage.

Discussion of semidry, new condition sausage may be continued with one of the methods suggested by Merck (Anon. 1938) for preparing a thuringer-type sausage. These suggestions undoubtedly came from the American Meat Institute.

The degree of tang, amount of smoke flavor, size of chop, meat formulation, size and type of casing, choice of species, and the finishing temperature are variations encountered in processing these products. The first basic formula consists of 30% cow chucks, 30% bull meat, and 40% regular pork trimmings. Pork and beef hearts are suggested in another basic formula. The beef hearts, if used, are ground through a $\frac{3}{16}$–$\frac{1}{2}$ in. plate, while the pork trimmings are ground through a 1 in. plate. The beef and pork are then mixed by regrinding through a ⅛–³⁄₁₆ in. plate. The spices and cure are added before regrinding. The spice and cure for 100 lb of meat consist of sodium nitrate, 2 oz; sodium nitrite, ⅛ oz; salt, 2–3 lb; dextrose, ¾–1½ lb; sucrose, 2 lb; black pepper, 4–8 oz; red pepper, 3 oz; and optional ingredients, mustard seed, 1–3 oz; nutmeg, ½–1 oz; coriander, ½ oz; allspice, ½ oz; garlic, ¼ oz; and sodium isoascorbate, ½–¾ oz. The culture of *Pediococcus cerevisiae* (ACCEL) is suspended in water, 1–3 oz, agitated, and distributed over the mixed meat, spice, and cure, and mixed for 1–2 min. The blend is stuffed into the casing, the product is allowed to warm up to room temperature, 65°–70°F, and then moved into the smokehouse. Variations in the smokehouse play an important part in affecting the final tang or acidity of the product. It is during this period that active fermentation occurs. The extent of fermentation will be dependent upon the temperature and the length of time of holding the sausage in the smokehouse. For a sharp tang with total acidity of 0.8–1.2% acid and pH 4.5–4.8, the temperature is held at 80°–90°F with relative humidity of 85–90% for 12–16 hr. This period allows for drying of the product and the initiation of growth of the starter culture. The temperature is increased to 100°–105°F at 85% RH for 24 hr during which time smoke is introduced to produce the desired smoke color and flavor. The culture will produce its maximum lactic acid during this period. The temperature is then raised to 150°–155°F and held 4–8 hr to insure a minimum internal temperature of 137°F for at least 10 min. The product is finally removed from the smokehouse, allowed to dry and cool at room temperature, and then stored at 40°–45°F to age or mellow.

A noncooked product may be obtained by holding at 100°–115°F for 28–32 hr after which it is dried and cooled to room temperature and placed in the cooler at 40°–45°F. A milder product with 0.6–0.8% acid and pH of 4.9–5.2 may be prepared by reducing the time held at 100°–105°F to 16–20 hr.

Many variations of these processes may be used. In one procedure used, the sausage maker held the sausage blend in the cold room at 36°–38°F for 24 hr before filling into casings and then smoking it at 95°–100°F for 24 hr. The culture was blended immediately before filling into casings.

No attempt will be made to describe all of the individual types of sausages prepared; a few, however, are of special interest. Mills and Wilson (1958) prepared experimentally pork rolls using 1 oz of starter per 100 lb of pork. These attained an acidity of about 0.95% acid as lactic acid with a pH of about 4.3 in 48 hr. Processing at 85°–100°F for 36 hr was recommended when acidity of about 0.90% and pH of about 4.65 were attained. Pork roll was prepared in New Jersey in colonial days.

Lebanon bologna was originally produced in Lebanon, Pa. It was smoked 5 days to 2 weeks in a large unventilated smokehouse with a cool, heavy smoke. This sausage is interesting in that Jensen (1942) cites the failure of a large packer to duplicate this product by chance inoculation. Jensen states that after isolation of the *Lactobacillus* strain responsible for the characteristic flavor, color, and texture, the bologna could be prepared elsewhere.

The native chorizo-type sausage prepared in the Philippines was studied by Sison (1967) under the direction of the writer (Table 7.1). Sison observed the presence and growth of considerable numbers of strains of the genus *Micrococcus* when the blends were held at 50°F prior to smoking, but a fermentation almost entirely due to *Pediococcus cerevisiae* resulted whether or not it was inoculated with pediococci. Cultures of *Leuconostoc mesenteroides* and a *Streptococcus* strain closely related to *S. faecalis* were frequently isolated. Acidities as high as 2.40% acid as lactic acid and pH readings as low as 4.00 were recorded. The fermentation of sausage blends inoculated with *Pediococcus cerevisiae* or *Lactobacillus plantarum* was controlled almost entirely by these species while *Leuconostoc mesenteroides* and *Lactobacillus brevis* persisted in sausages inoculated with the latter cultures.

The most obvious change during curing is the conversion of sugar to lactic acid as evidenced by the development of tangy flavor, by the increase in total acidity to about 1.00% or more, expressed as lactic acid, and by a reduction in pH from levels of 6.2–6.6 to 4.2–5.0. The degree of acidity attained depends upon the extent of the fermentation, the particular blend of ingredients, and the rate and extent of drying. Most sausage makers try to attain sufficient acid to preserve the meat and to impart the desirable flavor without making the sausage too sour. Excessive sourness is sometimes regarded as spoilage. The primary end product of a homofermentation, a fermentation effected by organisms such as *Lactobacillus plantarum* or *Pediococcus cerevisiae*, is lactic acid. In contrast, heterofermentative lactic acid bacteria, such as *Leuconostoc mesenteroides* and *Lactobacillus brevis*, convert slightly less than 50% of the sugar fermented to lactic acid and a similar amount to a combination of ethyl alcohol, acetic acid, and carbon dioxide. Some of the species of lactic acid bacteria produce acetoin that imparts a nutty flavor and aroma, and other by-products in small amounts. The lactic, acetic, pyruvic, and other acids produced in fermentation may be altered; for example, they may combine with alcohols to contribute the typical desired flavor. Little change occurs among the protein constituents other than those required to furnish nutrients for growth of the organisms. Certain lipid constituents, particularly phospholipids, are altered to furnish lipid nutrients for the growth of bacteria (Niven, 1961).

As previously noted, the heterofermentative lactic acid-producing bacteria produce carbon dioxide. Under certain conditions that favor development of these heterofermenters, carbon dioxide may be produced more rapidly than it can diffuse through the casing. Gaseous swelling may occur under such conditions. Some sausage makers puncture the casings to allow escape of excess gases. Genoa salamis are believed to be fermented in part by heterofermentative lactic acid bacteria.

The lactic acid bacteria ordinarily do not reduce nitrates to nitrites. Nitrates are essential to sausage for the proper development of color. Species of the genus *Micrococcus* are the primary organisms responsible for reduction of nitrates to nitrites. The micrococci are generally indigenous to the skin of animals although they may also occur elsewhere. They produce a slight acidity in broths but seldom produce enough acid to curdle milk. While they are not truly pathogenic, they are closely related to the staphylococci that are frequently associated with skin lesions. Before the use of nitrite in cures became standard procedure, the micrococci were essential in the early stages of fermentation for nitrate reduction; however, if their activity was not inhibited by growth of the lactic acid bacteria, excessive nitrite was produced, resulting in a condition known as nitrite burn (Deibel and Evans 1957; Niven 1951). Micrococci even suppressed the growth of lactic acid bacteria. Today, addition of too much nitrite in the cure may also cause this type of spoilage. The addition of the proper amount of nitrite and inoculation with the starter minimize the danger of nitrite burn.

ABNORMAL PRODUCTS

If a food is inclined to spoil, methods must be devised to prevent such abnormalities. It is remarkable that ancient man observed and developed methods of preserving foods when he possessed no knowledge of the causes of spoilage. Today, the relationship of environmental factors such as moisture content, temperature, osmotic pressure, acidity, and the presence of microorganisms to the growth of microorganisms are well established. Nevertheless, these factors were also quite well controlled in the preparation of fermented foods long before the first observations of microorganisms were made. Presumably, during the early period of development of fermented sausages spoilage did occur with subsequent losses, illness, and death resulting from their consumption.

Properly prepared fermented sausage will not support the growth of pathogenic bacteria because of its characteristic high acidity with pH of 4.0–4.6, its quantity of salt and curing agents, its dryness, and content of spices, herbs, and smoke ingredients with also possibly the presence of antibiotics. Surface mold and slime, sourness, green discoloration, rancidity, gassiness, nitrite burn, and other deleterious conditions, although infrequent, are at times troublesome and costly. At present, rancidity is considered a spoilage condition that, although not necessarily rendering the food unfit or causing illness, still decreases the palatability with some of the high pork-containing sausages. While lipid changes occur in normal lactic acid bacterial fermentations, they do not progress to the degree considered rancid. Since a reduced condition is produced in a lactic fermentation, rancidity is rarely encountered.

Present-day spoilage conditions may be termed nuisance problems that cause minor financial losses. The greenish discoloration occurring on the surface and within sausages is associated primarily with unfermented types of sausage. The discolorations result from changes in blood pigments caused by growth of certain strains of lactic acid bacteria. They are a direct reflection of malpractice in the holding of the finished product. In contrast, fading, browning, and surface discoloration are chemical in nature. Certain types of nitrite burn occurring in fermented sausages are associated with excessive nitrite either through addition in formulation (Deibel and Evans 1957) or excessive growth of nitrate-reducing organisms (Deibel *et al.* 1961A). The excessive growth of nitrate-reducing organisms may result from a significant lag in growth of the lactic acid bacteria.

Excessive production of acid may yield a product that is too sour for the average consumer. This may occur by prolonging the fermentation. For this reason, the lower acid-producing pediococcus is preferable to the higher acid-producing lactobacillus. Case-hardened sausage may not permit proper drying in the center area of dried sausage and, thereby, continued fermentation may produce a sour core. Excessive souring should not be confused with putrefaction of meats occasionally referred to as sour.

Surface growth of mold, yeasts, or bacteria that may cause a moldy or slimy condition may occur when sausage is improperly cured and dried. Moisture may accumulate when the temperature of storage is changed, particularly from a low to a higher temperature. Surface slimes and molds are the massive accumulation of microbial cells that may cause little change in the product and often can be removed mechanically.

Botulism is a very serious condition that has caused numerous deaths in the past; however, the organisms causing botulism, strains of *Clostridium botulinum*, cannot grow at the low pH of semidry sausages produced under commercial conditions. While it is true that the term, botulism, and the specific name for the causative organism is derived from the Latin word, *botulus*, meaning sausage, the organism will not grow in properly made fermented sausage. It should be emphasized that there are no short cuts in the recommended methods for sausage making.

NUTRITIONAL VALUE AND USES

The nutritional value of sausage is about equal to the meats and ingredients from which it is prepared. Fermentation effects little change in nutritional values. Since fermented sausages may have as much as 60% of their water evaporated during drying and smoking, these sausages are actually concentrated meat products. It is a well-known fact that meats are excellent protein food, especially rich in the vitamins, thiamine, riboflavin, niacin, and others.

The American public has not fully utilized sausages, except for wieners or hot dogs, in the many ways used by Europeans particularly in Germany, Hungary, and Italy. Church (1967) has presented about 200 recipes for using sausage. For many of these recipes the fermented sausages are specified.

DEVELOPMENT OF THE PROCESS

It has been stated that the development of fermented sausage processes occurred by a series of steps or advancements. The truth is obvious. These advancements or steps may be outlined as follows: (1) Drying of meats. (2) Chopping and, much later, mincing, grinding, and emulsifying. (3) Salting. (4) Spicing. (5) Casing and, later, the preparing of casings. (6) The use of nitrate. (7) Smoking. (8) Adopting of sanitary procedures. (9) Realizing the importance of working in special rooms with the same equipment. (10) Chilling before casing. (11) Semidrying. (12) The increasing knowledge of the fermentation. (13) Controlling temperature and humidity. (14) Heat processing. (15) Inoculating with starters.

One can only surmise that this may have been the order of developments resulting in the preparation of the present-day, semidry, fermented sausages. These advances obviously occurred in different areas at different times and, certainly, not necessarily in the order outlined. Drying of meats came first. The American Indian dried his meats for centuries and later he cut up the dried meat, added dried berries and nuts, and stored it for consumption in leaner days. Salt was not available in many areas of the world and much of it was impure. People of the Mediterranean shore areas preferred heavily spiced meat, while people further north in Europe preferred the more highly smoked meats. Temperature and humidity control were gradual developments that have culminated in the present-day elaborate smoking, drying, and pasteurizing of equipment. Finally, a gradually increasing knowledge of microbiology and chemistry culminating in the use of inoculation with starters has occurred within the past century.

Commercial Production

In the commercial production of fermented foods manufacturers strive constantly to exercise maximal control of the process to produce a consistently uniform product of high quality. Sausage may be defined as a food prepared from chopped and seasoned meats and molded to a symmetrical shape. The methods of sausage making practiced in the United States and Europe vary considerably. The formulations and methods of various companies also are extremely diverse, not only for the several varieties of sausage, but also for those of a single type. Beef and pork are the most frequently used meats. Some companies use all of the edible portions of the carcass including even the most desirable cuts. Shanks, chucks, and bull meat are commonly used because they have good binding properties. The amount of trimmings and other cuts used vary for different companies. The proportion of fat to lean, the amount of connective tissue, the fineness and uniformity of cut or grind, the spicings, as well as the fermentations themselves are variable and important quality factors. By careful practices and in some cases by government control, the products of certain companies are recognized for their superior quality. In these cases, substantial businesses have been developed. Sausage manufacture is an important part of the meat industry whereby the sausage maker can exercise his individuality and establish a singular reputation.

Sausage may be classified as cooked, or fermented, or cured types, either dry or semidry. Discussion here will be confined to the cured or fermented sausage such as salami, cervelat, and other so-called summer sausages in which bacteria play an important part in the curing, the development of flavor, and the inhibition of growth of undesirable microorganisms with their undesirable chemical changes. Bacteria have little or no part in the preparation of other types of sausage; for them it is advantageous to keep the bacterial load to a minimum. Curing, development of tangy flavor, and inhibition of growth of undesirable microorganisms in fermented sausages are the result of the growth and fermentation by bacteria. Previous to the use of nitrite in the meat formula, reduction of nitrate to nitrite for proper color fixation required the growth of nitrate-reducing bacteria, particularly the micrococci. In such cases there existed a delicate balance between the nitrate-reducing and the lactic acid bacteria. The latter are responsible for the major acid production. They ordinarily do not reduce nitrates.

The fermentation of sausage is sometimes compared to the fermentation of cheese; there are hundreds of varieties of sausage, many of which are native to and preferred by the people of a particular area. The possibilities for variation in sausage formulation are greater than for cheese since different blends of lean and fat meat from different animal species, various salt concentrations, spicing formulae, and degrees of drying are possible. Some sausage manufacturers use all of the meat of a carcass in their blends; others use only those cuts that cannot otherwise be sold to advantage. Beef and pork trimmings, chucks, bull meat, hearts, and other pieces are most frequently used. Sugar, salt, sodium-nitrate, sodium nitrite, and spices are the most commonly used additives.

The added sugar is fermented by bacteria that produce the fermentation products contributing to the tangy flavor. In the wide variety of fermented sausages, sugar serves indirectly as a preservative as a result of its fermentation to lactic acid and other by-products. The resulting reduction of pH in combination with salt in a product of relatively low moisture content produces a high degree of stability. Sugar aids in improving the color of cured meat by helping to establish reducing conditions when nitric oxide reacts with hemoglobin. This tends to prevent the oxidation of ferro- to ferri-hemoglobin derivatives. Sugar tends to exert a protein-sparing action during fermentation. The sugar and the desirable fermentative and reducing microorganisms, the lactic acid bacteria, are essential in this type of meat product. Any of the forms of sugar could be used. Dextrose and levulose, derived from the hydrolysis of sucrose, are readily metabolized by microorganisms to produce an increase in acidity and the desired lower oxidation-reduction potential. Common sugar, sucrose, and dextrose are the most frequently added sugars.

Salt is added primarily as a flavoring ingredient; however, because of its osmotic effect it also improves the texture of sausage and acts as a partial preservative. In sufficient concentrations, salt will inhibit the growth of microorganisms as a result of increasing osmotic pressure; however, the amount of salt used in sausage is far below these concentrations. The added salt and the nitrate-nitrite blend act to inhibit undesirable bacteria, and together with the acids produced and the semidry conditions in the finished sausage, exert the desirable preservative effect.

The literature concerning the action of nitrate as a preservative is voluminous, but its function is somewhat obscure. It can be reduced to nitrite by nitrate-reducing bacteria, but the numbers of such bacteria must be high to accomplish this. The number of micrococci attain these high levels in some sausage fermentations. Nitrite serves a specific purpose as previously mentioned. Since its use is permitted in sausage making, the need for the presence of nitrate-reducing bacteria is obviated.

The essential oils and other flavoring ingredients present in spices are added primarily for their flavor. Although they contain bacteriostatic substances, the quantity of spice used in any of our food products is far below the level required to exert a preservative effect alone. The historical background of ancient travels for procurement of spices is not only interesting but important to the development of sausage of particularly characteristic flavors.

After sausages are stuffed into casings they are frequently smoked at about 90°F for varying lengths of time. Smoking imparts flavor, helps to develop the permanent, desirable red color, and to retard fat oxidation. During this period, an active fermentation by lactic acid-producing bacteria occurs. Smoke is generally produced by slow combustion of sawdust derived from hardwoods. Until recent developments in equipment designed to control temperature, humidity, and degree of smoking, smoking was uncontrolled. It was produced by burning the wood sawdust beneath the meat. Improvements in equipment have resulted in more consistent quality, not only in the degree of smoking and the amount of drying effected but also the rate and extent of fermentation. The composition of wood smoke is very complex, and it contains many acids, aldehydes, and phenolic substances. Major components are formic, acetic, butyric, caprylic, vanillic and syringic acids, formaldehyde, acetaldehyde, furfural, diacetyl, acetone, methyl glyoxal, ethanol, methanol, and phenols as well as many other minor components. Smoke inhibits the growth of various microorganisms; however, the many species of bacteria, yeasts, and molds vary in their tolerance to the smoke ingredients.

As a result of the combination of ingredients added to the meat, the products of fermentation, the relative dryness of the sausage, and the smoke components, dry sausages are essentially sterile. The semidry sausages may be rendered sterile, if they are not already so, by low temperature pasteurization.

Until recently, sausage was prepared in the winter months in order to use natural conditions to keep the meat cold. Refrigeration with accurate humidity control is now used to keep the meat reasonably dry as well as cool. Storage temperatures of approximately 35°–40°F will permit the growth of only a few species of the more psychrophilic types of microorganisms. These are primarily aerobic species. After meats have been comminuted and blended with curing agents and spices, they are then packed tightly in shallow pans and covered with parchment pressed firmly against the meat and then held for 1–3 days or more for curing. The meats are, therefore, not only cold but contain little air. Since the psychrophiles are aerobic, they cannot develop readily.

The meat blends are packed in casings and placed in drying or smoking rooms operated at temperatures ranging from 55°–90°F, temperatures at which many organisms could grow if it were not for the salt, the other curing agents, and the more or less microaerophilic condition. This environment is favorable for the lactic acid bacteria that develop rapidly and produce metabolic products including acids, creating an environment inhibitive to many other organisms.

Dry Sausage.—Dry sausage was probably the first type of sausage prepared for storage. It undoubtedly was developed in a series of steps as a result of the experiences of sausage makers. Since dry sausage is processed by drying and is eaten without cooking, it is particularly essential that the meat used in its manufacture is very carefully selected and free from gross contamination. The age of the meats and the proportions of beef to pork, particularly fat pork, are important factors determining the binding properties. Excess fat, sinews, gristle, and bone must be removed. The method of chopping varies for different kinds of sausage. Salamis are generally coarsely chopped while cervelats are usually finely chopped. The type of chopping or comminuting depends upon the experience of the sausage maker. After the meats have been cut to the proper size, they are blended with spices and curing materials and transferred to the coolers, ordinarily maintained at 36°–38°F. To facilitate quick cooking the blend is spread on pans about 8 in. deep for coarsely cut meat and more shallow pans for finely cut meat.

The meat is packed down or kneaded to exclude as much air as possible and then covered with a piece of parchment to protect the surface from contact with air. Air will not only tend to oxidize the meat pigments but will also permit growth of certain psychrophilic types of microorganisms. The meat is held in the cold room for 1–3 days before it is packed into casings. During this period some enzymatic changes are undoubtedly effected by the natural enzymes of the meat. Some individuals believe that this holding period is essential for blending and development of flavor, others believe a favorable nitrate-nitrite relationship is established then. If the temperature of the blend is reduced rapidly to 36°–38°F, increase in growth of lactic acid bacteria is doubtful. Psychrophilic microorganisms could grow slowly if not inhibited by lack of oxygen and presence of sausage ingredients. Chilling insures a better filling of meat into casings. It may also tend to retard growth of microorganisms.

The chilled meat blend is packed firmly into casings in preparation for drying and smoking, with special care exercised to exclude air pockets. A variety of casings are used for sausage. Casings are prepared with great care by first soaking them in water, cleaning, salting, and then holding in salt brine for a month or more. Care is taken to remove traces of fat. If some microbial growth occurs, it tends to loosen the fat and mucous materials. The casings must be pliable and strong enough to withstand the pressure of the meat. After the casings are stuffed the ends are tied securely and corded. They are then transferred to the green sausage room for preliminary drying and thence to the drying room.

In the manufacture of dry sausage the temperatures in these rooms are carefully controlled at 55°–60°F. Humidities are also carefully controlled at 55–65% to prevent too rapid drying and yet provide a sufficient rate of drying to evaporate moisture on the surface of the sausages. It is during these periods that the desirable bacteria should develop and initiate fermentation. Moisture on the surface of the sausage will permit growth of bacteria, yeasts, and molds. Sausage must dry from the inside outward and uniformly so that case hardening does not occur and the dried sausage will contain no hollow areas or moist interiors. Moist interiors may permit growth of detrimental organisms.

Drying may require from 1–6 weeks depending upon the type of sausage. Loss of moisture may vary from about 20–45% or more. At the low temperatures, growth of bacteria is slow, acid is produced slowly, and there is less chance for excessive acid production than when fermented at higher temperatures. With the excellent drying rooms and equipment available today to regulate temperature and humidity, the problems of slimy surfaces, case hardening, too rapid growth of desirable bacteria with production of excessive acid, and growth of undesirable microbes should be prevented.

With the inadequate equipment used in the past it is not at all surprising that serious spoilages occurred on occasion. The role of the lactic acid-producing bacteria in producing acid and other by-products that inhibit growth of undesirable bacteria is now understood. The temperatures now used do not permit rapid growth of species of the genus *Lactobacillus*, but they delay, at the same time, rapid growth of nonacid-producing species. The species *Leuconostoc mesenteroides* as well as strains of the genus *Streptococcus* will, however, grow more rapidly at these lower temperatures than will the higher acid-producing lactics. The writer has observed rapid growth of *Leuconostoc mesenteroides* at these temperatures in other food fermentations, and there is every reason to suspect this species will grow in sausage at these low temperature conditions. The fact that casings are often punctured quite thoroughly to permit escape of gases from among the particles of meat strongly suggests that *Leuconostoc mesenteroides* initiates these low temperature fermentations. In 2 experimental preparations this species was isolated for nearly 48 hr in a choriza-type sausage smoked at the higher temperature of 102°F and for approximately 72 hr in an inoculated pack. It certainly would develop more rapidly at the lower temperature which approximates the optimum temperature for growth of this species.

Smoking.—Dry sausage for smoking is prepared in the same general manner as dry sausage until it is filled into the casings. The cased sausage, which may be held in a cooler at 38°–40°F overnight, is placed in a smokehouse producing smoke at 85°–90°F. Successful smoking is dependent on proper conditions of temperature, humidity, and uniformity of smoke distribution. These temperatures are near the optimum for growth and fermentation by species of the genus *Lactobacillus*. Marked changes in bacterial populations and acidity occur in 24 hr. The length of smoking time required will depend upon the size of the sausage, the degree of smoking desired, and the tang or tartness desired in the sausage. Excessive smoking or use of higher temperatures may yield a product too acid or sour, and high temperatures may also yield a product of inferior color and of soft texture. After the product is smoked to the degree desired, it is transferred to the drying room for further drying.

Some smoked sausages are smoked at temperatures lower than 85°–90°F. This will alter the rates of growth and fermentation as well as the rates of drying and curing. The proportionate numbers of the various species of lactic acid bacteria may differ at temperatures lower than 85°F.

Semidry Sausage.—The semidry or new condition sausages differ from dry sausages in that the dried as well as the smoked and dried sausages are finished by heating at a relatively high temperature of 138°–140°F after smoking. The product is subjected to a shorter drying period than dry sausage. Products included in this group are sausages such as the newer thuringer, cervelat, and some salami such as kosher salami and others. This method of preparing sausage has become more generally used. Furthermore, many sausage manufacturers are using starters or pure cultures of bacteria in the preparation of this type of sausage.

The improvements in equipment for preparing sausage, particularly refrigerated storage facilities and accurately controlled temperature and humidity, have resulted in great improvements in the uniformity and quality of sausage available on the market. Before the days of adequate refrigeration, the various types of dried fermented sausage had to be prepared during the cold winter months and stored until they were offered for sale during the summer months. The aging and curing that occurred during this period were significant in the development of character, but they were by no means uniform. The variations of temperature and humidity obviously affected quality. Also, preparation of semidry sausages presented greater problems than preparation of dry sausages.

BIBLIOGRAPHY

ANON. 1938. Technical service ACCEL for the production of thuringer, summer sausage, cervelat, Lebanon bologna, and pork roll. Merck and Co., Rahway, N.J.

BREASTED, J. H. 1938. The Conquest of Civilization. Harper & Row, New York.

CHURCH, R. E. 1967. Mary Meade's Sausage Cook Book. Rand McNally & Co., New York.

DEIBEL, R. H., and EVANS, J. B. 1957. "Nitrite Burn" in cured meat products, particularly in fermented sausages. Am. Meat Inst. Found. Bull 32.

DEIBEL, R. H., and NIVEN, C. F., JR. 1957. Pediococcus cerevisiae: a starter culture for summer sausage. Bacteriol. Proc. of the General Meeting Soc. Am. Bacteriologists, 14–15.

DEIBEL, R. H., NIVEN, C. F., JR., and WILSON, G. D. 1961A. Microbiology of meat curing. III. Some microbiological and related technological aspects in the manufacture of fermented sausages. Appl. Microbiol. 9, 156–161.

DEIBEL, R. H., WILSON, G. D., and NIVEN, C. F., JR. 1961B. Microbiology of meat curing. IV. A lyophilized Pediococcus cerevisiae starter culture for fermented sausage. Appl. Microbiol. 9, 239–243.

EAGLE, R. F. et al. 1938. Sausage and ready to serve meats. Inst. Meat Packing, Univ. Chicago.

FRAZIER, W. C. 1958. Food Microbiology. McGraw-Hill Book Co., New York.

FRAZIER, W. C. 1967. Food Microbiology, 2nd Edition. McGraw-Hill Book Co., New York.

JENSEN, L. B. 1942. Microbiology of Meats. Garrard Press, Champaign, Ill.

JENSEN, L. B., and PADDOCK, L. 1940. Sausage treatment with Lactobacilli. U.S. Pat. 2,225,783.

MILLS, F., and WILSON, G. D. 1958. Use of Pediococcus cerevisiae starter in pork roll. Am. Meat Inst. Found. Bull. 38.

NIINIVAARA, F. P. 1955. The influence of pure cultures of bacteria on the maturing and reddening of raw sausage. Acta Agral. Fennica 85, 95–101. (German)

NIVEN, C. F., Jr. 1951. Sausage discolorations of bacterial origin. Am. Meat Inst. Found. Bull. 13.

NIVEN, C. F., JR. 1961. Personal communication. Del Monte Corp., 215 Fremont St., San Francisco, Calif. 94119.

NIVEN, C. F., Jr., DEIBEL, R. H., and WILSON, G. D. 1958. The AMIF sausage starter culture. Am. Meat Inst. Found. Bull. 41.

SISON, E. C. 1967. Microbiology and technology of native fermented sausage. M.A. Thesis, Univ. Philippines Coll. Agr., Laguna, Philippines.

TANNER, F. W. 1944. The Microbiology of Foods, 2nd Edition. Garrard Press, Champaign, Ill.

CHAPTER 8

Some Cereal Foods

ORIGIN OF BREAD

Of all the plants, the grasses are the most important to man. All of our breadstuffs, wheat, rice, corn, oats, and barley, are grasses. Sugar cane and bamboo are also grasses. Man has been eating cereals for at least 10,000 yr without growing tired of them. Primitive people ate the seeds of wild grasses and obviously relished their flavor. Man discovered early that cereal grains could be sun dried and stored for months and even years without spoiling. The Biblical account of the storage of grain for the seven lean years by the Egyptians upon Joseph's recommendation is evidence for this early practice. Man learned to grind the grains into a crude powdery flour that could be blended with water and baked to improve its character. The Swiss lake dwellers mixed the ground grains, wheat, barley, and millet, with water, rolled the ground mash into sheets, and baked these on hot stones. These early breads were unleavened. Jensen (1953) traces certain practices to Neolithic man of 6000 to 7000 B.C. who migrated from the area west of the Nile River. Jensen stated further that Ceton-Thompson in 1934 found 165 silo pits in this area containing remarkably well-preserved grain.

A large amount of literature has been developed on the origin of bread baking. One of the earliest devices devised by man was the stone quern and crushing cylinder used to grind wheat and barley. Flour sieves came into use before 2000 B.C. Pliny stated that good bread depends upon the goodness of wheat and the fineness of the flour.

The forerunners of modern baking may be the practices started in Egypt about 3500 yr ago. The Egyptians made a great step forward when they observed that if they allowed the bread dough to cure for several hours, the dough expanded when baked and they obtained a spongy light loaf of bread. Triangular loaves of leavened bread have been found in Egyptian tombs. Without doubt, the more observant bakers noted that the dough was somewhat sour, and they may have observed the relationships between souring and quality characteristics. Bread was the principal food of the Egyptians; they were often referred to as wheat eaters; and bread was often used in lieu of wages.

The ancient Greeks prepared a leavened bread which contained barley flour. Women were the bakers as they are in homes today. With the development of leavened bread, the use of barley declined because it does not produce the light airy leavened bread typical of wheat bread. Baking was an important industry in ancient Rome, and it is estimated that there were more than 250 bakeshops in Rome as early as 100 B.C. The Romans of a later period collected the yeast from wines and used it for leavening bread dough. The Germanic peoples of that early period were nomadic and even Attila, years later, had not learned the use of leavened bread.

These early breads were quite different from the light airy breads we enjoy today. The flour was coarsely ground and some of the bread doughs were prepared only with flour and water. The baker's art consisted, in part, in carrying over the fermented dough, that is, the yeast-bacterial culture, from day to day. This was basically a problem of suppressing growth of spoilage organisms, bacteria, yeasts, and molds, the development of which they could not comprehend. The self-rising breads of antiquity, like the sourdough breads of today, depend upon a mixed flora of bacteria and yeasts to produce the alcohol, organic acids, and other substances responsible for the flavor and texture characteristics of bread.

The science of microbiology began several thousands of years later. Some of the early bakers may have observed a relationship between souring of the dough and lesser losses from unexplainable spoilage, but no one understood the nature of changes that occurred. The bakers' guilds of the period undoubtedly kept secret certain practices of the art. There is little doubt that the practice of carrying over a small amount of the dough from one baking to blend with the new dough was a later development in the art of breadmaking. Whenever this practice was first introduced, it must have been observed that if this carryover dough soured, the loaves of bread were not only light and spongy but were more uniformly delicious and less subject to spoilage. Originally, the leavening process was used for raising the dough to produce the light spongy loaf, but it is well known now that the leavening also improves the flavor of the bread.

CEREALS USED IN DOUGHS

Wheat, *Triticum vulgare,* was an established crop at the beginning of recorded history. Its origin is unknown but evidence indicates that it may have been developed from einkorn, *Triticum monococcum,* and/or emmer, *Triticum dicoccum,* by hybridization. Einkorn may have developed from the wild grasses native to the arid lands of Asia Minor and Southeastern Europe. Emmer is a superior wheat and resembles the wheat found in the mountainous regions of Syria and Palestine. Brace, the wheat peculiar to the Gauls, was a spelt wheat and was particularly well-adapted for bread baking as well as beer fermentation.

Wheats are sometimes classified into five types: (1) hard red spring, (2) durum, (3) hard red winter, (4) soft red winter, and (5) white. The hard spring and winter wheats are used for bread doughs; therefore, about 70% of the annual acreage sown in the United States is of these types. They contain a high percentage of protein of good quality. They mill well and yield good quantities of flour. The separation of bran from endosperm depends upon two principles related to the construction of the wheat berry. The first is the fact that when wheat is soaked with water, the bran becomes tough and rubbery while the interior remains soft and friable. Secondly, when the wheat berry is sheared by corrugations of the rolls and breaks in milling, it will split open releasing some of the endosperm and flour. Most of the pieces will flatten to expose the remaining endosperm.

When the hard wheat flours are mixed with water in correct proportions, the gluten forms an elastic dough or sponge capable of retaining the carbon dioxide produced by fermentation. When baked, the dough will set to a spongy structure. These, then, are the important characteristics that make wheat flour breads so acceptable to consumers and the reason why wheat flour is so frequently blended with other flours for other types of bread. No other cereal flour has these characteristics in the degree that the flour of the hard wheats has.

Rye, *Secale cereale*, is a major grain crop in Northern Continental Europe where its flour is used in the dark breads of these areas. Rye may have originated from the wild ryes or rye weeds growing among the wheats or barleys. Rye flour, although produced in a number of grades, does not contain protein with the unique glutinous character of wheat protein. Bakers of Russia, Poland, Germany, and other European countries are continually increasing the content of wheat flour in their bread doughs. Flat breads such as the knäkebröd of Scandinavian origin are still prepared with 100% rye flour and are now usually leavened with yeast. The sourdough rye breads containing wheat flour are increasing in popularity in some areas of the United States.

Rice, *Oryza sativa*, is one of the oldest and most important cereal crops. It is the staple food for over ½ of the world's peoples. The earliest records of rice production date back to about 2800 B.C. in China and 1000 B.C. in India. It was cultivated in the Euphrates Valley in 400 B.C. Today, over 90% of the rice is grown in Asia; and Burma, Thailand, and Indochina have usually been the only exporting nations. Among the 8000 or more botanically different varieties of rice, only a few have protein of the glutinous character that makes them suitable for producing satisfactory leavened products.

Oats, *Avena sativa*, may have originated with the ancient Slavonic peoples of Western Europe. The first authentic records of cultivated oats appear at the beginning of the Christian era. Writers of classical antiquity refer to it as a weed used for medicinal purposes. Oatmeal has often been blended with wheat flour in making breads. This was commonly practiced in the United States during the period of 1914–1918 when wheat flour was in short supply.

Barley, *Hordeum vulgare, H. sativum*, is one of the world's oldest cultivated cereal grasses. The ancient Greeks used barley flour for their unleavened bread, and in some areas such as Syria and Lebanon barley flour is still blended with wheat flour in preparing bread doughs. It has been used primarily in porridge and in the beverage industries.

Corn, *Zea mays*, was systematically cultivated by the American Indians. It may have originated from a subtropical plant, teosinte, an annual grass native to Mexico and Central America. It has little use in leavened bread but is used in many other fermented products.

Millet, *Panicum miliaceum*, another annual grass sometimes called barn grass, was found among the sites of the Swiss lake dwellers. It is used to a limited extent in fermented products.

Adlay, *Coix lacryma Jobi*, Job's tears, is a cereal grass whose seeds have a high protein content and have been used in breads. The grain requires a long growing season.

Flours are obtained from other plants but the above named are the more important. In some areas of the world cassava is commonly used in the high carbohydrate diets.

INGREDIENTS OF BREADS

Although bread can be prepared with flour, water, and yeast, most of the breads available today contain in addition several other ingredients. The principle ingredients now used are flour, water, salt, yeast, sugar, shortening, and various additives. Wheat flour is the basic ingredient of most of the breads. Wheat is unique among the cereals in that the milled product, flour, is capable of forming a dough that will retain the gas evolved during fermentation and when baked, yield a well-aerated bread. This has been ascribed to an orientation of gluten molecules. When properly blended, the dough acquires a uniformly smooth character, and the dough has the greatest degree of elasticity of doughs of any of the grains.

Water, an essential ingredient of doughs, has excellent dissolving properties and dissolves to some degree all ingredients of a dough. With these dissolved ingredients and the natural mineral content of most waters, the liquid furnishes an excellent growth medium for microorganisms in addition to producing the proper texture of the dough.

Milk as fluid milk or as dried milk is now used in many bakery products. It is the basic means of supplying liquid of high quality.

Wheat contains about 1% of lipids. Butter, lard, beef fats, corn, soybeans, or cotton seed oils and other lipids are used in most bakery products, the amount varying with the particular product. An antioxidant is usually used with lipids. Fats, oils, and shortenings impart richness and tenderness, improve eating qualities, contribute to flavor, provide better texture, lubricate gluten, act as emulsifiers, and provide aeration and better leavening. In some baked products such as Danish pastry they develop flakiness.

Flour contains about 1% of fermentable sugar. Sugar is essential for the growth of microorganisms and production of carbon dioxide, alcohol, and other fermentation products. The amount of sugar used in commercial baking has been increased in recent years. Sugar imparts flavor, color, and texture characteristics as well as furnishing the fermentable carbohydrate. Sugars, in the form of sucrose, maltose, lactose, corn syrup, or malt syrup are used. The latter syrup is often rich in enzymes capable of hydrolyzing starches to dextrins and maltose.

Eggs, flavors, spices, coloring materials, and other ingredients add flavor, aroma, and color to various baked products. Vitamins and minerals are added to improve nutritional qualities.

The function of yeasts in bread making is to lighten the dough and to impart the characteristic flavor and aroma. Four kinds of yeast are available but the ordinary compressed yeast or active dry yeast is most commonly used.

The chief use of yeasts in the household is not for producing fermented drinks but to raise bread. The raising of bread by means of yeasts has been brought to a state of great perfection. The methods of producing a desirable fermentation are now extremely simplified. It has taken many centuries of experience and experiment to understand the subject sufficiently to bring the fermentation under control.

PANARY FERMENTATION

The change which occurs in the flour and the other constituents of dough before baking into bread may be termed panary fermentation. It consists essentially of the production of alcohol and carbon dioxide gas. If the dough is left in a warm place for a number of hours, it will rise due to the production and retention of gas. An understanding of the changes that occur is essential to the consistent production of a good loaf of bread.

After thousands of years of bread baking and its development to a universal art in homes and in commercial bakeries, the knowledge of the fundamental principles of the process must be considered a comparatively recent acquisition that is still far from complete. To many peoples the history of bread making began in mother's kitchen where the generous slice of warm bread liberally spread with butter is remembered. In some homes a part of the raw dough from the previous day's baking was blended with the new batch of dough to serve as a starter or leavening agent. This is still practiced in homes and in bakeries in many areas of the world, even though good quality commercial yeasts are available and dependable. Although some variation exists in practices and in ingredients used in many homes, basic procedures are quite well standardized. The general methods used in the home were adapted to small bakery operations. As the small bakeries grew in size, more and more mechanical equipment was introduced. Today, the process in some bakeries is automated and continuous. Automation and equipment as well as their use is fully discussed in various books on the subject.

The principal ingredient in bread is flour. Although the incorporation of several other ingredients varies, the general procedure is to dissolve or blend the salt, sugar, and other ingredients in part of the water to be used, and then add some of the flour and yeast to this to give a somewhat thin dough or sponge. In the home this was done in a large bread pan. The dough was allowed to set for several hours, or overnight, or until it had increased in volume several times. Occasionally, the housewife punched down the sponge to retain it within the pan. When a proper volume was attained, the sponge was transferred to a bread board and more flour was added as the dough was kneaded. The board was continuously sprinkled with flour so that the thin sponge would not stick to the board. When the dough was compact, it was again placed back in the bread pan and allowed to rise again. Sometimes this operation was repeated and the dough would again rise. After an hour or more the dough was placed on the board again, kneaded, and divided into loaves large enough to fill the baking pan $\frac{1}{3}-\frac{1}{2}$ full. These were pressed carefully into the pan and the surface was wiped and smoothed with a small amount of water. The practices varied according to the experiences of the housewife. In some homes the surface was coated with butter or sprinkled with flour. In Europe, instead of the flat bread board, a hollowed, usually oblong board was used. These are more or less museum pieces today, and small decorative boards are made to simulate the large boards. In many homes the bread board was scraped, if necessary, to remove adhering dough, and then sprinkled lightly with flour and hung up for the next baking. In other homes the equipment was washed. By either method, the board retained on its surface and in its pores the yeasts and bacteria that inoculated the next batch of dough.

This is a brief description of panary fermentation. Relatively few people understand the nature of the microbiological changes that occur. To a microbiologist there are numerous opportunities seen during these operations for introducing a number of foreign microorganisms, bacteria, yeasts, and molds in addition to those added in the yeasts or dough starter, and numerous opportunities for contaminants to grow. These organisms may originate from the equipment, the ingredients, the hands of the baker, or the air. It is surprising that greater contamination and spoilage do not occur.

Generally, these methods were adopted in small bakeries centuries ago. Today the large bakeries are using the same general procedures but have mechanized the processes for speed and efficiency of operation.

The baking industry knew little or nothing about scientific control principles, even for centuries after pioneer work had begun. With the increased knowledge and improvement in equipment and control, acceptance of bakery products has increased rapidly since 1900. The advances in biochemistry, physical chemistry, microbiology, and technology have revolutionized the baking industry. The improvement in wheat culture and knowledge of the physical and chemical properties of wheat and other cereals has standardized flours and has been largely responsible for greatly improved quality. Knowledge of fermentation, which had to await the early discoveries of von Leeuwenhock, Pasteur, and others, has advanced to such a degree that fermentation is now standardized and controlled.

The sponge and dough method of mixing is commonly used because of the greater flexibility of control of fermentation and better adaptability to mass production schedules. The principle involved in the sponge and dough process is the mixing of part of the flour with other ingredients to a fairly stiff dough called a sponge and allowing the fermentation to proceed 3–4 hr or until the gluten of the flour has become well-conditioned to a soft web-like structure capable of retaining a large volume of gas. These gases, air, and water vapor together with alcohol vapor produced by fermentation increase the volume of the sponge many times. The mass is allowed to rise until the extensible limit is reached and the sponge drops.

MICROORGANISMS IN BREAD FERMENTATION

The flavor, texture, and other characteristics of bread depend to a large extent upon the microorganisms involved in the leavening process. Leavening depends upon the activity of the microorganisms, the yeasts, and bacteria. The leavens responsible for rising of bread dough may be divided into three

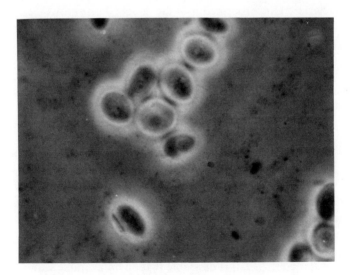

FIG. 8.1 MICROPHOTOGRAPH OF A STANDARD BREAD YEAST

Note the comparatively small bacteria, presumably lactic acid species, among the yeast cells.

types: the relatively mixed microbial leaven that develops spontaneously in mixtures of flour and water under appropriate conditions; the leavens prepared in the home or in bakeries in solutions of various kinds; and the relatively pure yeasts secured in the form of yeast cakes (Fig. 8.1). In many areas of the world, yeasts in several forms are available for bread making.

Cereals and the flours prepared from cereals are always heavily seeded with microorganisms. They may have been introduced onto the cereals from soil, from water, from air, or equipment during harvesting of the cereals, or subsequent handling of the grain and flour. A large majority of the microorganisms are generally harmless and do not develop in the dry cereals or flour, or even in the doughs prepared from the flours. Since yeasts, bacteria, and molds are ever present, one cannot prepare a dough without incorporating these organisms. The dry cleaning and washing of grains and the milling of wheat to flour reduce the microbial content from several million organisms per gram to several thousand per gram of flour. This is, in part, the effect of bleaching of flour by oxidizing agents. These remain as a heterogeneous flora of yeasts, bacteria, and molds that are primarily aerobic and which will not develop in the dry flour. Mold species of the genera *Aspergillus* and *Penicillium* can develop in grain or flour if the moisture content exceeds 15%. Yeast and bacteria require a higher moisture content, possibly 17–20%, before they can grow. Thus the activity of each species of microorganism is influenced by water content as well as other environmental conditions.

In order to establish proper leavening action the yeast and/or acid-producing bacteria are incorporated into the dough. Their growth will establish in the dough a partial anaerobic and acid environment unfavorable to the growth of the undesired aerobic and usually acid-intolerant flora such as mold and aerobic spore-forming bacteria. In spite of these conditions spoilage does occur at times.

Yeast, and to a lesser extent bacterial fermentations, are the cornerstone of the baking industry. Yeast manufacture is concerned principally with one species of yeast, *Saccharomyces cerevisiae*. There are a great number of strains or races of this species which may differ from each other in minor characteristics. The bakers in the past have selected their strains which were kept alive in bread doughs or other media and usually guarded them carefully as do the wine and brewing interests today. The manufacturers of yeast select for propagation the strains most suitable for bread making. These strains are isolated and preserved in pure culture, and today continuous study and research are conducted to improve the characteristics of various strains. Under certain conditions yeast may pass through a series of mutations that modify its character; also, it is desirable that no unfavorable variations occur in the culture. The culture used in preparation of dried yeast cakes is cultivated under somewhat different conditions than those used for the moist yeast cakes. Similar study and care should also be applied to the bacteria involved in fermentations.

Commercially produced yeast is a relatively recent innovation in the art of bread baking. The relative roles of the yeasts and the acid-producing bacteria in the art of baking, as practiced by the ancient Egyptians and all subsequent bakers, are unknown. Two extremes in leavening action in respect to yeasts and bacteria are practiced. At one extreme we have a pure yeast fermentation in which relatively no lactic acid bacteria are present and the primary product is alcohol and carbon dioxide. At the other extreme we have leavenings due entirely to the heterofermentative lactic acid bacteria. This has been observed in the fermentations of idli, puto, and pumpernickel breads. The yeasts produce sufficient carbon dioxide to result in a great increase in volume and a resulting light spongy bread. Since the heterofermentative lactic acid bacteria produce relatively small amounts of carbon dioxide compared with that produced by yeasts, the increase in volume of the doughs is relatively small and the bacterial leavened breads are, therefore, much more compact than yeast-leavened doughs. Since the homofermentative lactic acid bacteria produce very little carbon dioxide, their value in bread baking is entirely due to the acidity and the flavor they may impart to the dough and bread. For preservation of sourdoughs, in which yeasts and bacteria both play a role, the homofermenters may play an important role. There are a number of lactic acid-producing bacteria, both homo- and heterofermenters, that have been identified with the leavening activity. These will be discussed subsequently.

The majority of dough fermentations are intermediate between these extremes and are due to various mixtures of yeasts and bacteria. In fact, since environmental conditions conducive for growth of both these yeasts and bacteria are so similar, it would be difficult to create a leavening action exclusively due to one. The commercial yeast cakes available usually contain some lactic acid bacteria; and the intermediate changes that occur in both yeast and lactic acid bacterial fermentations are so similar that it would be difficult to determine which of the substances produced by fermentation are the products of one rather than the other. The major end products, carbon dioxide and ethyl alcohol, are usually attributed to the yeasts while lactic acid is considered a product of bacteria. This is not entirely true. The heterofermentative lactic acid bacteria produce some carbon dioxide and alcohol, and the yeasts produce some acid. Certain studies have indicated that the presence of bacteria in yeast cultures result in improved bread quality. On the other hand, others do not believe that the bacteria in comparatively small numbers and of small size could have much effect upon flavor, particularly in commercial operation.

Originally, it was assumed that the exclusive purpose of leavening was the loosening of the dough and, therefore, production of a light spongy bread. An important role is the production of flavor and aroma. Bread flavor is derived from the components of the formula, the by-products of microbial action, and the thermal effects of baking on the components and the products of fermentation. Although the yeast cells in commercial yeasts far out-number and outweigh the bacteria, research has indicated that the presence of these bacteria with the yeasts is essential to superior bread flavor. The pithy and sour taste of rye bread, although pleasing to some consumers, is unsatisfactory to others. In the same way, the bland flavor of yeast-leavened bread is not entirely satisfactory to everyone.

Considerable research has been conducted concerning identification of the microorganisms of breads, both the desirable types and the spoilage organisms. *Saccharomyces cerevisiae* is the yeast used almost entirely for leavening doughs. The yeast *Candida utilis* is used in some rye breads primarily as a flavor supplement.

A great number of species of lactic acid-producing bacteria of the genera *Lactobacillus, Streptococcus, Pediococcus,* and *Leuconostoc* have been listed in the results of various researches as active agents in dough fermentations. The true lactic acid-producing bacteria can be divided into two types: the homofermenters which include the streptococci, the pediococci, and the majority of the species of lactobacilli; and the heterofermenters that include the leuconostoc and some of the lactobacilli. The former produce lactic acid as the primary end product of fermentation, and, therefore, their contributions are primarily flavor and acidity production. The heterofermenters may, in addition, produce leavening action due to production of carbon dioxide as well as lactic and acetic acids and alcohol. The lactic acid bacterium first isolated and described by Henneberg as *Bacillus panis fermentati* is generally considered a synonym of *Lactobacillus brevis*, a heterofermenter; however, from his drawings it may well have been a mixed culture of this species with *Leuconostoc mesenteroides*. At the time of Henneberg's studies the latter species was associated only with spoilage in sugar solutions.

Nearly every species of the genus *Lactobacillus* as well as species of *Leuconostoc, Pediococcus,* and *Streptococcus* have been isolated from bread doughs by many scientists. Spicher (1959) reviewed bacteriological studies of sour bread doughs, and he isolated 120 strains of lactic acid bacteria rep-

TABLE 8.1

SPECIES OF GENUS *LACTOBACILLUS* ISOLATED FROM SOURDOUGHS

Group	Number of Cultures	Species Designation
1	4	*Lactobacillus delbrüeckii*
2	2	*Lactobacillus leichmannii*
3	36	*Lactobacillus plantarum*
4	12	*Lactobacillus casei*
5	32	*Lactobacillus brevis*
6	14	*Lactobacillus fermenti*
7	11	*Lactobacillus pastorianus*
8	5	*Lactobacillus buchneri*

Adapted from Spicher (1959).

resenting 4 hetero- and 4 homofermentative species (Table 8.1). Over ½ the strains were identified as either *Lactobacillus plantarum* or *Lactobacillus brevis*. This is not surprising since the various doughs in which sugar is present furnish a nutritious medium not only for the fastidious lactic acid bacteria but also for great numbers of other microorganisms. When the lactic acid bacteria grow and produce acid or when the yeasts produce alcohol and carbon dioxide, an environment is created that is inhibitory to great numbers of other microorganisms. The heterofermentative species *Leuconostoc mesenteroides, Lactobacillus brevis, Lactobacillus fermenti,* and *Lactobacillus buchneri* can grow in dough and produce leavening action as well as acidification of the dough. It is even conceivable that *Lactobacillus pastorianus* could grow in the beer bread (Øllebrød) prepared in Scandinavia. If sufficient free sugar is present, *Leuconostoc mesenteroides* could produce dextran; stringy bread doughs have been described. Most of the higher acid-producing and homofermentative strains of lactic acid bacteria could grow well and, if milk is used in the doughs, the high acid-producing fastidious milk fermenters would find the dough medium satisfactory. The apparent confusion concerning the lactic acid-producing bacteria isolated by microbiologists from bread doughs is not surprising. In the bread doughs prepared for the steamed breads, idli and puto, *Leuconostoc mesenteroides* has initiated the fermentation (Fig. 8.2).

FIG. 8.2. EQUIPMENT USED FOR BAKING PUTO BY STEAMING

Undesirable Microorganisms

Because bread doughs, as previously stated, furnish a medium in which any number of other microorganisms may grow, it is not at all surprising that bread spoilage has been ascribed to various species. Excessive or extensive souring of doughs by the true lactic acid-producing bacteria may be quite undesirable in certain breads and should then be considered a form of spoilage. The facultative species *Escherichia coli, Aerobacter aerogenes,* and *Serratia marcescens* are acid-producing and acid-tolerant and have been isolated at times from doughs. Souring of bread doughs has been attributed to the first two species and red or "bloody" bread to the latter. Red or "bloody" bread rarely occurs and is striking in appearance. *Serratia marcescens* often produces a brilliant red pigment. Some of the early "miraculous" appearances of drops of blood on communion breads or wafers after storage in damp areas have been attributed to the growth of this organism. Other shades of red bread have been caused by growth of molds such as *Monilia sitophila.* Ropiness has been caused by *A. aerogenes* as well as *Bacillus mesentericus* and other species. Blackening of bread doughs may be due to growth of other species of the genus *Bacillus.* This should not be confused with the normal so-called black breads made with coarse wheat and rye flours. Rancidity and other off-flavors may be the result of growth of several types of bacteria, including strains of species of the genera *Micrococcus, Flavobacterium, Achromobacter,* and *Pseudomonas.* Species of the latter are aerobic and acid-intolerant. Under aerobic conditions it is conceivable that *Acetobacter* species could develop. Under certain conditions proteolytic organisms of certain species may develop and destroy some of the glutinous property of doughs. The rare chalk-like spots that occur in so-called "chalky" bread are ascribed to the growth of the yeast-like fungi *Trichosporon variable* and *Endomyces fibuliger.*

Moldiness and ropiness have been the major types of spoilage of baked bread. The former has been observed by nearly all older people at some time or another. These spoilages are seldom observed today in commercially made bread, but moldiness remains a problem with homemade breads.

Since molds are aerobic and their spores are ubiquitous, their presence can be assumed in all doughs. During baking, the surface crust attains a temperature sufficiently high to kill the spores; however, spores within the loaf may survive. They can also contaminate the surface of a loaf after baking and may be particularly disastrous if the bread is stored in a warm, humid place. The surface of sliced bread is particularly subject to mold growth. Spoilage is evidenced by black, green, yellow, and brown cottony but usually white growths with colored sporangial or conidial dots. Species of the genera *Penicillium, Aspergillus, Rhizopus, Monilia, Mucor,* and others have been implicated in mold spoilages.

Ropiness of bread is still observed on occasion with home-baked bread. The temperatures attained in baking are high enough to kill all vegetative bacterial cells. The spores of the spore-forming bacteria can withstand the 100°C baking temperature. Heavy spore contamination of the bread doughs is a major cause of spoilage, but suitable conditions must occur before the spores can germinate. A warm, humid condition is most favorable. This may occur when bread is cooled slowly or when stored in a humid, warm atmosphere. The common spore-forming organisms *Bacillus mesentericus* and *Bacillus subtilis* usually associated with this type of spoilage are inhibited by acid and, therefore, do not develop in sourdough bread and many of the breads with a lower degree of acid.

Although staling of bread may be considered spoilage, it is a physical change that is not related to microorganisms.

Prevention of Spoilage.—The types of dough and bread spoilages mentioned are seldom observed today in commercially baked bread but still may be troublesome. Research to determine causes of spoilage and methods of prevention of such spoilage has had a marked effect in reducing losses to a minimum.

The prevention of spoilage that is due to growth of molds or to germination and subsequent growth of spore-forming bacteria that cause ropiness has been a serious problem of the baking industry. Flours will always contain some mold and bacterial spores. During baking, the molds' vegetative growth and most of the spores are killed. The bacterial spores are, however, resistant to the temperatures attained within a loaf of bread during baking. These spores and the organisms introduced in handling of the baked bread may cause subsequent spoilage. Preventative measures are designed, first, to obtain flour of low microbial content; second, to prevent recontamination by means of cleanliness, filtering air, and the use of ultraviolet radiation or electronic heating; and, third, by proper cooling of baked bread and storing at reasonably low temperatures in low humidity atmospheres. Use of inhibitory or mycostatic substances, such as sodium and calcium propionate, has become quite common in the industry. Sorbic acid has been used as well as acetic, lactic, citric, and phosphoric acids to lower the pH so that spores of bacteria cannot germinate.

THE FERMENTATION

The changes that occur in the dough and its constituents before baking of bread are often referred to as panary fermentation. Although this consists essentially of the production of carbon dioxide gas as a result of yeast fermentation, the changes in constituents of the dough and the production of alcohol, acids, and other fermentation products by yeasts and bacteria are very important in imparting characteristic flavors to the baked product.

The leavens used for bread making may be divided into three groups: the relatively pure yeasts obtained as compressed or dried yeast, the leavens propagated in the bakery or home in various dough solutions, and the spontaneous leaven that develops in mixtures of flour and water.

Fermentation starts as soon as the sponge or dough is prepared and continues until the dough loaves reach the oven where the heat of the oven kills the microorganisms and inactivates their enzymes. During fermentation, the enzymes not only produce the fermentation products but also condition the dough so that it will be in a proper state to retain the maximum amount of gas.

The yeast cake is broken up and suspended uniformly in a liquid portion of the mix. In commercial operation, an emulsifier is used for this purpose. This liquid suspension of leaven is propagated in doughs and blended uniformly with the mix until a uniform mixture of the ingredients is obtained. Since heat is generated in blending the ingredients, cooling is sometimes essential in commercial operation. In the small amounts of dough prepared in homes or in small bakeries, the heat retained in the dough may be just adequate for raising the temperature of the sponge to about 75°F, a temperature usually considered desirable for the rapid growth and fermentation in the sponge. The sponge at this stage should be a smooth glutinous mass and the microorganisms should be multiplying and fermenting the sugars. Within an hour or more the sponge will increase in volume several fold. The length of time required for adequate fermentation and increase in volume will depend upon a number of factors such as temperature, amount and condition of the yeast used, the strength and hardness of the dough, and upon other ingredients. In commercial operation in which temperature is held at about 75°–80°F and the relative humidity is about 75%, sponge fermentation time is predetermined. Rapid fermentation can occur in the relatively moist sponge. The size of gas spaces may vary. At this stage the doughs are kneaded with additional flour to produce a uniform dough. This dough may be fermented again for 15 min to 1 hr before the fermented dough is cut into the proper sized pieces and placed in containers for baking. Each loaf is individually rounded up to form a smooth skin over the dough that will aid in retaining the gas during subsequent fermentation. As fermentation continues, this loaf will again increase in volume. In commercial operation the fermentation may continue for about 10–15 min in a so-called proofing room at temperatures of 94°–98°F with a higher relative humidity of 85%.

During the first few minutes of baking, as the temperature is increasing, enzyme action continues at a rapid rate. The dough continues to rise rapidly, not only due to the increased enzyme activity and rapid production of gas but also due to the increasing temperature and expansion of gases previously formed. When the temperature reaches 140°–150°F, the yeasts die, their enzymes are inactivated, and the free alcohol has been evaporated. Proteins coagulate and starch is gelatinized in part. As baking continues, the exterior of the loaf becomes very hot and dextrins are formed from starch on the outer surface to form the crust. The heat causes a series of chemical reactions sometimes referred to as caramelization that produce a brown flavorsome crust. During baking, molds, vegetative cells, and some spores of bacteria as well as the yeasts are killed. The presence of too many bacterial spores may cause spoilage of improperly cooled and stored bread.

CLASSIFICATION OF BREADS

From a microbiological viewpoint breads may be classified in several ways. The first, and most obvious, is separation into unleavened, chemically-leavened, and microbiologically-leavened breads. A second classification can be on the basis of microorganisms responsible for leavening, that is, the yeasts and bacteria. A third separation based upon the relative influence on flavor could be included.

Unleavened Bread

The breads prepared from ground grains in the early history of the development of man were unleavened breads. They were ordinarily prepared from flour and water with salt added if available, and baked on the hot surfaces available. Unleavened breads or bread-like foods include the hosts used for Holy Communion, the very thin Matzah prepared by the Jewish peoples for their Passover, flat unleavened chapaties common to areas of India, some of the breads prepared by the American Indians, some of the steamed rice flour preparations of the Far East, and the various macaronis and spaghetti.

Chemically Leavened Bread

Chemical leavening is commonly used in the preparation of cakes and cookies, but it is also utilized in the preparation of certain breads, biscuits, and bread-like products. Since about 1890, prepared packaged pancake and waffle mixes have increasingly replaced the microbiologically leavened mixes. Some of the doughnut mixes are also leavened by chemicals. Since these various products are not leavened by fermentation, they will not be considered further.

FIG. 8.3. PHYSICAL DIFFERENCES BETWEEN A WHITE BREAD (TOP RIGHT) LEAVENED PRIMARILY BY YEAST, A SOURDOUGH RYE BREAD AND A PUMPERNICKEL (TOP LEFT AND CENTER) BOTH LEAVENED BY YEASTS AND BACTERIA, AND PUMPERNICKEL (BOTTOM) PRIMARILY LEAVENED BY BACTERIA

TABLE 8.2

COMPARATIVE NUMBERS OF YEAST AND BACTERIA IN COMPRESSED YEAST AND
EFFECT ON FLAVOR

| | Number of Cells × 10,000 | | |
	Bacteria	Yeasts	Flavors Score
Pure yeast		30,000	6.0
Compressed yeast	1,100	25,000	7.2
Pure yeast and			
bacteria	32,500	28,500	6.5

Adapted from Carlin (1958).

Microbiologically Leavened Bread

The leavening effect of the more common types of breads is almost entirely the result of activity of the yeasts, while leavening of some types of sourdough breads is almost entirely effected by the lactic acid bacteria (Fig. 8.3). There are, however, few leavenings that are caused by pure cultures of either yeasts or bacteria (Table 8.2), and distinctive quality of many breads is imparted by the activity of both yeasts and bacteria. Most of the sourdough and salt-rising breads available on the market today are leavened by a combination of yeasts and lactic acid-producing bacteria, and it is generally recognized that bacteria play some role in imparting character to yeast-leavened breads.

Many of the sourdough rye bread recipes available today call for a mixture of rye and wheat flours and the use of yeast as well as the "sours" or starter. Some of the dense, compact, dark and black rye sourdough breads and pumpernickels are leavened entirely by the heterofermentative lactic acid bacteria and may acquire an acidity of pH 4.0–4.5, in contrast to a pH of about 5.3 for ordinary rye bread or 5.7 for white bread. The sweet rye breads in which sugar or molasses is added favor the development of yeasts. The idli of India and puto of the Philippines are often leavened entirely by the heterofermentative lactics, usually *Leuconostoc mesenteroides*. Puto is neutralized during leavening to reduce the acidity. It should be stressed here that the homofermenting lactic acid bacteria do not produce sufficient carbon dioxide to play a major role in leavening and, furthermore, their development inhibits growth of the heterofermenters. The sourdough bread prepared with yeasts and bacteria are less dependent upon the heterofermentative lactic acid bacteria since the yeasts produce the carbon dioxide essential for leavening. The lactic acid provides the acid condition and inhibits growth of undesirable bacteria.

FLAVOR AND TEXTURE OF BREAD

Food flavors are associated with the presence of certain chemical compounds and bread is no exception. There are great variations in flavor among the hundreds of types of bread products, and breads are prepared to suit the desires of nearly all people. Basically, the flavors of breads are derived from the ingredients used in the bread, the changes produced during fermentation, and the flavors produced by the baking process. The flavors of dessert breads are in large part due to the ingredients used, which usually are of such intensity that they mask the flavors produced by fermentation or baking. On the other hand, the flavor of sourdough breads is in large part due to the fermentation and the changes that occur among fermentation products during baking.

Freshly baked bread has aroma and flavor that are subtle yet attractive to most individuals. Unfortunately, these appealing characteristics disappear too quickly. The functions of microorganisms in bread making are to lighten the dough and to impart a characteristic aroma and flavor. Microorganisms provide flavoring compounds. Many bakers feel that yeasts and bacteria contribute little flavor to bread under the rapid methods now employed. The professional baker adds a considerable quantity of yeast, but by the rapid methods he employs, growth of microorganisms is not encouraged and leavening is in large part due to enzyme action. There is little opportunity for growth of the bacteria that may impart a characteristic bread flavor. In contrast, the repeated kneading of dough practiced in the home lengthens the fermentation period and provides opportunity for the development of flavor by yeasts and bacteria. Many experts maintain that lactic acid bacteria growing in the dough can contribute considerably to flavor.

Numerous scientists have undertaken research concerning the fundamental chemistry of bread flavor and aroma. This subject is amply reviewed by Coffman (1967). In the many studies involving the identity of bread flavor substances, consideration has been given to the origins of the various flavor compounds. No doubt the presence of some of the chemical substances that remain in the loaf and contribute to flavor is due to fermentation. It is a well established fact that during fermentation, acids (particularly lactic and acetic), alcohols (including many besides ethyl alcohol), the esters formed by acids and alcohols, (acetoin, aldehydes, ketones), and various carbonyl compounds are formed. An important source of flavor and aroma is the changes and chemical substances produced by the Maillard reaction during baking. Little study has been given to the changes in the protein and lipid constituents and the formation of new chemical substances. Some 45 acids of the fatty acid series have been identified.

Comparative Influence of Leavening on Flavor

The simplest types of breads, those made primarily from flour, water, salt, and leaven, depend for flavor primarily upon the type of flour used, the products of fermentation, and the changes that occur during baking. At the other extreme are the dessert breads in which leavening is essential to yield the fine texture desired, and the true bread flavor is present but the flavors imparted by fruits, sugars, syrups, spices, and many other added ingredients are dominant.

Satisfactory white bread can be made from flour, water, salt, and yeast or sour dough. The Italian straight-dough bread is this type although malt is sometimes added to increase amylase activity to compensate for the lack of added fermentable sugar. When properly prepared and baked, the bread will have a firm, hard, golden brown crust and the bread-like flavor will be pronounced. French bread is quite similar but may contain some added oil. The Yemenites prepare a similar bread, kubani, in which rendered butter is added. Some of the hard crust sourdough breads made with different amounts of rye and wheat flours may have even more bread flavor. It is this bread-like character which appeals to people, and one can eat the bread day after day without seemingly tiring of it. The appeal of the many comparatively thin flat breads and rye crisps is due to their distinctive flavor. Some of the homemade breads also have this pronounced crusty character and bread-like flavor. The average bakery bread has a softer crust and usually a less developed bread-like flavor. Beginning with the comparatively simple breads one can sample the entire series of minor variations to some of the most delightful dessert breads (Fig. 8.4).

FIG. 8.4. SOME COMMON LEAVENED BREAD-LIKE PRODUCTS: PRETZELS, DOUGHNUTS, SALTINE CRACKERS, CINNAMON CRISP, RYE WAFERS, AND GARLIC BREAD STICKS

TYPES OF BREAD AND BREAD-LIKE PRODUCTS

Crackers and pretzels are prepared from comparatively simple doughs. Kryddskorper, the wheat or rye rusks of Sweden, flat brød or flat bread, rogbröd, the rye bread sometimes flavored with fennel, anise seed, or with syrup, and the khobaz arabec of Syria are dry breads in which much of the flavor is due to fermentation and baking. The latter is used on holy days in a bread-breaking rite. Irban is another bread used for religious purpose while kibiz is a bread used daily. In general, they also depend for flavor somewhat on the salt on their surfaces. The flavor of crusty toast appeals to people. Similarly zweiback and some of the sweet flavored packaged toasts rely for palatability upon the bake. Available now are prepared toasts flavored with various ingredients such as almond oil. Zweiback is also prepared from whole wheat bread. As the name indicates, it is baked twice.

There are a whole series of standard, low-density, soft crust breads, some containing flours from other cereals. The sweet rye bread, oatmeal, rice, wild rice, whole wheat, and graham breads are variations. Milk, potato, egg, honey, corn syrup, molasses, and fruits, including even cranberry, are common additions to impart flavor. Many of these are of unusual composition such as the so-called chuck wagon bread containing honey, the egg-twist bread containing egg yolk, and, of course, the old-fashioned raisin molasses bread. Øllebrød is a bread in which beer is used as the liquid. Other interesting breads include the sweet rye breads and pumpernickels served with syrup or molasses; the dark rye bread of Western Asia in which the sour milk, kumiss, is added; the khamari of India in which milk curds are used; and the sour cream bread, bereke. The Ethiopians prepare a slightly sour bread, injera, sometimes from wheat, barley, or millet, but ordinarily from the flour of teff, a type of grain.

The English crumpets, the French croissants, the Czechoslovakian poppy seed crescents, the Latvian klingeri shaped like a doughnut, bagels, and many others may be somewhat richer, but challah, the egg bread of Jewish origin, kanebuller or cinnamon rolls, buns, and breads, or the cinnamon ring vetekrans, and vetebrød, containing eggs and flavored with cardamom are usually somewhat richer. Kaffekaka, coffee cake with raisins, gugelhof, buchteln served with a sweet sauce, and some of the nut rolls like pecan rolls are other variations.

These breads cannot be placed in any order of richness and they vary considerably. Generally, the richer dessert breads, many of which are prepared only for Christmas, Easter, weddings, and other special occasions, include: Christmas breads or buns such as yulekakka; saffron bread flavored with saffron; babka, baba, and the larger savarin, usually glazed and served with special syrups; kulich; pannetto; ensaimanda with its rich sweet high egg content; pasha; yul pumpernickel; brioche, the French sweet bread baked to crisp character; strudel; stollen; and many other sweet breads. Often, they contain glazed fruit including apple, cherry, pineapple, and raisins; and others contain nuts, such as filberts, almonds, walnuts, and pecans, and various flavor extracts and powders. Of course, these are enriched with eggs, sugar, and butter or other shortening. Possibly the Danish pastries in which the typical flakiness is imparted by the shortening used and the handling of the dough may also be classified as dessert breads.

Danish pastry is one of the best known international delicacies usually prepared with a high content of butter as well as eggs and sugar, blended and handled to yield a delightful, flaky, and light dessert. Considerable care is required in the blending, fermentation, and baking. The many variations in shape, e.g., horns, rings, butterflies, twists, and sticks, and the different flavors such as almond, walnut, and coconut provide a considerable variety. There are a great number of similar sweet, rich breads or cakes such as the delicate crescent-shaped French croissant.

It is particularly interesting that these products are still prepared by yeast leavening even though admittedly they could be prepared at lower cost by chemical leavening. Even though the flavors due to fermentation and baking are so well masked, these flavors are important in the overall character of the products.

The origins of any of these breads are difficult to trace. Babka is of Polish origin, baba French, gugelhof and buchteln Austrian. The Cornish use saffron as a flavoring and coloring substance. Ordinarily the German term, strudel, is applied to cake-like breads covered with fruits such as apple slices, cherries, pineapple, or others, but a similar product is made in many countries other than Germany. Although the name streussel is sometimes applied to breads coated with streussel, actually streussel is a crumb-like mixture of flour, shortenings, sugar, and flavorings used to spread on cakes. Many are served hot and at times with special syrups such as a rum-flavored syrup.

The breads of America, like so many other American foods, are truly international. The French, Germans, English, Scotch, Cornish, Scandinavians, Austrians, Russians, Polish, Czechoslovaks, Latvians, Spanish, Italians, Swiss, and others have given us the many variations but possibly credit for the universality must be given to the Jewish peoples.

There are a great variety of bread-like products which depend upon fermentation to impart their light, airy, and spongy characteristics. Some of the best known are the various types of pancakes, griddle cakes, and waffles. Some of these fermented products have been replaced in many homes by prepared mixes containing chemical leavening agents. The use of chemical leavening agents began in the 1880's. They have not replaced entirely the sour dough pancakes, buckwheat cakes, or the sour milk pancakes still prepared by fermentation. The Ukrainian mlyntae, a wheat and buckwheat flour pancake made with milk, sugar, salt, and butter blended about 5 hr before baking and to which egg is added after rising, is typical of the more acceptable types. The Lebanese prepare a bread-like pancake in which they roll certain flavorsome fillings. Sour milk pancakes are common in India.

The steamed pancake-like product called idli is common in the southern

TABLE 8.3

BACTERIOLOGICAL AND PHYSICOCHEMICAL CHANGES IN IDLI BATTER

| | | | | Estimated Number of Each Species × 10[6] | | | | | Duplicate Batter[5] |
| | Volume of Batter Ml | Total Count per Gm of Dry Grains × 10[6] | | Microaerophilic Species | | | | | Total Acidity (Gm of Lactic |
Time Hr			Aerobic Species	Leuconostoc mesenteroides	Streptococcus faecalis	Pediococcus cerevisiae	pH	pH	Acid per Gm of Dry Grains)
0[1]		0.0372	0.015	0.006	0.016		6.1	6.2	0.06
0[2]		0.0117	0.003	0.002	0.007		6.2	6.2	0.04
4[2]		0.183		0.174	0.009		6.2	6.2	0.12
8[2]		120		114	6		6.25	6.2	0.24
8[3]		198		188	10		6.35	6.3	0.93
9[4]	505	326		326			6.00	6.1	0.77
12	505	1,260		1,170	190		5.70	5.75	1.2
16	645	2,520		1,890	630		5.20	5.10	2.23
20	705	3,000		1,500	1,500		4.85	4.70	2.50
24	745	2,550		1,530	1,020		4.65	4.55	2.71
28	745						4.72	4.50	3.14
32	525	2,640		1,320	1,190	130	4.70		3.14
36	450	2,410		846	362	602	4.55	4.35	3.14
40	380	2,130		852	852	426	4.50	4.30	3.21
44	380	1,650		990	330	330	4.45	4.30	3.28
48	380	1,780		980		800	4.45	4.25	3.43
52	380	2,090		420	1,250	420	4.50	4.20	3.43
56	380	2,200		880	330	990	4.40	4.15	3.57
71	380	770		460		310	4.35	4.10	3.70
460	380	213				213	4.00		

Source: Mukherjee et al. (1965).
[1] Unwashed black gram wash water.
[2] Washed black gram soak water.
[3] Ground black gram.
[4] Ground black gram, rice semolina, and salt.
[5] Data on a second batch.

area of India. It is ordinarily prepared with rice flour and black gram, dhal, or bean flour (Table 8.3). In some areas, wheat flour is used in place of rice flour. Other steamed breads include puto prepared in the Philippines, possibly of Chinese origin, and kanom-tan that is also made with rice flour in Thailand. Palm fruits and coconut are used with the latter.

Yeast-leavened products include not only those foods in which yeast serves principally as an aerating agent but also those in which the leavening action is secondary to the function of dough conditioning and flavor development such as in Danish pastry and crackers. It is not the purpose of this book to describe all of the breads made throughout the world from the simple flour, water, and yeast breads to the delicate rich flaky dessert breads. The great variety of rich sweet breads made from yeast-leavened doughs are prepared in a great variety of flavors, sizes, and shapes with various amounts of shortening, eggs, sugar, and flavoring ingredients.

Sourdough Breads

Possibly there are greater differences of opinion in regard to the meanings of sourdough breads and sours, starters, and barms than for any other type of bread. Sourdough breads are sometimes confused with salt-rising breads. Salt-rising breads may well become sourdough bread if the lactic acid bacteria develop rapidly. Sourdough breads were undoubtedly the first leavened breads prepared by the Egyptians. With the conditions existing at that time and for centuries afterward and the lack of knowledge in regard to bread microbiology, experience must have taught the ancient bakers that to produce bread of uniform quality they had to use a sourdough process. Sour dough is simply a self-perpetuating yeast-bacterial mixture. In America, the word, sourdough, is associated with the early prospectors who carried sourdough starter pots strapped to their packs. Many stories and experiences have been told about sourdough breads. Sourdough breads are increasing in popularity in many areas of the United States. One company in the West turns out 75,000 loaves daily.

The starter still holds a place of honor in many households throughout the world. It is kept as cool as possible. If tightly covered, it will blow up. Small portions are passed on to daughters at marriage. If bread, pancakes, waffles, biscuits, or doughnuts are to be prepared, a cupful of starter may be removed the night before, placed in a warm place, and next morning added to the mix of flour, water, and other ingredients. A cupful of flour and sugar-water are added to the starter to replace the cupful removed. If a dough or sponge starter is used, it is handled similarly. If by chance the starter acquires a bad flavor, new starter is obtained from some neighbor. This, then, is the background of the sourdough breads. With our present day knowledge of microbiology better control of starters is possible.

Sourdough starters may vary from a mixture of yeasts and lactic acid bacteria in which yeasts predominate to a starter consisting exclusively of lactic acid bacteria (Tables 8.4 and 8.5). In the presence of sufficient yeasts to leaven the dough the function of the lactic acid bacteria is primarily the acidification of the starter to inhibit undesirable microorganisms. In the absence of yeasts, lactic acid bacteria of the heterofermentative type are essential to produce leavening. Today, rye breads including pumpernickels are the more common sourdough breads.

TABLE 8.4

EFFECT ON MICROBIAL CHANGES DURING PROOFING

Added Bakers' Yeast (% of Flour)	Cell Counts per Gram of Bread Dough[1]					
	Bakers' Yeast		Sourdough Yeast		Sourdough Bacteria	
	0 Hr	7 Hr	0 Hr	7 Hr	0 Hr	7 Hr
0	0	0	26	156	64	1027
0.10	86	1	28	118	73	1132
0.25	202	46	25	88	59	700
0.50	395	178	25	83	62	708

Source: Sugihara et al. (1970).
[1] × 100,000 for yeasts; × 1,000,000 for bacteria.

TABLE 8.5

MICROBIAL CHANGES DURING SOURDOUGH BREAD PROOFING

	Millions of Microorganisms per Gram of Bread Dough	
	At Start of Proof	After 7 Hr Proof
Sourdough yeast	3	18
Sourdough bacteria	86	1650

Source: Sugihara *et al.* (1970).
Note: Composition of starter sponge: yeast = 15 to 28 million per gram; bacteria = 600 to 2000 million per gram.

The microorganisms involved in sourdough fermentations have been studied extensively in Europe. Henneberg (1926) summarized knowledge up to the time of publication of his book. More recently, Spicher (1960) characterized the breads produced by various species of lactic acid bacteria which he had isolated from sourdough breads. He had previously observed that *Lactobacillus plantarum* and *Lactobacillus brevis* were the most common species.

Rye Breads

Rye breads may have originated from the wild ryes of Europe and Asia, and for centuries rye flour was the basic flour used in making the dark breads of Europe. Rye bread, Ry-krisp, Swedish health bread, Munich rye, the Scandinavian knäkebröd, and the black breads were originally made solely with rye flour. Nearly all were sourdough breads and were relatively flavorsome, but solid type breads that were distinctly acid (pH 4.4). Since rye flour does not have the glutinous character of wheat flour, it does not hold the carbon dioxide to yield a light airy bread. Many of the old rye breads were made without a starter and relied upon the bacteria present in the mixture of flour and water for leavening action. The addition of sponge from the previous day's baking became common. This type of bread is still available in some areas; however, it has been generally replaced by a rye-wheat flour blend in which as much as 80% wheat flour is used. The latter is light in color and has the spongy characteristics of white breads. Furthermore, they are made with yeast or with sour doughs in which yeasts are present in high proportions.

Pumpernickel bread originated during the wheat famine of 1443 and was supposedly the idea of a Swiss baker, Pumper Nickel. It was a true sourdough bread containing a high proportion of rye flour. Like rye bread, most of the pumpernickel breads available today contain a high proportion of wheat flour, and leavening action is brought about by yeasts and bacteria. The old type sour, very flavorful pumpernickel bread is still made. This is a compact dark bread with a course texture and with uniform small holes throughout. The leavening dough prepared by one baker when examined had a pH of 3.98 and was characterized by millions of small cocci which apparently were strains of the species *Leuconostoc mesenteroides*. No yeasts were present. This type of bread was apparently the first to be successfully canned. Its highly acid character would not permit germination of bacterial spores if they survived baking. Some pumpernickel breads are made with rye flour alone but others include wheat flour, bran cereal, molasses, and even cold mashed potatoes and unsweetened chocolate.

When one considers some of the unusual breads prepared by the American Indians and early white settlers, one can surmise that microbiologically they may be classed as sourdough breads or possibly more closely related to salt-rising breads. Acorns from some species of oaks have been used in breads by the Indians for centuries. The dried or roasted acorns were ground to a coarse meal usually after they had been leached to remove the bitter tannins. Acorns from some species of oaks are low in tannins. Shelled acorns were also allowed to mold and eaten as nuts. The Indians of the Southwest have used piñon nuts in piñon bread and bread sticks. The nuts are obtained from the cones of a low spreading conifer. The plains Indians have used sunflower seeds in bread. Like all seeds, they are highly nutritious and when roasted and ground, they impart a unique flavor to bread. The starchy roots of the prairie turnip, prairie potato, or breadroot, a member of the pea family, were relished by the early plainsman and may have been used in bread. A wild rice bread prepared with wild rice flour, wheat flour, yeast, and skim milk is available in certain areas of the midwest. It is assumed that such breads were made by the Indians, most likely by a sour dough fermentation. Little information is available in regard to jamin-bang, a coarse bread prepared by the Indians of Brazil. Chufa, a sedge of the cyperus group, has tubers that can be ground to a powdery flour that when blended with wheat flour produces a distinctive bread. Millet, adley, oats, and other cereals have been used in bread. Some of these may have been sourdough breads. Oatmeal breads have always been favorites among many peoples and were used to a considerable extent in the United States during the period 1914–1918 when the demand for wheat flour was so great. The use of rice in the steamed bread preparation puto and with black-gram beans in idli has been discussed previously.

Crackers

The cracker industry in the United States is very highly mechanized to produce uniform, high grade products consistently. After the sponge and then the dough are fermented, the dough is rolled into long thin sheets which pass through cutting machines for shaping before entering the baking ovens. There are few secrets in the cracker industry. The yeast leavening process contributes a characteristic aroma and flavor. Even though they are more expensive than chemical leavening, the yeast-leavened crackers are more acceptable.

Crackers are prepared with relatively strong gluten, unbleached flours of low protein content. Fermentation of the dough tends to mellow the gluten. Either compressed or dry yeast is used. As in similar fermented doughs, carbon dioxide and alcohol are the primary products of fermentation, but side reactions provide small amounts of flavor substances. The small number of lactic acid-producing bacteria present in yeast may contribute a small part of this flavor. The diastatic malt imparts flavor and color to the baked products. The amylolytic and proteolytic activity of diastatic malt produces relatively small amounts of fermentable sugars. Since starch is nonfermentable, sugars serve as the substrates for microorganisms for elaboration of carbon dioxide and flavoring compounds.

There is a great variety of crackers other than the well-known soda crackers and saltines. The automatic machinery, the extruders, sheeters, laminators, and cutting and stamping machines are fully described by Matz (1968).

Cheese crackers, oyster crackers, and others are prepared with the same basic doughs. The imparting of color and flavor requires specific practices. For example, the color of cheese crackers may be intensified by the addition of paprika.

Pretzels

Similar basic doughs are used for making pretzels. The typical flavor contributed by yeast-leavened dough is characteristic and desirable. This industry is also highly mechanized (Matz 1968). The pretzel dough is forced through an extruder, rolled to desired thickness, the dough strip twisted and pressed into knots, passed through a caustic bath, salted, and baked by mechanically-devised equipment. The homemade pretzel is rare but very flavorable when freshly baked. Such pretzels often become very dry and hard when aged.

Doughnuts

The old-fashioned yeast-raised doughnuts, in contrast to fried cakes, were prepared from an enriched bread dough. Beside flour, sugar, salt, yeast, and liquid they may contain milk, eggs, spice, margarine, and jam. After dispersing the yeast in liquid with sugar, salt, and some flour, the liquid mixture is set aside to rise. Later, other ingredients are added, the dough is kneaded, rolled out on a board, and cut into circles. If the doughnuts contain jam, this is placed within the circles after frying. Danish cheese doughnuts and the Austrian carnival doughnuts, faschingkrapfen, are also variations.

Grape Nuts

Grape nuts is a mixture of wheat flour, malted barley flour, yeast, salt, and water. The dough is fermented 4½–5 hr at 80°F. Some starch liquefaction occurs. The fermented dough in pans is baked for a longer time than bread, then crumbled; the crumbs are then rebaked or dried, and graded. Some grape nuts are sold in flake form. The density, sweetness, and other characteristics of grape nuts is the result of proper control of amylase activity.

BIBLIOGRAPHY

BUCHANAN, E. D., and BUCHANAN, R. E. 1938. Bacteriology, 4th Edition. Macmillan Co., New York.

CARLIN, G. T. 1958. The fundamental chemistry of bread making. Part I. Proc. Am. Soc. Bakery Engr. 55–61.

CATHCART, W. H. 1944. The production of bread and bakery products. In The Chemistry and Technology of Food and Food Products, M. B. Jacobs (Editor). Interscience Div., John Wiley & Sons, New York.

COFFMAN, J. R. 1967. Bread flavor. In Chemistry and Physiology of Flavors, H. W. Schulz, E. A. Day, and L. M. Libbey (Editors). Avi Publishing Co., Westport, Conn.

FRAZIER, W. C. 1967. Food Microbiology, 2nd Edition. McGraw-Hill Book Co., New York.

HENNEBERG, W. 1926. Handbook of Fermentation Bacteriology, Vol. 2. Paul Parey, Berlin. (German)

JENSEN, L. B. 1953. Mans' Foods: Nutrition and environments in food gathering time and food producing times. Garrard Press, Champaign, Ill.

LINDEGREN, C. C. 1949. The Yeast Cell, Its Genetics and Cytology. Educational Publishers. St. Louis.

MATZ, S. A. 1959. Cereals as Food and Feed. Avi Publishing Co., Westport, Conn. (Out of print.)

MATZ, S. A. 1960. Bakery Technology and Engineering. Avi Publishing Co., Westport, Conn.

MATZ, S. A. 1968. Cookie and Cracker Technology. Avi Publishing Co., Westport, Conn.

MATZ, S. A. 1969. Cereal Science. Avi Publishing Co., Westport, Conn.

MATZ, S. A. 1970. Cereal Technology, Avi Publishing Co., Westport, Conn.

MUKHERJEE, S. K. et al. 1965. Role of Leuconostoc mesenteroides in leavening of the batter of idli, a fermented food of India. Appl. Microbiol. 13, 227–231.

POMERANZ, Y., and SHELLENBERGER, J. A. 1971. Bread Science and Technology. Avi Publishing Co., Westport, Conn.

SPICHER, G. 1959. The microflora sourdough. I. Studies on the species of rod-forming lactic acid bacteria (Lactobacillus Beijerinck) isolated from sourdough. Zentr. Bakt. Parasitenkunde Infect. Hygiene II Abt. 113, 80–105. (German)

SPICHER, G. 1960. The stimulators in sourdough fermentations. Brot Gebaeck 14, 27–32. Chem. Abstr. 55, 9708b, 1961.

SUGIHARA, T. F., KLINE, L. and McCREADY, L. B. 1970. Nature of the San Francisco sour dough French bread process. Baker's Dig. 44, 51–53, 56–57.

SULTAN, W. J. 1969. Practical Baking, 2nd Edition. Avi Publishing Co., Westport, Conn.

TANNER, F. W. 1944. The Microbiology of Foods, 2nd Edition. Garrard Press, Champaign, Ill.

CHAPTER 9

Alcoholic Beverages

In nearly all areas of the world some type of alcoholic beverage native to its region is prepared and consumed. Fermented beverages of the world represent independent discoveries in the many societies and, therefore, vary considerably. Beverages produced from cereals are usually referred to as beers; those produced from fruits are classified as wines. Still other beverages are prepared by fermentation of a variety of foods or blends of fruits, cereals, milks, saps, honey, molasses, and/or other foods containing fermentable carbohydrates.

The origin of many of these beverages is lost in ancient, unrecorded history. Many foods undergo fermentation to produce alcoholic beverages. The fermented drinks of the world represent independent observations or discoveries. Many of the fermentations are considered natural phenomena, and early man was unable to understand the changes that occurred. As man required food, he consumed the available food regardless of changes that it may have undergone. The mild fermentation with effervescence produced in some products during the early stages of fermentation resulted in beverages pleasing to many peoples and comparable to our present-day apple ciders. The general adoption of these beverages cannot be explained merely by their exhilarating effect upon the consumer; the alcohol content is too low in the early stages of fermentation to produce an intoxicating effect. The fully fermented products, clear and pleasing in flavor to many peoples, resulted in their general acceptance and preference at times to the unsatisfactory waters available. In spite of some of the temporary illnesses resulting from drinking orgies, the illnesses were less severe than those resulting at times from the consumption of contaminated water. Today, it is known that the nutritive value of some of the fermented preparations was not impaired appreciably during fermentation and in certain respects may have been augmented.

A considerable amount of research has been conducted, and numerous articles and books have been published concerning wines and beers. It seems almost superfluous to include even a brief discussion of these and the beverages in the following chapters; however, there are innumerable beverages prepared from a variety of food substances that undergo fermentations. It is apparent from the brief descriptions available that these have not been carefully studied and that they are not necessarily merely yeast fermentations brought about by the species *Saccharomyces cerevisiae*. Many are described as acid as well as alcoholic. If a clearer understanding of some of these fermentations is obtained from this brief discussion of wines and beers, and the beverages in the following chapters, their inclusion will be justified.

WINES

Wines have been an integral part of the diet of the human races throughout history. Their use is as old as recorded civilization. They were the beverages of nearly all agricultural peoples and in many societies are still considered as essential as bread in the diet. The actual birthplace of wine is unknown. Presumably, grapes were cultured for wine production since man first settled in the Tigris-Euphrates Valley. Grape culture spread with advancing civilization from Asia Minor to the shores of the Mediterranean. Wine was known to have been prepared by the Assyrians by 3500 B.C. It is evident by figures in mosaics that wine was produced in Egypt about 2400 B.C. and in China about 2000 B.C. Hesiod, writing in Greece during the 8th Century B.C., gave directions for the care of the vine. Later, a Greek handbook on wine appeared. The Romans undoubtedly borrowed practices from the Greeks. They also adopted the Greek god of wine, "Dionysus," renamed "Bacchus." The Romans left eloquent testimony in praise of wine and established grape production as an important agricultural pursuit. The Phoenicians are sometimes credited with introducing the vine into France at about 600 B.C. Wines have played an important role in the religious practices of primitive peoples as well as those of the more highly developed societies. Reference to grape growing and use of wine in Biblical times is found in both the Old and New Testaments. In fact, the early Christian Church was involved in much of the development of both the wine and beer industries. Throughout the Middle Ages the church was the important factor in maintenance and spread of vineyards and the development of wine. Some of Europe's present-day vineyards were established by religious orders centuries ago.

No other drinks, except water and milk, have earned such universal acceptance and esteem throughout the ages as has wine. Most of the world's wines have been consumed with food or as a food itself. Wines reputedly aided in maintaining health, not only because of their own nutritive value, but also because they replaced inadequate, impure, or otherwise unsatisfactory water supplies. Wine has always been considered a safe and healthful beverage. Wine has been used to welcome guests, to minister to the sick, to perform church rites, and even to launch ships.

Wine is the naturally fermented juice of fresh ripe grapes; however, the term is also applied to beverages produced by the fermentation of many other fruits as well as vegetables. Most of the world's grape wines are prepared from varieties of a single species of grape, *Vitis vinifera*. This probably originated somewhere in the Caspian Sea area, the Tigris-Euphrates basin, or in Egypt several thousands of years before our era.

Wines are produced in all of the countries along the Mediterranean,

Europe, Africa, and Asia Minor as well as inland areas favored with proper climatic conditions for growing grapes. France, Italy, and Spain are considered the leading wine producing countries of Europe, but wine is produced in nearly all others as well as in South Africa, Australia, New Zealand, and Central and South American countries, led by Argentina. Vineyards are present wherever suitable climatic conditions permit grape culture. California is by far the largest wine-producing state in the United States; however, appreciable quantities of wine are produced in the Eastern United States and Canada. The wines of the northeastern areas are produced primarily from the *Vitis lubrusca* varieties while those of the southeast are produced from the *Vitus rotundifolia* varieties.

One would hesitate to list the named varieties of wine. Even during the time of Christ, Pliny distinguished 50 different types. Today, names are applied based not only upon the grape varieties and method of preparation but also upon the area in which the wine is made. The differences are real in that numerous varieties of grapes are used and their characteristics are influenced by climatic factors, soil, rainfall, temperature, methods of production, and other factors. A general classification of wines into red table, white table, sherry, dessert, sparkling, and flavored wines will include most of the types.

The preparation of wines, care of vineyards, varieties of grapes and wines, and types of equipment used are discussed in numerous publications cited and summarized by Amerine *et al.* (1967). Discussion herein will relate primarily to the microbiology of fermentation.

Red table wines are made by fermenting crushed colored grapes before separation of the skins and seeds from the juice. Some of the anthocyan pigments, tannins, and other substances localized in the skins and seeds are extracted during fermentation, imparting to the wines flavors so extracted. White wines, generally prepared from white grapes, are crushed and pressed to separate the juice prior to fermentation. These wines have a more delicate flavor since they have a low content of extractives; however, the effect of deteriorative flavor and aroma defects are more readily discerned. Such wines are often sweetened slightly for general table use. Dessert wines may be prepared from either red or white wines by sweetening and fortifying with brandy. These are frequently given further aging treatment to impart characteristic flavors and bouquet. Sherry wines and some port wines, usually classified as dessert wines, are subjected to an aging or baking process to impart characteristic flavors through chemical changes. Flavored wines such as vermouth are prepared by blending extracts of aromatic materials with grape wine. Sparkling wines are so prepared to retain some of the carbon dioxide. Some are artificially carbonated.

Grapes on the vine and after harvesting contain a diverse microflora consisting of numerous species of several genera of yeasts, molds, and bacteria. The flora is influenced by environmental conditions and will change as the fruit ripens. The flora is also affected during preparation for fermentation; however, unless the juice is pasteurized it will contain a mixed flora that may influence the fermentation. Mass inoculation by a single species or mixture of strains will usually control the fermentation.

In grape musts the first evidence of growth of microorganisms is a slight haziness. This is followed by distinct clouding and later, the formation of a sediment. Carbon dioxide gas is evolved which will agitate the liquid so that the yeasts tend to remain in suspension. A froth or foam develops on the surface (Fig. 9.1). Although there is considerable variation in practices, usually in wineries, after the initial fermentation the liquid is racked or separated from the sediment of yeast and insolubles. This racking process may be repeated several times as the wine ages. Microorganisms left in the wine will slowly autolyze, thereby imparting flavor to the wine and sometimes causing a slight haziness. In many wines the yeast fermentation is followed by a malo-lactic fermentation in which the lactic acid bacteria convert malic acid to lactic acid to reduce the total acidity.

Wines are the products of an alcoholic fermentation by yeasts. A good wine yeast is one which will impart a vinous or fruit-like flavor, will ferment sugar to a low content producing 14–18% alcohol, and is characterized by remaining in suspension during fermentation and then agglomerating to yield a coarse granular sediment that settles quickly and is not easily disturbed in racking (Fig. 9.2). Many strains of the yeast species *Saccharomyces cerevisiae* fulfill these requirements and, therefore, this is the most common

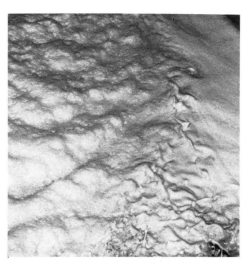

Courtesy of Taylor Wine Co.

FIG. 9.1. FERMENTATION IS FIRST OBSERVED AS A SLIGHT HAZINESS FOLLOWED BY DISTINCT CLOUDING AND THEN FORMATION OF A BUBBLY FROTH OR FOAM ON THE SURFACE

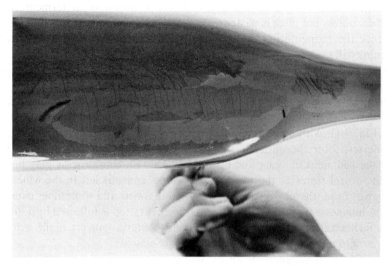

FIG. 9.2.   YEAST SETTLING IN A BOTTLE OF CHAMPAGNE

A good yeast is characterized by remaining in suspension during active fermentation and later agglomerating to yield a sediment that settles readily and is not easily disturbed.

species used. Strains of *Saccharomyces cerevisiae* var. *ellipsoideus* often referred to as *Saccharomyces ellipsoideus* are generally used in wineries where pure culture inoculation is practiced. Pure culture inoculation generally yields more consistent results. Some wine makers use a mixture of pure cultures to give the wine a distinctive bouquet.

A great variety of yeast strains of different species as well as bacteria and molds have a role in producing certain wines. Many of the alterations and the variety of microorganisms causing them are due to the conditions existent. Until the development of methods for isolating microorganisms and the use of pure culture inoculation (Fig. 9.3), the wine maker relied upon chance inoculation or upon massive inoculation from a previous fermenting wine for introduction of the particular yeast strain or strains. Usually strains of *Saccharomyces cerevisiae* dominated such fermentations, but in many instances proliferation of other strains of other species resulted in the development of a distinctive type of wine. Since such conditions generally prevailed in a specific area, a wine typical of the area became common. True sherry wine is one of this type.

Sherry wines prepared in the Jerez de la Frontera region of Spain are fermented by special "Jerez" strains of yeasts. The wines are fermented in partially filled barrels and aged for a long period of time to produce the distinctive character. The yeast species first classified as *Saccharomyces cher-*

FIG. 9.3. THE YEAST INOCULUM IS EXAMINED MICROSCOPICALLY TO DETERMINE ITS STATE OF GROWTH, THAT IS, THE NUMBERS OF CELLS AND THE EXTENT OF BUDDING

esienses, later as *S. beticus*, is now generally called *Saccharomyces oviformis*. During fermentation, the yeasts rise to the surface of the wine to form a continuous film or "flor" after the sugar is fermented. During aging, the "flor" may settle and rise again several times for a period of months or years of aging. *Saccharomyces oviformis* produces a dry wine of 14–15% alcohol. The yeast used in this process in California is considered to be a strain of the species *Saccharomyces chevalieri*. During aging, among other chemical changes involved in flavor development, the aldehyde content increases as the acidity decreases. The flavor developed may be due in part also to autolysis of yeast. Similar oxidative changes and development of flavor are now effected by chemical methods.

Bordeaux and certain Italian wine fermentations are initiated by strains of nonspore-forming yeasts of the genera *Kloeckera* and *Torulopsis* and continued by strains of the genus *Saccharomyces*. In a similar manner, other wines produced by natural fermentation are subject to fermentations by different species of yeast. Sometimes the initiating yeasts are referred to as "wild" yeasts and sometimes as apicula yeasts. *Kloeckera apiculata* is the specific name applied to one species.

Sauterne is one of the most generally accepted and imitated wines. The grapes used for sauternes of the Bordeaux area of France are generally attacked by *Botrytis cinerea,* a fungus that is inclined to develop in the cool and humid areas of Europe. The growth of *Botrytis cinerea* tends to loosen the skin, causing excessive moisture evaporation resulting in concentration of sugars and other ingredients of the grapes. The high sugar-content retards the rate of fermentation and yields a relatively sweet wine. The cracking of grapes sometimes permits the growth of bacteria and yeasts that may actually reduce the sugar content. Botryized grapes are used in production of other wines including sweet table wines.

Mold species of the genera *Penicillium, Aspergillus, Mucor,* and *Rhizopus* do not grow in wines; they may, however, grow on equipment and damage the quality of wines by imparting moldy flavor. Such molds are responsible at times for considerable losses to the industry.

Grapes vary considerably in their acid content and the higher acid-content grapes yield a wine that may be too sour for the average consumer. Fortunately, the excess acid is often reduced during fermentation. The titratable acidity of grapes decreases during ripening. Both tartaric and malic acids decrease. Tartaric acid, the major acid of grape juice, when excessive, is decreased by deposition as potassium acid tartrate. Malic acid is naturally reduced during alcoholic fermentation and often by a malo-lactic fermentation by bacteria, in which lactic acid is produced. Citric acid is slowly decarboxylated during aging. Other acids are present in grapes while still others such as succinic, lactic, and acetic acids are products of alcoholic fermentation.

Possibly no other aspect of the effect of lactic acid bacteria on wine has elicited so much interest as the so-called malo-lactic fermentation. The malo-lactic fermentation is a bacterial fermentation in which malic acid is decarboxylated to lactic acid. Although it occurs in many wineries, it is often unrecognized. It is considered desirable in wines of excessive acidity since it mellows and softens the harsh taste. Although it is considered indispensable in many red wines, it also occurs in some white wines. Following the primary yeast fermentation, particularly with wines that are lightly racked or those left on the lees longer than usual, lactic acid bacteria develop and convert malic acid to lactic acid. It is presumed that autolysis of the yeasts releases the protoplasmic nutrients, particularly vitamins and amino acids essential for growth of the lactic acid bacteria. The conversion of malic acid to lactic acid lowers the total acidity and raises the pH. Diacetyl and acetoin are often increased. With these changes there is a mellowing of flavor and a change in color that are considered desirable in wines of high acidity; however, if the malolactic fermentation continues or occurs in wines of low initial acidity, a wine lacking in characteristic flavor results. This may be considered spoilage.

This reduction of acid and mellowing of flavor were recognized as early as 1897, but the organisms causing these changes were not recognized until several years afterward. Even then, the apparent confusion between orga-

TABLE 9.1

CHANGES DURING MALO-LACTIC FERMENTATION OF RED SEIBEL WINE

Time (Days)	pH	Volatile Acid (Gm/100 Ml)	Malic Acid (Gm/100 Ml)	Lactic Acid (Gm/100 Ml)	Reducing Sugar (Gm/100 Ml)	Log. of Bacterial Count
0	3.40	0.030	0.24	0.01	0.16	2.00
7	3.35				0.17	3.48
14	3.33	0.027			0.16	4.95
21	3.40	0.030			0.18	6.11
28	3.45	0.030			0.17	7.25
35	3.60	0.036			0.16	7.77
42	3.80	0.042			0.17	7.86
49	3.80	0.045	0.01	0.19	0.17	7.63

Adapted from Rice and Mattick (1970).

nisms causing spoilage in wines and those causing a mellowing of wines was difficult to comprehend. Pasteur recognized four types of spoilage: *tourne* and *pousse,* bitterness or *amertume,* sliminess, and acetification. Pasteur considered *pousse* a gassy form of *tourne.* These conditions are due to the growth of lactic acid bacteria of the same species that are associated with malo-lactic fermentation and in part with mannite formation. The *tourne* of Pasteur may be applied to the homofermentative species and *pousse* to heterofermentative species of lactic acid bacteria. The extent of growth and the carbon dioxide produced by heterofermenters sometimes is insufficient to produce a gassy condition. The heterofermenters also produce mannitol from levulose and, therefore, are associated with the ferment mannitique. The relationship between malo-lactic fermentation and spoilages is complex and is further complexed by the fact that since the acid grape juices and wines are not ideal media for growth of the lactic acid bacteria, growth is frequently atypical of the species. The organisms, when transferred to media ordinarily used in laboratories, revert to the more common characteristics associated with the species.

Following the Pasteur studies, Carles, Gayon and Dubourg, and Laborde recognized mannitol production in wine. Müller-Thurgau (1908) and Müller-Thurgau and Osterwalder (1917) isolated a series of organisms from wines and described several strains of bacteria that may be associated with known cultures available today. Nearly every species of lactic acid bacteria that may be expected to grow in wines and cause a malo-lactic fermentation has been isolated by investigators; see Amerine *et al.* (1967). Among the homofermentative species *Pediococcus cerevisiae, Lactobacillus plantarum,* and *Lactobacillus delbrüeckii;* and among the heterofermentative species *Leuconostoc mesenteroides, Leuconostoc dextranicum, Leuconostoc citrovorum, Lactobacillus brevis, Lactobacillus hilgardii, Lactobacillus buchneri, Lactobacillus fermenti, Lactobacillus pastorianus,* and even the alcohol-tolerant *Lactobacillus trichodes* have been isolated. *Micrococcus variococcus* described by

Müller-Thurgau and Osterwalder and *Micrococcus malolacticus* described by Siefert are undoubtedly strains of the genus *Pediococcus;* and *Bacterium gracile,* also described by Müller-Thurgau and Osterwalder, is obviously a strain of the genus *Leuconostoc.* The heterofermentative species are all mannitol formers, and many strains will produce the slime, dextran, from sucrose. Many of the fruit wines discussed in the next section may be subject to a malo-lactic fermentation. Apples have a high content of malic acid and some of the changes in apple wines may be due to such bacterial changes.

Differences between a desirable malo-lactic fermentation and one that may be considered quality spoilage in wine are slight. A mellowness produced in a harsh acid wine by a malo-lactic fermentation is desirable, but when the reduction of acid is excessive, the wine is inferior. Malo-lactic fermentation should be prevented in wines that are excessively low in acidity. Any of the above species can metabolize sugars if present in the wine and actually increase the total acidity by production of lactic and acetic acids. The true acetic wines, however, are generally the result of growth of acetic acid bacteria. Other types of microbial spoilage also occur in wines, some caused by spore-forming bacteria, others by various species of yeasts.

Sparkling wines may be defined as wines that have more than 1.5 atmospheric pressure at 50°F. This definition may include champagne and champagne-type wines, carbonated wines, and other wines in which an excess of carbon dioxide remains in the wine. Champagne in France is the sparkling wine from the Champagne region. Other wines produced in France by a similar process are known as Vin mousseaux. In Germany they are known as schaumwein, in Italy as spumante, and in Spain and Portugal as espumante. The United States Internal Revenue Service defines champagne as a "sparkling light wine which derives its effervescence solely from a secondary fermentation of the wine within the containers of not greater than 1 gal. capacity and which possesses the taste, aroma, and other characteristics attributed to champagne as made in the Champagne district of France."

The grapes to be used in making champagne are fermented in the usual ways to about 11.0–11.5% alcohol, cold stabilized at about 26°F, and then varieties are blended to give a desirable blend of about 0.7–0.8% acid. The second fermentation is made in bottles at temperatures below 60°F. Sufficient sugar must be present to yield the carbon dioxide required to produce the desired pressure. The type of yeast used is important. A champagne-type yeast is an agglomerating or granulating yeast that forms a coarse heavy sediment in the bottle and will not rile during the disgorging process. After complete fermentation, which may require 1–3 yr, the bottles are placed upside down in a rack to allow the yeasts to settle on the cork. The bottles are turned and jolted daily to loosen any yeast that may settle on the sides. The yeasts will eventually settle on the cork and are then removed by disgorging. After disgorging, a small amount of sugar and brandy solution is added, and the bottle is sealed for the consumer market.

Occasionally, difficulty is encountered in completely clarifying the wine by racking. This may be due to an unsatisfactory type of yeast, to bacterial contamination, to autolysis of yeast liberating protoplasm into the wine, or other causes.

As previously stated, sparkling type grape wines or wines prepared from other fruits may also be made in other ways. The Charmat process has come into common use. The secondary fermentation is conducted in tanks and after fermentation the yeasts are removed by a filtering process. The same type of pure yeast culture is required as in the typical champagne process. Sparkling wines can also be produced by carbonation. Some wines are sparkling because the carbon dioxide produced in the natural fermentation is retained in the bottled wines.

WINE-LIKE BEVERAGES

The term wine is ordinarily associated with the fermented beverage prepared from grapes. Fruit juices as well as juices from the vegetable portions or saps of plants are subject to alcoholic and acid fermentations. They are also referred to as wine.

The most important of these is cider, cyder, cidre, or apfelwein, an alcoholic drink produced by a vinous fermentation of apple juice. For many years, cider was consumed in larger quantities in the United States than any other fruit beverage. The beverage was generally produced on the farm or in neighborhood cider mills. Until the general use of chemical preservatives, cider was subject to an alcoholic fermentation by any of the yeasts, usually resulting in an effervescent drink of comparatively low alcohol content. In fact, the ciders often were held for a day or more by the consumer. Under the conditions used in pressing the fruit, the juice contained great numbers of microorganisms, usually in their logarithmic stage of growth and, therefore, ready to initiate a fermentation. Since apples are not high in sugars, a relatively low alcohol content beverage was produced. When fermentation was complete, the containers were filled to exclude air and inhibit acetic fermentation. In many instances, these hard ciders were allowed to freeze, resulting in concentration of alcohol and solids to the center of the container. This concentrate was often recovered to yield a brandy-like hard cider. If the containers were improperly filled, an acetic fermentation occurred resulting in production of vinegar of various degrees of acidity.

In Europe, special varieties of apples, distinguished by their characteristics and usually unsuitable for table use, were used for making cider. The cider orchards of Normandy and Western England were particularly notable. Germany and Switzerland produced considerable amounts of cider. As in the United States, cider making was confined to farms and small mills, and a natural fermentation by wild yeasts occurred. Later, production was industrialized and research in several institutions was conducted to improve and standardize quality. Pure yeast culture inoculation is now practiced by some processors, particularly in France, but a natural fermentation is still more common practice. Alcohol content, sweetness, acidity, and astringency are carefully considered by the better producers. A malo-lactic fermentation occurs in some ciders, particularly if held for a period. The clear ciders are separated from the lees by methods similar to those in use in the grape wine industry. The clear cider when stored at low temperature, 40°F, may undergo a second fermentation. After final separation of the clear cider, it is bottled, but even then a carbon dioxide pressure may develop to impart a desirable effervescence. Sparkling ciders are produced by the champagne process or an adaptation of this process referred to as the Charmat bulk fermentation. Hard ciders similar to those produced in Europe have been produced experimentally and to a limited extent, commercially, in the United States; however, wine still has the greater appeal. At present, apple products are being investigated intensively in Japan.

Cider spoilages or disorders occur, the most common of which is acetification by the acetic acid bacteria. In such instances there is little that can be done except to convert completely to vinegar. Obviously, microbial spoilages occur characterized by off-flavors, ropiness, and other conditions. By washing fruit, pasteurization of juices, use of preservatives, and by pure culture inoculation such conditions are reduced to a minimum.

Perry, made from pears, is a product bearing the same relation to the pear as cider to the apple. This is also produced in Normandy, Brittany, and in areas of England. Pears are generally more astringent than apples and somewhat lower in acidity; these characteristics are imparted to the fermented products.

Sour cherries are frequently used in production of wines in Europe in preference to the more bland sweet cherries. The highly flavored cherry wines are used extensively in brandies or liqueurs such as kirchwasser, maraschino, and ratafina. A cherry flavored vodka, vishnyovaka, is made in Russia and apparently a rice beer, rusiviina, prepared in Finland is flavored with cherry. The rice wines, sool and yagjoo, made in Korea might better be classed as beers.

Cherry, raspberry, strawberry, loganberry, blackberry, Boysenberry, currant, elderberry, and apple have all been fermented to produce wines on the west coast. Since some of these are low in essential nitrogenous foods for yeasts, the rate of fermentation of blackberry, currant, loganberry, and raspberry have been increased markedly by addition of urea. Bananas, bigney, duhat, casoy, pineapple, and tomato wines have been produced experimentally and on a small scale, commercially. Casoy is the fruit that carries the cashew nut, duhat is a purple berry. Pineapple is fermented readily by yeast, but much of the fermented wine is converted to a flavorful vinegar. Reports are available of the use in the Middle East of pomegranates, lemons, tangerines, oranges, dates, and figs for making wines. Boukha is the name applied to a fig wine. Recently an Austrian patent has been granted on making elderberry wines. Elderberry wine, like many other wines, have been produced in the homes in the United States for many years.

It is generally assumed that these wines are all produced by yeast fermentations; in fact, baker's yeast and pure culture wine yeasts have often been used as inocula in their production. Many of these juices will support the growth of yeasts as well as species of lactic acid bacteria. The natural acidity of some would inhibit growth initiation of lactic acid bacteria, but they may develop later during the fermentation. The rate of fermentation may be increased in some cases by addition of nitrogen sources, minerals, and vitamins. Unfortunately, too little information is available in regard to nutritional requirements.

A number of so-called wines have been prepared from vegetables. Dandelion wine is one of the best known. Parsnip wine is made in England. Carrots and other vegetables apparently have been used. Vegetable materials in general are subject to fermentation by lactic acid bacteria, and even though the vegetable extract may be heated and inoculated with yeast, it would be difficult to exclude *Leuconostoc mesenteroides* and *Lactobacillus plantarum* from such fermentations.

Tomatoes are actually fruits but are often referred to as vegetables. Like vegetables they are very subject to fermentation by lactic acid bacteria, and some acid alcoholic drinks are distinctive in character. One small company produced an acid alcoholic product for a short time, but it was difficult to control the relative growth and influence of bacteria and yeast and, therefore, hard to standardize the product. A fruit similar to tomatoes, naranjilla, is fermented in Central and South America.

Wine-like beverages are produced from the fruits and saps of varieties of palms, agaves, and cacti. Pulque, a popular drink in the Southwest and in Mexico, is made from the juice of agave. Sánchez-Marroquín (1953) found that the chief organisms isolated in the fermentation of pulque are homo- and heterofermentative species of the genus *Lactobacillus*, species of *Leuconostoc*, and the yeast *Saccharomyces carbajali*. Faparusi (1969) obtained strains of *Leuconostoc mesenteroides*, *Lactobacillus plantarum*, and a so-called sarcina from palm wine fermentations. The so-called sarcina is undoubtedly a strain of the genus *Pediococcus*. The Ecuador Indians prepare chontaruru from the fruit of the chonta palm. Tuba is prepared from sap of the coconut palm. The Apaches of the Chiricahua use laurel berries with the banana-like fruit of the yucca in a fermented beverage. The addition of sugar is necessary in some of these products in order to obtain a satisfactory fermentation. One of the more common sweetening agents was honey.

In order to ferment satisfactorily, it was observed centuries ago that the substance to be fermented had to have a high sugar content. Honey was one of the first sweetening agents known to mankind. Honey is mentioned in the Old Testament. It was used by the ancient Assyrians, Egyptians, and others. The Roman mulsum was prepared from grapes sweetened with honey. The Romans introduced wine into Britian but ale and mead were the common beverages. Wine was a luxury item. Honey has been diluted and fermented alone for centuries. The beers of ancient Egypt were sweetened with honey. Some products called honey were apparently saps from trees. Beverages were produced by the early Britains from wheat and honey, and honey was an important ingredient of the meads or metheglins prepared by the ancient Nordics. Honey must be diluted to allow growth of yeast and in dilution, essential nutritive substances as well as sugar are diluted. Some of the early meads were prepared with blends of other substances such as a blend of malt, honey, and water. In preparation of beverages today, honeys are fortified with essential nutritives, nitrogen sources, vitamins, and minerals, and are acidified, usually with citric acid.

It seems logical that the experiences observed in grape wine fermentations should be reviewed carefully and serve as a guide in study and preparation of any of these types of wine.

BEER

The first evidence of beer manufacture has been traced to ancient Babylonia, possibly dating as far back as 5000 to 7000 B.C. Crude pictures of brewing painted on stones have been found of the period the pyramids were built. One ancient relic depicts a Babylonian king drinking a beer-like beverage through a straw. Some 18 varieties of beer, called bousa, were said to have been prepared in Babylonia as long ago as 2200 B.C. By 1000 B.C., the Egyptians had produced hopped beer and spiced beers were prepared. Taverns managed by women were common. The brewing of beer and the baking of bread were co-associated and were women's work. The sediment from brewing was consumed by the poorer classes. Certainly, the solid sediment was nutritious. Barley was the predominant cereal used, and not always malted. At first, malting was apparently a method of improvement of flavor. Later, it was generally practiced in baking and brewing. At some time barley was moistened and when germination began, it was crushed with a pestle, roughly ground, and made into loaves with sour dough or leaven. They were baked sufficiently to form a crust without cooking the interior. When beer was required, the loaves were broken, mixed with water, and allowed to ferment. The liquid was pressed, separated from the dough, and when fermentation was completed, the resulting acid-alcoholic beverage was called boozah or bousa. One can appreciate the fact that the solids given to the poor were nutritive.

Beer was an important beverage in the daily life of the ancient Egyptians. An Egyptian beer known as hek or heqa was brewed from malted barley and later from malted barley and fruit. The Egyptians may have learned the art from the Babylonians. Beer was used as a medicine in both Babylon and Egypt. The Egyptian beer was variously attributed to Ra, the sun god, or to Osiris, god of the netherworld. Beer played an important part in sacrifices to propitiate the gods as well as in ceremonials. Rameses III was said to have offered 466,000 jugs of beer for sacrificial purposes. Bread baking and brewing were closely allied and were the duties of women. The art of brewing was passed on to the Greeks and Romans and later to the Celts, Germans, and Scandinavians. The ancient Hebrews prepared a bread beer called machmetzeth from stale bread, malt flour, and water.

The Chinese may have been the first peoples to prepare a beer-like beverage from cereal grain. Samshu was brewed from rice. A beer-like beverage, kin, was supposedly produced about 2300 B.C. The Chinese also may have been the first to produce distilled liquors. Santchoo was prepared by distilling the rice and millet beer, tchoo. Hung-chu, called a red wine, is a beer colored with ang-kak.

The brewing of beer was and continues to be considered an art in many circles. Formerly, no part of the brewery was more mysterious than the fermentation room. The brewing industry is, moreover, another distinctive example of the great advances accomplished in an industry, developed step-by-step as a result of observation and research. It is sometimes stated that this began with the notable research of Emil Christian Hansen. One cannot, however, discount the careful observations of numerous other individuals throughout history who were responsible for earlier developments. The advances involve not only microbiology, but also chemistry, physics, engineering, agronomy, and other sciences. It has been stated that "malt is the soul of beer." Malting has been practiced for centuries; hops are believed to have been used about 3000 B.C. The important studies by agronomists concerning varieties, properties, and the suitability of barley and other cereals have been responsible to a large extent for quality improvement. The many research papers relating to barley and hops reported in the numerous articles and abstracts summarized in the *Wallerstein Laboratory Communications* and by Hind (1940) on brewing are ample evidence of the importance of these ingredients.

Beer is a beverage obtained by alcoholic fermentation of malted cereals. Barley, spelt, and other cereals were used, and during the Middle Ages, oats were commonly used by the poorer people. Hops were used when available, and the use of aromatic and bitter herbs was common.

Germanic tribes made their bior or peor with germinated and steeped barley, and it was usually sour in taste. Even today some of the beers have a slightly sour taste, indicating lactic acid bacterial fermentation. The Celts fermented wheat and honey. Honey and grain beverages were made in the British Isles in the 4th Century. Columbus reputedly was presented by friendly Indians with a beer made from corn. Beer was made in Virginia by its settlers as early as 1548 and later in New England. A brewery was erected in Manhattan in 1612 by Dutch settlers and in Pennsylvania in 1683 by William Penn. Other notable figures in the early history of brewing in America include Washington, Jefferson, Samuel Adams, Madison, Henry, and others. The Germans and Dutch, however, apparently exerted the greater influence on brewing in later years.

The application of scientific instruments in breweries occurred comparatively late. It was not until about 1760 that brewers appreciated the value of a thermometer and 1785 before the first saccharometer was used. Fermentation was formerly a mystery. Several theories were offered before 1800 to explain the changes, some of which sound almost fantastic today. For example, vinous fermentation, acetous fermentation, and putrefaction were considered revelations of the same process. Malting was regarded as a vegetable degree of fermentation. According to Hind (1940), Dubrunfant in 1830 converted starch to sugar by means of an extract of malt. Three years later, Pagen and Persoz precipitated and dried an active substance from similar extracts and called it diastase. The'nard in 1803 (Hind 1940) stated that yeasts were the cause of fermentation but believed the yeasts were of animal origin. In 1836 Cagnard de la Tour and Schwann and Kützing independently declared that yeasts were of vegetable origin. This started the dispute between those led by Liebig and supported by Berzelius and Wöhler who believed fermentation was strictly a chemical process and those who believed it was a biological process. Liebig's theory, stated in 1839, was that fermentation was brought about by an unstable body called a ferment, and was essentially a chemical process.

A new era opened in 1860 when Pasteur published the results of his study of yeasts. He found that sugar was not decomposed exactly according to Gay-Lussac's equation but that glycerol and succinic acid were formed at the same time. The fundamental outcome of this was the doctrine that fermentation was coincident with the life of the microorganism. This was a universal doctrine for all fermentations, whether by yeasts, lactic acid, or acetic acid bacteria. Pasteur's Etudes sur le Bie're published in 1876 had a far-reaching influence on brewery practice.

When in 1883, Emil Christian Hansen introduced his pure culture of yeasts at the Carlsberg brewery, a system of fermentation control was set up that proved to be of inestimable value. Hansen had previously shown that brewery yeasts were a mixture of several races of yeasts and that defects were caused by wild yeasts. The results of Hansen's pure yeast culture were so successful in the Carlsberg brewery in Copenhagen that the method was adopted permanently in 1886. Pure cultures maintain regularity in lager breweries. It was not until 1897, however, that Buchner demonstrated the connecting link between the ideas of Liebig and those of Pasteur. He extracted the enzymes from yeasts and demonstrated that fermentation was possible in the absence of the living cell. This gave impetus to the vast amount of scientific investigation that followed that also has been of such great value in many other branches of knowledge, e.g., medicine, nutrition, chemistry, and other sciences.

BREWING

The fundamental basis of the liquor industry is the conversion of sugars to ethyl alcohol and carbon dioxide by the action of enzymes. The process of conversion of cereal starches to alcohol is not a direct one and involves many steps that are capable of modification. Furthermore, the simultaneous conversions of nonsugar components may have a marked effect upon the ultimate product. In addition, nonenzymatic chemical changes, such as those involved in aging, occur and affect quality. For centuries brewing has been conducted on the basis of experience. Centuries ago, beer was made by women. Then the master-brewer became the key man. He represented experience; but of late years the capable master-brewer accepted the results of the laboratory and tried to utilize them.

The significant difference between the beer and whiskey industries and the wine industries is the composition of the raw material. Beer and ale are malt beverages made with malt, malt adjuncts, hops, yeasts, and water. Malt is prepared from barley grains, germinated, and dried after removal of the sprouts. Hops are the dried flowers of the hop plant. Malt adjuncts are starch or sugar-containing materials, such as corn and corn products, rice, wheat, barley, sorghum grain, soybeans, cassava, potatoes, sugars, and syrups.

The steps in brewing include malting, mashing, fermentation, separation of the alcoholic beverage from the mash, and rectification.

Malt is prepared from barley or other cereal grains by soaking or steeping the grain in water at 50°–60°F and allowing it to germinate at 60°–70°F for 3–4 days or more. Malting activates the enzyme systems. The grain is then dried to about 5% moisture to preserve enzymatic activity; the sprouts are removed and the malt is ground. The malt is ground in order that it may come into more uniform contact with the starches, proteins, and other constituents of the wort to convert the starches to sugars and hydrolyze other constituents more efficiently.

Mashing is the process in which the ground malt and adjuncts of the corn, rice, and other cereals are mixed with water and extracted to obtain the fermentable liquid known as wort or sweet wort. Different blends of barley malt, roasted barley, oat malt, caramel, sugar, and other materials are used for distinctive types of beer or ale. The malt and adjuncts are mixed with water at 100°–120°F then the temperature is slowly raised to 150°–158°F, during which time starches are hydrolyzed to fermentable sugars, principally maltose, while proteins are converted in part to peptides and amino acids. The blend is finally reheated to inactivate the enzymes. The solids settle out leaving the soluble extract, the wort, that is filtered from the spent grains and pumped into brew kettles where it is boiled for upwards of $2\frac{1}{2}$ hr with hops to impart the characteristic flavor of beer. The boiling of wort and hops concentrates the liquid, sterilizes and inactivates enzymes, extracts the hop solubles, precipitates proteins, and caramelizes sugars. The bitter extracts from the hops enhance flavor, stability, and removal of tannins. The wort is cooled, inoculated with brewer's yeast, and fermented.

The composition of the wort will depend upon the composition of the blend and the extent of enzyme action. At 100°F enzyme action is very slow; proteolytic enzymes become most active at 122°–129°F and saccharifying enzymes at 140°–150°F. During mashing, the color will change to a light to dark amber. Adjustment of acidity is essential because the normal pH of 5.8 is too high for optimum enzyme activity. Of course, there is considerable variation in methods used within the brewing industry. Reduction of pH to 5.1–5.2 is commonly practiced, but the desirable pH varies with the specific type of beer. Some ales are adjusted to as low as pH 4.2–4.3; it is apparent, however, that there is ample opportunity for bacterial growth, notably during the process outlined but also in its variations.

Acidity adjustment is accomplished by the addition of acid, preferably lactic acid, or by biological action, either by microorganisms or enzymes. The heat-tolerant *Lactobacillus delbrüeckii* is considered to be the most suitable organism for this purpose because it converts sugars almost entirely to lactic acid; it grows and produces acid at temperatures of 107°–111°F and continues to grow at 124°F. This is important because the purity of the culture can thereby be fairly well assured. Few other organisms will grow at 124°F. An acidity of 1.0% acid expressed as lactic acid at pH 3.2 is attained by this species. With this sour mash, a blend with the larger volume of mash can be attained at almost any acidity from pH 3.2 to 5.8. Since biological acidification by this method requires constant supervision, enzyme acidified malt may be used.

After the desired enzymatic changes have been attained, hops are added and the mash is boiled. During boiling, evaporation of water occurs and temperatures of about 215°F may be attained. The wort becomes essentially sterile, enzymes are inactivated, flavor and preservatives are extracted from the hops, coagulation of proteins and caramelization of sugar occur, and various chemical reactions occur among various ingredients. Boiling plays an important part in the stability of beer.

The wort is then cooled. Wort cooling would be simple if it were not for the possibility of microbial infection and the changes in physical state of some of the constituents. At about 150°F the clear wort will become cloudy, but there is little chance for microbial contamination until the temperature falls below 140°F. The open cooling kettles used years ago afforded ample opportunity for microbial contamination. The closed cooling systems now commonly used obviate this hazard but present problems because of solubility changes in the wort. The wort is now ready for yeast inoculation and fermentation.

The fermentation rooms are special rooms that can be kept clean and at a constant and uniform temperature and humidity (Fig. 9.4). Many brewers still prefer the wooden vats, but stainless steel, aluminum, and glass-lined tanks are common. Similarly, some brewers prefer the old style oak vats in their lager cellars to the modern air-conditioned lager cellars with glass-lined steel tanks. Because of the need for carbon dioxide, completely enclosed tanks or tanks with domed lids that can be lowered onto the tank are employed.

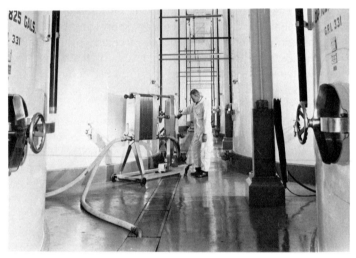

Courtesy of Taylor Wine Co.

FIG. 9.4. HEAT EXCHANGER USED FOR COOLING WINE DURING ACTIVE FERMEN-TATION

Lager beers are generally bottom fermentations conducted at a relatively low temperature by strains of *Saccharomyces carlsbergensis*. Strains of *Saccharomyces carlsbergensis* are considered to have stronger fermentative capacity but weaker respiratory activity than strains of *Saccharomyces cerevisiae*. Ales are generally top fermentations conducted by strains of *Saccharomyces cerevisiae*. Although yeasts ferment most rapidly at the higher temperatures of 77°F for bottom yeasts and 95°F for top yeasts, the use of these higher temperatures impairs the vitality of yeasts and also stimulates development of bacteria. Lager fermentation temperatures from 41°–46.4°F to 42.8°–50°F and even as high as 53.6°F are used. The main fermentation may be completed within 5–15 days. The lower temperature fermentations are usually thought to yield beers of finer quality and fuller flavor and supposedly retain their character over a longer period. Pale beers are often fermented at the lower temperature. The higher temperatures that are often used for dark beers result in more vigorous fermentations. Also, there is then less tendency for wild yeast development. There is a trend toward use of the higher temperature fermentation, particularly in the United States. Ales are fermented at the higher temperatures of 68°–77°F.

The rate of fermentation is extremely important, particularly in its effect upon growth of contaminants. A rapid change in pH during early stages is suitable for yeast growth because it is unfavorable for many bacteria. During fermentation the pH drops from about 5.2 to about 4.1–4.2 and in some top fermentations to pH 3.85. These low pH values are unfavorable for bacterial growth. Temperatures may increase by as much as 8°F and of course specific gravity and wort solids will decrease.

Since yeasts must be collected from one fermentation for use in another, it is necessary to protect the yeasts from contamination. Yeasts remain remarkably constant in some breweries through many fermentations. The lag phase can be shortened considerably by selection of yeasts from one fermentation for inoculation into the next by selecting yeasts during the most active stage in fermentation. The middle crop of yeasts are selected since the yeasts that separate first may be more heavily contaminated with hop resin and protein. The most active fermentation proceeds while yeasts remain suspended in the wort.

Regardless of whether the beers are fermented by bottom or top yeasts, there are certain other desirable characteristics of yeasts beyond their ability to produce good flavor and alcohol content. A good brewery yeast should remain in suspension during active fermentation, then flocculate and settle without haze and the sparkling beer may be readily separated from the yeast sediment. The character of the rate of settling is therefore important.

Since the classical studies of Hansen, considerable research has been conducted to obtain the most suitable yeasts. The genetic possibilities, for example, are illustrated in the studies of Lindegren in which he has obtained 1600 different strains from the original 12 yeast strains by genetic crossing. Some taxonomists favor simply using the generic name, *Saccharomyces,* to include all species and types.

After fermentation is completed, beers may be aged for an indeterminate period of months at temperatures as low as 32°F but usually 33°–36°F. During this period a slow fermentation may continue, colloidal changes occur, the beers clarify by precipitation of yeasts, proteins, and other suspended solids, and chemical reactions occur that impart a clear, mellow, and smooth flavor. Lager beers, as the name implies, are generally stored for a longer period of time than ales.

Several types of beer are produced in various areas. In general, beers can be grouped into ales and beers. The ales that are characteristic to England are produced by a top fermentation. The ordinary ale has a pale color, a pronounced hop flavor, and a somewhat tart taste with an alcohol content of 4–5%. It is fermented at 68°–75°F in 3–5 days and stored at 40°–46°F. Porter has a darker brown color with a full-bodied malt flavor and sweet taste. It contains less hops than ale and attains about 5% alcohol. Stout is still darker in color with a rich, sweet, malt flavor and a strong, bitter hop taste. Alcohol contents of 5.0–6.5% are attained in this. Weissbeer is a special pale ale made from wheat and barley; it has a tart taste and rich foam characteristics.

The lager beers that have become common in the United States are prepared by bottom fermentation. The fermentation is slow, and after completion the beers are stored 1–2 months or more at 32°–34°F to acquire the mellowness desired. Pilsener and Dortminder are characterized by their pale color, medium amount of hop flavor, and their relatively low alcohol content of 3.0–3.8%. Vienna has a more amber color with a mild hop flavor. Munich beer has a dark brown color, full body, and a sweet, malt-like slight hop flavor, and contains 3.0–5.0% alcohol. Bock beer is a special heavy beer with a slightly sweet malt-like flavor. The Faro and Lambic beers of Belgium are somewhat acidic in character.

These statements are generalizations and, of course, are subject to variation. Some beers that are sometimes called tonics have a high nutritive value, low alcohol content, and are often enriched with minerals and vitamins. So-called "near" beers, which are similar to Pilsener-type beers, have alcohol contents as low as 0.4%.

BEER-LIKE BEVERAGES

Beer-like beverages are produced in nearly every country in the world where basic ingredients can be grown. One of the most important beverages of Japan is sake, a rice wine or beer which dates back to ancient times.

Sake, prepared by the Japanese, is the popular beverage manufactured on a large commercial scale. It is a light yellow beverage with a flavor somewhat resembling sherry. The starch of steamed rice is saccharified by the mold, *Aspergilus oryzae*. The koji thus produced is added to a thin paste of fresh boiled rice. Fermentation by the yeast, *Saccharomyces sake,* is then initiated and may continue for 30–40 days. More rice and koji may be added to continue the fermentation. It is finally filtered, pasteurized, and bottled. Sake contains from 12 to 15% alcohol, about 3% solids, and about 0.3% lactic acid. Shaw-shing-chu is closely related to sake but has a lower alcohol content.

Japanese scientists have conducted considerable research on the production of sake. Kitahara *et al.* (1957) have isolated two types of lactic acid bacteria known for their ability to sour the beer. The names *Lactobacillus homohiochi* and *Lactobacillus heterohiochi* have been applied to the homofermentative and heterofermentative strains of the genus.

The Javanese also prepare a rice beer in which the starch is saccharified by species of the genera of molds, *Mucor* and *Rhizopus*. Oo is a home brew prepared in Thailand from uncooked rice, husks, and yeasts sealed in earthen jars and buried for several weeks. Water may be added from time to time. The sweet alcoholic liquor is sipped out by straws and more water may then be added. Bananas are sometimes added for flavor. Another preparation is made from steamed rice in a similar manner; however, this is distilled to yield a strong drink known as lao rong.

Binuburan is the name applied to a rice beer prepared in the Philippines. The sediment from the fermented product is pressed into cakes that are used as inocula for the next batch. The cakes consist of yeasts with a smaller number of bacteria. Kanji is a sour rice beverage prepared in India in which the starch is hydrolyzed by a strain of *Aspergillus*. It is interesting that the sake´ of Japan, the Javanese beverage, and the Indian pachwai should be prepared from rice saccharified by species of three different genera of molds.

Pachwai is a high alcohol rice beer made in India by inoculation with a preparation known as bakhar also known as murcha, ramu, and u-t-iat. The bakhar is prepared in small cakes. The beer is prepared by adding the powdered bakhar to half-boiled rice and allowed to incubate 24 hr. The whole mass is transferred to earthern jars where alcoholic fermentation proceeds. Species of *Mucor* predominate in a mixture of other molds of the genus *Rhizopus* and yeasts and bacteria. The molds saccharify the rice. Atsumandie and saraimandie are similar preparations used as inocula. Sonti is another rice wine or beer prepared in India in which the *Rhizopus sonti*, mold, and a yeast are the fermenting organisms. Torani is a sour beverage prepared from rice in India, supposedly fermented by molds of the genera *Aspergillus* and *Penicillium* and by yeasts of the genera *Hansenula, Candida,* and *Geotrichum*. It would seem that lactic acid bacteria would be involved in such a fermentation. The sour beverage is cooked with vegetables and spices to make kanji. The people of India commonly use sour milk preparations with vegetables.

Arrack, arak, rak, or rack is a generic name applied to a variety of beverages and distilled liquors prepared from rice beers in the Far East, including India, Ceylon, Siam, Java, Batavia, China, and Korea. Molasses or juice of certain palms are sometimes added to the rice. Ragi is an inoculum. Toddy, the fermented juice from the unexpanded flower-spathes of various palms including the cocoa palm, is distilled in Ceylon to produce arrack. The Indian arrack is prepared from rice toddy and the refuse from sugar factories. Another type of arrack contains coconut milk. The yeasts obtained from a previous fermentation are probably a mixture of wild yeasts, bacteria, and strains of *Saccharomyces*. These preparations were supposedly made as early as 800 B.C. Okelehao, another type of arrack, is prepared in Hawaii by fermentation of the rice, taro, and molasses mash or by fermenting rice, molasses, and coconut milk.

Rakshi is the name applied in Nepal to the high alcohol content rice beer used by the Lamas in religious observances. The rakshi is poured on boiling butter and the flames blaze high, symbolically consuming the prayer and lifting it to heaven.

Kvass is a general Russian name for the sour type of alcoholic beverage prepared with rice and barley or rye that has been allowed to sprout before milling. A Ukrainian drink, kvass, is described as having a mild, sour flavor; usually, beets are used in its preparation. Still another type of kvass is prepared with dry rye bread, sugar, raisins, and added orange flavor. The word kvas, kvass, or kwass refers to leaven. Teekvass is a tea beverage produced in Russia, supposedly by a symbiotic fermentation of a tea sugar solution by *Schizosaccharomyces pombe* and *Acetobacter xylinum*. A similar product is prepared in Indonesia from an aromatic tea infusion to which 10% of sugar has been added. This is supposedly fermented by *Saccharomycodes ludwigii* and *Acetobacter xylinum*. A study of these fermentations seems essential to determination of the microorganisms involved. *Acetobacter xylinum* should produce a heavy matted film on the surface similar to that produced in "Nata."

A Finnish beer made with rye malt is called kalja.

The early settlers in America met with many types of beverages prepared with corn as the main ingredient. Sir Walter Raleigh was introduced in 1587 to pogotowr, an Indian beer made from corn. Later, the Pilgrim Fathers partook of nohelick, considered a great delicacy. Apples and grapes were used with corn by the Indians in beverages. The Creek Indians flavored corn beer with the ashes of hickory wood and/or leaves of a species of holly. Although the early settlers brought beer with them, many were obliged to use corn with other grains in the preparation of more beer. The brewing of beer by the usual methods practiced in Europe developed rapidly in America.

The South American Indians used corn in their preparation called chica or chichara. It is said that the older Indian women chewed the red corn and collected the chewed corn for fermentation. This is a common practice among primitive peoples. Obviously such a practice would not only introduce diastatic enzymes, but also both yeasts and lactic acid bacteria. Chicha, a light beverage, and sora, a heavy beverage, were names applied by the Incas for corn beers.

The Mexicans and Indians of the southwestern United States prepared several alcoholic beverages. Relics of prehistoric brewing antidating by centuries the arrival of the white man have been discovered in the Great Bend region of Texas. Zaca is a sacred corn gruel of the Mayas. Tesquino is a corn beer of the Tarahumares and was used by the Indians in rituals, festivals, and dances. Tiswin or tulpi are other names of beers.

Wheat is seldom used in making beer. However, the ancient Egyptians did prepare a beer, zythum, from wheat and malt; and, of course, the beers made by fermentation of breads soaked in water contained wheat.

The Bantu tribes of South Africa prepare kaffir corn beers called mpoho and machewa or magon. The latter has been described as an acid nonalcoholic beverage brought about by species of the genus *Lactobacillus*. It is fed to babies as a porridge-like beverage and is held in such high esteem that it is used in religious practices.

Kaffir beer is an important part of the diet of the natives of Nyasaland. According to Platt (1964) the term kaffir beer should be confined to that prepared with kaffir corn, a type of sorghum. The fermentation is complex, involving the molds *Aspergillus flavus* and *Mucor rouxii* as well as lactic acid bacteria and yeasts. Other starchy foods, e.g., cassava and plantains, are sometimes used in its preparation. Platt (1964) emphasizes that the beverage supplies a definite nutritional requirement for the people. Fermentation almost doubles the riboflavin and nicotinic acid content of the grain.

The practice of chewing or masticating the grain to induce fermentation is practiced quite generally by primitive peoples, not only in Africa, but also in South America and elsewhere. The chewing of foods other than grains by older people in many societies to feed directly to infants is quite common. The chewing is supposed to influence favorably the active spirit in the drink by sharing the magical power of the body. As previously stated, bacteria, yeasts, and diastatic enzymes are thereby introduced to the food.

Thumba is an alcoholic beverage produced from millet in West Bengal. The millet seeds are boiled, cooled, and inoculated with a yeast and fermented for ten days in sections of bamboo. The yeast has been identified as *Endomycopsis fibuliges*. The yeast is sold as small cakes in the market places. Chang is a millet beer brewed in Sikkim, and braga is a fermented drink prepared from millet in Romania. Busa is an acid alcoholic drink prepared by the Tatars from millet and sugar and often with rice. Species of *Saccharomyces (S. busae asiaticae)* and a *Lactobacillus* named *Bacterium busae asiaticae* by Tschekan (1929) are responsible for the fermentation. The latter is believed to be synonomous with *Lactobacillus plantarum*, but it may be closely allied to *L. delbrüeckii*. Millet beers are also prepared in Africa. The people of Mali call their beer konya and drink it from a gourd-like vessel.

The Mexican pulque rivals in historic antiquity the Babylonian bousa. Pulque is prepared by fermenting the juice obtained from a number of species of agave. The cultivation of agave or maguey for pulque manufacture constitutes a considerable industry. One plant yields a few quarts of juice daily for about six months. The juice is collected and carried to the tinacal where it is fermented to yield pulque. The method is sometimes credited to the Toltecs. Pulque is a light colored, thick, slightly alcoholic drink. Among the Aztecs, pulque was sacred and reserved for a ceremonial beverage. Cleanliness and prayers were stressed for its successful fermentation. It was used as an offering at weddings and funerals and for medicinal purposes; drunkenness was frowned upon and for it severe punishment was accorded. Pulque is sometimes sweetened with honey or fruit juice. The blend with pineapple is called tepache. Pulque is distilled to yield tequila, a clear, potent, alcoholic drink with a slightly bitter taste. When aged, it becomes amber in color.

Mezcal, an intoxicating drink from Oaxaca, Mexico, is prepared from the cores or hearts of the century plants. The core is stripped of its leaves and cooked to yield a sweet juicy pulp. This is crushed and extracted to yield the juice, acquamiel. Some of this juice is fermented naturally to yield a product, Madre pulque or Mother of pulque. If the product is good, a small quantity of this is added to fresh acquamiel; a rapid fermentation ensues and the product is ready to drink in a few days. The slightly brown liquid has a heavy flavor resembling sour milk and is esteemed by the natives. It has an alcoholic potency equivalent to beer.

Sánchez-Marroquín (1953) observed that the important organisms involved in pulque fermentation are the homo- and heterofermentative species of the genera *Lactobacillus* and *Leuconostoc* and the yeast *Saccharomyces carbajali*. Strains of *Leuconostoc mesenteroides* are specifically emphasized. This fermentation was at one time ascribed to high acid-producing yeast.

Balche is a ritual drink of the Mayas of Yucatán, fermented with the bark of a tree. The desert sotol lily is used in preparing another alcoholic beverage.

Tuwak or toowak is an alcoholic beverage prepared from the liquid which comes from the flower stalk of the aren-palm. It is supposedly a bacterial fermentation. It would seem more likely to be a mixed bacteria and yeast fermentation.

Masata is a beer made from the yucca plant in Mozambique. The older women chew the yucca plant, spit it out, and allow it to ferment. Shima is a palm wine of Mozambique. Chonta is a palm wine in Ecuador. The Shoshone Indians boil mesquite beans to a mush and dry this to sulfur-colored cakes that can be kept for long periods of time. These are mixed with wild honey, honey comb, and water to ferment and yield a mildly intoxicating drink. A beer, kuva, is prepared in South America from beans. Brazilians chew starchy roots of cassava to produce a beer called kaschiri.

Tein mein chang is a steamed bread product prepared from wheat flour. The steamed bread is allowed to mold for 3 weeks and then placed in jars with 5% salt and some water and allowed to ferment 6–8 months. This second fermentation would appear to be alcoholic. Water is added daily. The original steamed bread, if it is similar to the Philippine puto, may have been leavened by lactic acid bacteria or yeast or both.

Awamori is an alcoholic drink made in Japan from sweet potatoes. Uri is a similar preparation.

Koumiss and kefir are alcoholic, acid beverages prepared from milk by the symbiotic fermentation of lactic acid bacteria and yeasts. The milk from mares, camels, goats, sheep, and cows are used for these preparations. The koumiss prepared from mares' milk is distilled in some areas of Western Asia to produce a potent beverage called ariki, araka, or arrack. Similarly, kefir is distilled in Caucasia to produce an equally potent beverage called skhow.

The heather beer of the Scots and the ancient Picts and the sprossenbier of the Germans are highly flavored with heather and spruce respectively. A number of beers are flavored with substances such as sassafras, ginger, pepper, licorice, cinnamon, spruce as well as fruits. Spruce beers apparently originated in Scandinavia. Ginger beer or tibi is a weak alcoholic carbonated beverage that possibly originated in Switzerland. It is interesting because the tibi grains consist of a capsulated strain of *Lactobacillus brevis* var *vermiforme* and a yeast *Saccharomyces intermedius*. Lactic acid, carbon dioxide, alcohol, and dextran are produced. A solution such as 15% dried fig and/or raisin, lemon juice, and ginger are inoculated with the tibi grains to produce the beverage.

The origin of the practice of producing strong alcoholic beverages by distillation is lost in antiquity. The sautchoo prepared by the Chinese was apparently produced before the Christian era, and arrack was produced before 800 B.C. Similarly, ariki and skhow distilled from koumiss and kefir have ancient origins. Mead made from honey was distilled in Britain about 500 A.D., brandy in Italy by 1000 A.D., and cognac in France by 1300 A.D. Whiskies were produced at an early date; in fact, when the English invaded Ireland in 1170 A.D., they found that the Irish were distilling fermented mashes to produce whiskey. The Scots were also producing whisky at about this time.

Spirits are produced from a number of starchy materials such as barley, rye, oats, wheat, corn, potatoes, and cassava roots. The Russian vodka, vodki, or wodky was originally made almost entirely from rye and barley malt, but at present, potatoes and corn are also utilized in its preparation.

MICROBIOLOGICAL ASPECTS OF FERMENTATIONS

With some notable exceptions the majority of the alcoholic beverages discussed have been developed more or less naturally with little or no knowledge of the microorganisms involved other than yeasts. Many of them are described as acidic-alcoholic drinks. The basic substrates used in the preparation of most of these beverages are adequate media for the growth of many types of microorganisms, e.g., molds, yeasts, and bacteria. There is little doubt, therefore, that considerable variation existed among the various preparations depending upon the flora other than the yeasts. For example, the people of the Far East learned early in the development of their beverage production that the growth of molds would alter the character of their rice preparations so that they would ferment more readily. They also observed that an aerobic environment was favorable to the growth of molds. Mold species of the genera *Aspergillus, Rhizopus,* and *Mucor* have been mentioned. Although alcoholic fermentation is ordinarily attributed to *Saccharomyces cerevisiae* or to related species of this genus, obviously under the conditions of fermentation, species of many other genera of yeasts, particularly those grouped as wild yeasts, could also be involved in many of these fermentations.

It was possibly among the fermentations involving bacteria that many products acquired distinctly undesirable characteristics. Spoilage attributable to the growth of spore-forming bacteria must have been fairly prevalent. In the early studies of bread making, such organisms frequently caused undesirable changes, not only in texture but also in flavor. Many species of nonspore-forming bacteria could also develop. Among the various bacteria, the lactic acid bacteria would grow well in most of these substrates. Since the lactic acid bacteria are microaerophilic or do not require air for growth, they could develop simultaneously with the yeasts during fermentation. In some cases, such as the koumiss and kefir fermentations, the simultaneous growth of yeasts and bacteria is a truly symbiotic relationship. As in bread making, the early brewers must have observed that the flavor of an acid-alcoholic beverage was superior to that in which only acid was produced, and this type of fermentation was therefore encouraged. It seems obvious that an acidic character was desirable; however, this should not be confused with the acetic character developing in later stages of fermentation resulting from the growth of acetic acid bacteria and destruction of alcohol.

The production of lambic, the highly hopped beer produced in Belgium, may be a symbiosis between lactic acid bacteria and yeasts. Lambic is beer made in a slow fermentation of barley malt and wheat with the yeasts *Saccharomyces bruxellensis* and *Brettanomyces lambicus* and the lactic acid bacterium called *Lactobacillus viscosus bruxellensis.* The fermentation is initiated by the *Saccharomyces* and continued by the high viscosity acid-producing *Lactobacillus.* The acid favors the growth of *Brettanomyces lambicus* responsible for the typical flavor.

The organisms involved in fermentation grow in response to environmental conditions. Pure culture fermentations are almost nonexistent. In acid foods such as the fruit juices, fermentations are frequently initiated by wild yeasts, followed by the higher alcohol-producing species. Lactic acid bacteria are often involved, particularly those that convert malic acid to lactic acid. In the fermentation of cereal products, it is frequently essential to hydrolyze starches to readily fermentable carbohydrates. Molds are generally present on dry cereals, ready to grow when the cereal is moistened. Certain species of bacteria may grow, particularly at raised temperatures. The microbial activities are quite well elucidated for wines and beers in reference books such as those by Amerine *et al.* (1967) and Hind (1940) as well as in many scientific articles. Ault (1965) in discussing beer fermentation states that the whole taxonomic situation is complicated by mutations and even goes so far as to suggest that species differentiation might be discarded and deal only with genera. He further states that the spoilage situation outlined by Shimwell in Hind (1940) is still apropos.

Many species of microorganisms are involved in some of the lesser-known beverages. There is little doubt that the flora is varied in most of these and even that there is considerable variation in the fermentations of a specific beverage. Each beverage must be considered as a distinct entity, a product of environment and subject to variation with change in environment.

Spoilage in Beer

A wide variety of microorganisms are capable of growing in liquid substances. In beers, hops will inhibit many species of bacteria, but there has been too great a tendency to rely solely on hops for inhibiting all types. Bacteria in breweries were among the earliest microorganisms described by Pasteur in Etudes sur le Bière in 1876, but their study was neglected for a number of years thereafter.

Hansen used the term, wild yeast, to denote any yeast that may cause trouble in brewing or produce defects in beer. Hansen observed that brewing yeasts were a mixture of several races with different properties and suggested use of pure cultures. Five species of wild yeasts were described by Hansen as *Saccharomyces ellipsoideus* I and II, and *S. pastorianus* I, II, and III. Later these were referred to as *S. cerevisiae* var *ellipsoideus, S. cerevisiae* var *turbidans,* and *S. pastorianus* var *intermedius* and var *validus.* Many other yeasts have been described and named since then that produce in beer various defects such as sweet taste, rotten-fruit flavor, cloudiness, haziness, acetic flavor, and others. Species of a number of genera of yeasts have been described as spoilage organisms in beer.

The subject of microbial spoilage is reviewed by Shimwell in Hind (1940). In general, the bacteria that may cause defects in beer may be grouped as acid-intolerant and acid-tolerant types. They may also be separated into aerobic and microaerophilic types. Since the acid-intolerant bacteria, species of the genera *Achromobacter, Flavobacterium, Bacillus,* and others, are in general aerobic and cannot grow in acid beer, their spoilage is more or less confined to the nonacid mashes and worts. In contrast, the lactic acid bacteria are microaerophilic and acid-tolerant; therefore, they may grow in beer. In general, the species are intolerant to hops. The *Acetobacter* species are acid-tolerant but aerobic, and spoilage by these organisms is confined to beer exposed to the air. *Acetobacter* sp. may be detected in equipment that has been exposed to beer and subsequently to air, such as in returned empty barrels. The alcohol is converted to acetic acid. The slimy growth is difficult to remove. Shimwell has listed ten species of *Acetobacter* from beer.

Lactic acid-producing organisms have been associated with beer spoilage since the condition was first described by Pasteur in 1876. Unhopped, low-acid beers are excellent media for the growth of lactic acid bacteria. Both homofermentative and heterofermentative species have been implicated in spoilage. Although Pasteur did not apply names to the bacteria that he recognized, he demonstrated that yeasts contaminated with bacteria could be purified by growing in a slightly acid solution of sugars. Although the lactics may not be killed, their growth becomes attenuated. Two bacterial species, *Pediococcus cerevisiae*, a homofermentative coccus form, and *Lactobacillus pastorianus*, a heterofermentative rod form, are most frequently mentioned in reference to beer spoilage. Other species can grow in beer, and *Lactobacillus delbrüeckii* is frequently mentioned, particularly as a species that acidifies wort.

Pasteur observed that the cocci associated with the long rod bacteria were the cause of the so-called "beer sickness." The organisms were referred to as a sarcina by Hansen in 1879. The spoilage was studied by Balcke in 1884, and in a second paper he recognized that the organism was not a true sarcina. Balcke applied the binomial *Pediococcus cerevisiae* to the species. In a series of papers, Lindner described the organisms more completely and proposed new specific names for types producing somewhat different reactions. Mees (1934) recognized their relationship to the true lactic acid bacteria. In a series of papers, Shimwell and Kirkpatrick (1939) and Shimwell (1940) and later Pederson (1949) recognized the relationship of these organisms to the true lactic acid bacteria rather than to the sarcinae and included them in the tribe *Streptococceae*. They recognized that the peculiar flavor of the beer was due to diacetyl. The organisms are distinctly different from the species of the genus *Streptococcus* because they divide in two planes to produce tetrad groupings and they produce inactive lactic acid; therefore, the name *Pediococcus* proposed by Balcke is the acceptable generic name. Many names have been applied to strains of this genus (Pederson 1949).

Lactobacillus pastorianus was originally described by Van Laer in 1892 as *Saccharobacillus pastorianus*. It is heterofermentative lactic acid bacterium and is possibly related to the organism referred to as the tourne bacterium by Pasteur. The silky turbidity associated with this species is a characteristic of long rod-shaped bacteria. Henneberg later described two varieties of the species. This species can be controlled by using high amounts of hops and by acidification. Handling of the "pitching" yeasts is quite important in control of this spoilage. Since the "pitching" yeasts are collected from one fermentation for use in another, it is equally necessary to protect them from contamination to save only yeasts that are uncontaminated. Yeasts that separate first are contaminated by the hop resins and proteins. The middle crops only are used whether they occur in the middle layers of bottom yeasts or the middle heads of top fermentation. Different races of yeast are distinguishable by the rate at which they ferment and the degree to which they attenuate the wort. Active fermentation proceeds while the yeasts remain in suspension in the wort. Yeasts that separate rapidly from the wort perform like weak yeasts while those that show little tendency to separate are strongly attenuating. The difference depends upon the flocculating property of the yeast. While this property will remain constant through many fermentations, it may not necessarily continue to do so.

The numerous studies of spoilage in beer have been reviewed, not only by Shimwell in Hind (1940), but also in many papers and reviews abstracted in *Wallerstein Laboratory Communications* and the *Journal of the Institute of Brewing* during a period of many years. The subject cannot be adequately covered in this treatise. The sale of lagers and ales in Europe and America is so highly competitive that quality must be given first consideration in their manufacture. Too little is known in regard to the spoilages that have undoubtedly occurred in many other beer-like beverages.

Fermentation is carried out in breweries in two very distinct ways, known as top and bottom fermentations. As the terms imply, most of the yeasts rise to the surface in one case and/or sink to the bottom in the other. The guiding principles are the same, differing only in the nature of the yeast, the temperatures used, and the manner in which the yeast crop is removed from the beer. In whatever way the fermentation is conducted, the fact is that its successful issue depends not only upon the alcohol content of the beer but also upon its flavor and stability. The aim is to mix an active "pitching" yeast with the wort under conditions that will insure a rapid initiation of growth so that other microorganisms shall have little chance of contaminating the wort.

One must not be disturbed by the variety of scientific names applied to the microorganisms from various fermentations. Each investigator is privileged to apply a specific name to an isolate and as long as it is given the proper generic name, the species can be readily associated with other species in the genus. Newer names such as *Lactobacillus homohiochi* and *Lactobacillus heterohiochi* applied by Kitahara *et al.* (1957) are distinctive enough to associate them with the two physiological groups in the genus *Lactobacillus*. The species named *Lactobacillus trichodes* by Fornachon *et al.* (1949), *Lactobacillus hilgardii* by Douglas and Cruess (1936), *Lactobacillus malofermentum* and *Lactobacillus parvus* by Russell and Walker (1953), and *Lactobacillus frigidus* by Blandari and Walker (1953), all have been applied properly to strains that have characteristics somewhat at variance from species previously named. Some of the recent studies with the malo-lactic bacteria of the genus *Leuconostoc* have shown characteristics unlike the three species included in the genus in Bergey's Manual (Breed *et al.* 1957) (Table 9.1). Possibly the specific name *Lactobacillus gracile* should be used to distinguish these strains.

Finally, there still may be some question as to what constitutes spoilage in contrast to a desirable fermentation. It seems clear that this is a matter of customary usage. Certainly, the Mongolian accustomed to the acid-alcoholic koumiss would not consider yoghurt a substitute, while the consumer of yoghurt would consider koumiss a spoiled product. The English ale and German lager are distinct, but the changes that occur in preparation of lambic beer would be considered spoilage if they occurred in either. A malo-lactic fermentation is desirable in certain wines in mellowing a harsh flavor but in certain low acid wines the fermentation may result in a flat, tasteless product. Spoilage may be considered an abnormal fermentation in a product as it is ordinarily consumed by a certain group of people.

ALCOHOLIC SPIRITS

The term spirits is generally restricted to distilled liquors. The origin of the practice of distillation is lost in antiquity. Sautchoo was produced in China long before the Christian era; arrack was made in India as early as 800 B.C.; and later, distillation of a fermented grain liquor was practiced in Ireland and Scotland. The distillation of wine to produce brandy appears to have been practiced in the 9th Century and became a sizeable industry in France by the 14th Century. The production of brandy from wine or spirits from fermented molasses are simple processes compared with production of spirits from the grains, rye, barley, oats, wheat, corn, potatoes, and other starchy substances.

Brandy is the alcoholic spirit distilled from fermented grape juice and, in the United States, must contain not less than 54% alcohol by volume and must have been stored in wood for not less than 4 yr. When distilled, it is a colorless liquid, but upon aging it acquires an amber color and a characteristic taste. Brandy ordinarily contains organic acids, esters, higher alcohols, aldehydes, and furfural.

Whiskey is any spirituous liquor made from grain. Scotch and Irish whiskeys are primarily made from barley, rye whiskey from rye, and bourbon from corn, but all contain some barley malt. Scotch whiskey has a smokey taste because the barley malt is dried in kilns with a porous floor directly above the peat fires.

The first stage in the preparation of spirits, such as whiskey, from starch materials is the conversion of the starch into a dextrin-maltose mixture with intermediates. This is accomplished, as in the brewing industry, by the action of a diastatic ferment such as malted grain. Diastatic enzymes prepared by growth of mold have been used. The method of preparing the malt is similar to that used in brewing. Mashing is designed to produce the highest possible yield of alcohol; thus, it is done at relatively low temperature so that diastatic action will be maintained to a high level. However, the temperature must be high enough to restrict growth of a great variety of microorganisms while keeping it low enough so that diastatic action is not restricted. In addition, in order to avoid undesirable fermentative activity, it is necessary to allow the wort to sour naturally or to add acid. The mashes are generally soured by inoculating with a culture of homofermentative lactic acid bacterium. A thermoduric strain of the species *Lactobacillus delbrüeckii* is ordinarily used. It will grow at 122°F, produce lactic acid as a major end product, lowering the pH from about 5.6 to about 3.8 in 6–10 hr or more. Small amounts of volatile products are also produced which are considered to improve the flavor of the distilled spirit. When diastatic action and lactic acid production have reached a suitable level, the mash is heated rapidly to 165°–180°F, a temperature at which all microorganisms will be killed in the acid mash. It is then cooled to 75°–80°F for fermentation.

The mash is inoculated with an actively growing culture, in logarithmic growth phase, of a strain of *Saccharomyces cerevisiae*. This inoculum may consist of a portion of mash previously prepared. Vigor in the culture is extremely important and optimum environmental conditions, mash concentration, nutrients, pH, and buffer are all-important.

The factors governing mashing, fermentation by bacteria, and then by yeast must always be carefully controlled and are under continual study to improve the process. For example, the addition of a mold bran prepared by allowing mold to develop on bran has resulted in more vigorous yeast growth. Yeast fermentation may be started at temperatures as high as 80°F, but the fermenting mash is gradually cooled to about 60°F.

After fermentation is completed the liquid is distilled to recover the alcohol and other volatiles. These volatiles are diverse. The quality of the distilled product will be determined by the nature of the fermentation. Like distilled products prepared from other substances, chemical changes will occur during aging which will affect the character of the beverage. The character will also be affected by ingredients used in the mash. Different blends of substances such as barley, rye, corn, or other grain, molasses, potatoes, cassava, and mold bran produced by the action of a mold such as *Aspergillus oryzae, A. niger,* or a *Rhizopus* sp. will all affect the flavor of the spirits. When distilled, many of the volatile substances, acids, esters, aldehydes, and others present in the fermented mash remain in the final beverage.

There are a great number of alcoholic spirits produced from the various fermented beverages by distillation. The fermentations have been discussed. Brandy is one of the oldest of the alcoholic spirits. Brandy is ordinarily prepared from grape wine, but other wines are also distilled. Cognac is a brandy produced in the Cognac district of France. Apple brandy or apple jack, plum brandy or slivovitz or mirabelle, peach brandy or Southern Comfort, cherry brandy or kirsch, apricot brandy, or orange brandy, and raspberry brandy are just a few of the distilled products.

Rums are alcoholic spirits from fermented sugar substances such as molasses. A great variety is produced; for example, arak is a highly aromatic rum made in Java in which red rice is added. A similar preparation is made in Hawaii.

Aquavit produced in Scandinavian countries, vodka produced in Russia from grains and potato mashes, arika produced from the milk drink koumiss, skhow from kefir milk, tequilo produced from the palm wine pulque, and gin flavored with juniper berries and other aromatics are just a few of the types of distilled spirits. Since these are distilled products from alcoholic beverages, they will not be considered further.

BIBLIOGRAPHY

AMERINE, M. A., BERG, H. W., and CRUESS, W. V. 1967. The Technology of Wine Making, 2nd Edition. Avi Publishing Co., Westport, Conn.

AULT, R. G. 1965. Spoilage bacteria in brewing. J. Inst. Brewing 71, 376–391.

BLANDARI, R. R., and WALKER, T. K. 1953. Lactobacillus frigidus n. sp. isolated from brewery yeast. J. Gen. Microbiol. 8, 330–332.

BREED, R. S., MURRAY, E. G. D., and SMITH, N. R. 1957. Bergey's Manual of Determinative Bacteriology, 7th Edition. Williams & Wilkins Co., Baltimore.

BROTHWELL, D., and BROTHWELL, P. 1969. Foods in Antiquity. Thomas and Hudson, London.

DOUGLAS, H. C., and CRUESS, W. V. 1936. A Lactobacillus from California wine, Lactobacillus hilgardii. Food Res. 1, 113–119.

FAPARUSI, S. I. 1969. Effect of pH on the preservation of palm wine by sulfite. Applied Microbiol. 18, No. 1, 122.

FORNACHON, J. C. M., DOUGLAS, H. C., and VAUGHN, R. H. 1949. Lactobacillus trichodes nov. spec., a bacterium causing spoilage in appetizer and dessert wines. Hilgardii 19, No. 4, 129–132.

HIND, J. L. 1940. Brewing Science and Practice. John Wiley & Sons, New York.

KITAHARA, K. H., KANEKO, T., and GOTO, O. 1957. Taxonomic studies on the hiochi-bacteria, specific saprophytes of sake. I. Isolation and grouping of bacterial strains. J. Gen. Microbiol. 3, 102–110. (Japanese)

MEES, R. H. 1934. Studies with the beer sarcina. Diss. aus de Lab. A. J. Kluyver, Delft. (Dutch)

MÜLLER-THURGAU, H. 1908. Bacterial bubbles (bacterial cysts). Centr. of Bakt., II Abt. 20, 353–400, 449–68. (German)

MÜLLER-THURGAU, H. and OSTERWALDER, A. 1912. The bacteria in wine and fruit wine and the changes that occur. Centr. of Bakt., II Abt. 36, 129–338. (German)

PEDERSON, C. S. 1949. The genus Pediococcus. Bacteriol. Rev. 13 (4), 225–232.

PLATT, B. S. 1955. Some traditional alcoholic beverages and their importance in indigenous African communities. Proc. Nutr. Soc. 14, 115.

PLATT, B. S. 1964. Biological ennoblement: improvement of the nutritive value of foods and dietary regimens by biological agencies. Food Technol. 18, No. 5, 68–76.

RICE, A. C., and MATTICK, L. R. 1970. Malo-lactic fermentation in New York State wines. Am. J. Enology Viticulture, 21, No. 3, 145–172.

RUSSELL, C., and WALKER, T. K. 1953. Lactobacillus malofermentans n. sp. isolated from beer. J. of Gen. Microbiol. 8, 162. Lactobacillus parvus n. sp. isolated from beer. J. Gen. Microbiol. 8, 310–313.

SANCHEZ-MARROQUÍN, A. 1953. The biochemical activity of some microorganisms of pulque. Mem. Congr. Cient. Mex., IV Centenario Univ. Méx. 471–484 (tr) Chem. Abstr. 1955, 49, 2021i.

SHIMWELL, J. L. 1940. Brewing Science and Practice. H. L. Hind (Editor). John Wiley & Sons, New York.

SHIMWELL, J. L., and KIRKPATRICK, W. F. 1939. New light on the "sarcina" question. J. Inst. Brewing 45, 137.

TSCHEKAN, L. 1929. Microbiology of busa. Cent. of Bakt. II Abt. 78, 74–93. (German)

WARD, H. M. 1892. The ginger-beer plant, and the organisms composing it: a contribution to the study of fermentation-yeasts and bacteria. Phil. Trans. Roy. Soc., London 183, 125.

CHAPTER 10

Nutritious Fermented Foods of the Orient

Fermented foods are important components of the diets of peoples throughout the world. The orientals learned centuries ago the techniques of hydrolyzing the ingredients of many foods, particularly proteins and starches, by using the enzymes produced by molds. Although this might seem to be a wasteful procedure because these are aerobic fermentations in which some of the carbohydrates are destroyed, this disadvantage is more than compensated by the beneficial results. Van Veen and Steinkraus (1970) stated that fermentation does not increase the protein nutritive value but that digestibility and organoleptic characteristics of certain fermented foods such as tempeh, ontjom, and bongkrek are improved over the raw ingredients. The wholesomeness of some fermented foods is as good as, and in some cases, better than that of the substrate from which they are produced. Shallenberger *et al.* (1967) observed that during fermentation of tempeh, the content of stachyose decreases rapidly. The removal of stachyose and raffinose may reduce flatulent characteristics of soybeans. The keeping quality of fermented foods is generally increased and required cooking time is decreased. Some vitamins are increased, others decreased. The proteins in particular are partially hydrolyzed, made more soluble, and available nutritionally. The flavors of the foods undergo considerable alterations, and it is an accepted fact that such foods offer a change in monotonous high carbohydrate diets. Soy sauces, soy cheeses, misos, tempeh-like products, and all their closely allied products are widely used in the Orient, but only soy sauce has found general use in the Western world. The high salt content of so many of these foods make them unacceptable to those unaccustomed to them. The monotonous staple foods are eaten with greater relish when consumed with fermented side dishes, sauces, or relishes.

In all of these preparations, it must be appreciated that there is likely to be considerable variation. One may surmise that they developed in a manner similar to other classes of fermented foods, such as cheeses, wines, and others. First made in the various homes by methods primarily based upon experience of the individual family or group, some were later prepared on an industrial or semiindustrial scale and some standardization of practices followed. Although certain microbial species predominated, it is doubtful that general standardization existed until the preparations were industrialized; therefore, it would appear that many of the products described in the ensuing pages actually were mixed fermentations in their historical development and subject to considerable variation. The scientific names applied to the various species of microorganisms often were given to correlate the species with the food product. Later literature has tended to regard many of the species as salt-tolerant strains of well known species of bacteria, yeasts, or molds.

Much of the information presented in this chapter was obtained from the papers of Hesseltine (1965) and van Veen (1953), but considerable information also was obtained in personal conversations with Drs. Steinkraus, Fukushima, and others at Geneva, and in discussion with friends in the Philippines. A considerable number of scientific papers have appeared recently. Attention may be called to those appearing in the reports of the Noda Institute for Scientific Research.

Most of these food products are characterized by the marked hydrolysis of proteins and lipids that increases the soluble nitrogen content. The degree of hydrolysis is dependent to a large extent upon time, temperature, and the activity of the specific enzymes.

SOY SAUCE

The first recorded history of the soybean plant in China according to Lo (1964) was written in Shen Nan's Materia Medica in 2838 B.C. The soybean is a leguminous plant native to southeastern Asia. It is now grown in the United States, Russia, Italy, France, Egypt, South Africa, China, Korea, Japan, Manchuria, Indonesia, Malay, and elsewhere.

The preparation of soy sauce or shoyu is an ancient process used in China and Japan. It apparently has been produced in Japan for at least 1000 yr. It represents one of China's greatest uses for soybeans (Table 10.1).

TABLE 10.1

STEPS IN BREWING SOY SAUCE

Prepare koji
  Whole soybeans: soak, discard water, and autoclave ⎫   Blend together 50% of
  Wheat: roast in hot air   ⎬   each on basis of dry
                                                     ⎭   matter before treatment
  Inoculate with *Aspergillus soyae*
  Incubate 3 days in high humidity flowing air at 30°C. This is the inoculum, or koji
Soak koji in salt solution
  Mix 15 liters of saturated salt solution with 10 kb of koji to produce moromi
Ferment moromi
  Initial stage of fermentation: 3–6 months at 15°C gradually warming to 25°C
    Proteins and carbohydrates are hydrolyzed by *Aspergillus soyae*. Osmophylic bacteria
    gradually lower pH from about 6.7 to about 5.0. *Pediococcus soyae* is the predominant flora
  Alcoholic stage of fermentation: 3–6 months at 25°–27°C.
  Lower temperature gradually to 15°C over a period of 4–5 months.
    osmophilic yeast produce about 2.5% alcohol.
    Saccharomyces rouxii is the dominant yeast.
    Yeasts and bacteria continue fermentation. Final pH is 4.7–4.8. Salt
    concentration is 18%
Filtrate and pasteurize

Source: Fukushima (1968).

The average composition of soybeans is about 40% protein, 17% carbohydrate, 18% oil, and 4.6% ash. The ash ingredients include potassium, sodium, calcium, magnesium, manganese, phosphorus, sulfur, chlorine, iodine, iron, copper, zinc, and aluminum. Glutamic acid, leucine, arginine, lysine, valine, isoleucine, phenylalanine, threonine, histidine, tryptophan, and methionine are abundant. Sucrose, raffinose, stachyose, and pentosans are among carbohydrate constituents. Soybeans are rich in phospholipids, nucleic acids, and the vitamins, particularly thiamine, riboflavin, and niacin. Soybeans are used for their oil, for protein-rich foods such as soybean milk, soybean cheese, soy sauce, and numerous other preparations.

Soy sauce is a hydrolysis product of the soybean. Soy sauce is a dark brown liquid with a pleasant aroma, used primarily as a flavoring agent for meats, poultry, fish, vegetables, and rice. Its high salt content of about 18% and distinctive flavor makes it a useful adjunct for many of the bland food products on which it is used. Smith (1961) stated that the microbiological and chemical processes involved in soy sauce production are complex and not completely understood by even the best-informed scientists. Soybeans and wheat provide the essential substrate although in times of shortages of wheat, oats, rye, and sorghum are substituted. Smith (1961) stated that the preparation of soy sauce, e.g., brewing, is known to involve the action of molds, yeasts, and bacteria. He mentions *Aspergillus oryzae, Zygosaccharomyces soya, Zygosaccharmyas major,* and *Lactobacillus sp.* Yokotsuka (1960) stated that *Aspergillus oryzae* and *A. soyae* are used by manufacturers. A chemical process has been devised but that process produces an inferior flavored product that, therefore, will be excluded from this discussion.

The koji for making soy sauce is prepared by inoculating a blend of treated whole soybeans and wheat with *Aspergillus oryzae* or *Aspergillus soyae.* The koji is the actual inoculum. It will be discussed in a later section. To prepare this blend, soybeans are soaked in water, cooked in an autoclave, and mixed half and half on the basis of dry matter with wheat previously roasted in hot air. The inoculated mixture is incubated for 3 days at 30°C at high humidity. The koji so formed is mixed with sodium chloride in the ratio of 10 kg of koji to 15 kg of salt solution. This blend is called moromi. The moromi is fermented 3–6 months, first at 15°C and later at 25°C. The proteins and carbohydrates are hydrolyzed by the enzymes of the molds. Osmophilic bacteria develop slowly and lower the pH from about 6.7 to 5.0. Fukushima stated that *Pediococcus soyae* is the dominant lactic acid bacterium present during this stage (Table 10.1); however, Hesseltine (1965) reported that Lockwood isolated a strain of *Lactobacillus delbrüeckii.* Spore-forming bacteria and micrococci have also been reported. The final stage involves an alcoholic fermentation by osmophylic yeasts with *Saccharomyces rouxii* dominant. The osmophylic yeasts *Zygosaccharomyces soya* and strains of the genera *Hansenula* and *Torulopses* have been implicated. It

TABLE 10.2

ANALYSES OF FOUR REPRESENTATIVE SOY SAUCE SAMPLES

Item	Sauce-A Chinese Nanking	Sauce-B Japanese Noda	Sauce-C Chinese Old Process	Sauce-D Chinese New Process
Total solids	32.0	38.13	29.7	30.5
Mineral matter	. . .	19.70	20.2	16.4
Sodium chloride	16.0	18.02	18.8	14.0
Phosphoric acid (as P_2O_5)	. . .	0.48	. . .	. . .
Total nitrogen	1.0	1.51	0.76	1.01
Protein nitrogen	. . .	0.09	. . .	. . .
Nonprotein nitrogen	. . .	1.42	0.72	1.00
Amino nitrogen	. . .	0.70	. . .	. . .
Volatile acids (as acetic)	0.5	0.14	. . .	. . .
Nonvolatile acids (as lactic acid)	. . .	0.65	. . .	. . .
Total acidity	0.7	. . .	4.1	5.3
Sugar (as glucose)	2.0	5.99	3.06	6.74
Dextrins	. . .	1.06	. . .	. . .
Total carbohydrates (as glucose)	. . .	. . .	4.2	9.6
Viscosity	. . .	4.84	3.04	4.18
Hydrogen ion concentration (pH)	. . .	4.6	4.6	4.9
Specific gravity	1.2	1.2	1.195	1.19

Source: Smith (1961).

TABLE 10.3

AMINO ACID CONTENT OF SOY SAUCE FROM THE DIFFERENT SUBSTRATE COMBINATIONS

	Substrate					
	Wheat-soybean Free Amino Acid		Corn-soybean Free Amino Acid		Sorghum-soybean Free Amino Acid	
Amino Acid	Percentage	(Mg/100 Ml)	Percentage	(Mg/100 Ml)	Percentage	(Mg/100 Ml)
Basic						
Trytophan	0.25	30.62	Trace	Trace	0.38	40.82
Lysine	8.87	817.50	5.53	497.05	8.97	687.09
Histidine	1.69	155.15	1.63	155.15	2.00	162.01
Arginine	0.76	78.39	3.74	400.68	1.34	121.95
Acidic						
Aspartic acid	1.60	126.45	6.59	539.10	3.34	232.94
Glutamic acid	9.29	971.12	11.54	1044.69	8.59	622.13
Neutral						
Threonine	3.04	214.34	3.50	256.02	3.04	190.53
Valine	5.66	392.29	5.61	404.00	6.11	374.72
Methionine	1.86	164.14	1.63	149.22	2.39	186.53
Isoleucine	5.41	419.78	4.39	354.19	5.25	360.75
Leucine	7.60	590.31	8.54	688.70	8.97	616.55
Phenylalanine	3.80	371.70	3.66	371.70	4.00	346.92
Serine	4.31	267.90	2.28	147.08	4.10	225.88
Proline	3.89	264.68	3.25	230.16	2.48	149.60
Glycine	4.48	198.94	5.69	262.75	4.29	168.90
Alanine	8.67	458.66	6.59	360.69	8.59	400.77
Tyrosine	0.84	108.72	1.46	163.08	1.48	135.90
Cystine	Trace	Trace	Trace	Trace	Trace	Trace

Source: Noda Institute for Scientific Research, Noda-Shi, Chiba-Ken, Japan.

seems quite obvious that these organisms would grow, possibly symbiotically, during this slow fermentation period. For the next 4–5 months, the temperature is held at about 15°C after which the moromi is filtered to yield the raw soy sauce and a residue said to be used for animal feed. The sauce is heated to 80–85°C, filtered, and bottled. The finished soy sauce is high in soluble proteins, peptides, and amino acids and has a dark brown color and a pleasing aroma (Tables 10.2 and 10.3). The presence of alcohol and both volatile and nonvolatile organic acids as well as amino acids, may result in the formation of esters, but flavor has also been related to other aromatic constituents. Yokotsuka (1960) considered the ratio of amino nitrogen to total nitrogen an index of quality. He also discussed the great changes that have occurred in the industry during the past 60 yr. There are numerous references in Japanese literature in regard to the preparation and properties of soy sauce. Flavor is dependent upon salt, amino acids, organic acids, various volatile aromatic substances, and even substances such as vanillin, vanillic acid, and ferulic acid.

This is obviously a fermentation far more complex than described. Since ancient times the Japanese knew the techniques of enzymatically hydrolyzing certain protein foods to make them more appealing. The Orientals have enriched the flavor of fish and meat as well as vegetable materials by fermenting in high salt concentration brines. Such foods, formerly called soy, are believed to be the antecedents of the present day soy sauce as well as some of the other foods. In soy sauce the mold *Aspergillus soyae* is the source of the proteolytic enzymes in fish soys to be discussed later. The enzymes natural to the product are believed to furnish the digesting enzymes. The first soy sauce, called soy, sho, or mesho, supposedly resembled the present day miso or hard mash or the moromi and was consumed as such. Obviously the products prepared in various homes and semiindustrial plants varied and were prepared by variations of the general procedure. In some of the preparations, coloring matter, condiments, and /or preservatives were added. They include caramel, licorice, maltose, millet jelly, cloves, nutmeg, pepper, cinnamon, and other spices. A red pepper sauce is made by addition of red pepper to soy sauce. Wheat is an expensive cereal in the Orient and in many areas it is not available; therefore, various substitutes are used in part or in full in place of wheat. Other cereals and sorghum are common. The quality of those soy sauces is generally inferior. Tamari is a soy sauce made in China in which the proportion of soybeans is higher than regular soy sauce. Soy, shoyu, shoyon, sho-yu, soja japanais are all names for soy sauce. Ch'au yan is a Chinese name and toyo a Philippine name for soy sauce. In Indonesia soy sauce is made from black soybeans; for this the sauce is concentrated by slow boiling to a thick syrup that is sold as ketjap. Catsup (ketchup) supposedly originated in the Orient.

NATTO

Natto is prepared from soybeans by fermentation with *Bacillus subtilis* (*Bacillus natto*). It has a strong persistent flavor and is used as a side dish with rice. The soybeans are soaked in water 12–20 hr during which time they double their weight. They are cooked until fairly soft then inoculated with *Bacillus subtilis*. The beans are fermented in paper-thin sheets of wood at a temperature of 40°–43°C for 18–20 hr. As heat develops, aeration is essential to prevent overheating. The beans become covered with a viscous, stringy polysaccharide polymer and have a strong odor of ammonia.

KOJI, RAGI, AND SIMILAR INOCULA

The conversion of starches to yeast-fermentable carbohydrates may be accomplished in several ways. Malting is the common practice in brewing with cereal grains. The use of a variety of molds is common practice in preparing many of the foods prepared in the Orient. Koji, the best known term, is a general term for the molded masses of cereal or soybeans. The molded materials serve as sources of enzymes and in some cases as an inoculum. There are a number of types of koji, depending upon their use, but *Aspergillus oryzae* is the mold generally used. A specific koji is prepared for each type of product in order to produce the proper mixture of amolytic, proteolytic, and lipolytic enzymes. Tane koji is the inoculum used to start any type of koji. Raggi and Java yeast are used in Indonesia and consist of rice flour containing fungi, yeast, and bacteria. Certain strains of *Mucor, Rhizopus,* and other molds have been isolated from such preparations.

Hesseltine (1965) lists several other names for products of similar nature. Atsumandie and saraimandie are preparations fermented by *Mucors*. Bakhar, also known as murcha, ranu, and u-t-rat, also is fermented by a *Mucor* and is used in the preparation of the rice beer, pachwai, in northern India. Kyoku-shi, chiu-niang, and chou are Chinese terms for koji. Pek-khato is a Chinese term for ragi. Sherokoji and chiu-chu refer to Chinese yeasts and levain of Kahasia and levain of Sikkim are similar. Yeast is used in a general sense, and in some cases the preparation is mixed with molds and undoubtedly bacteria. Shero-koji contains a *Rhizopus* mold and yeasts; levain of Sikkim a *Mucor*. Men, meen, and mien are terms applied to Chinese yeast preparations. Men is used in Annam to produce the alcoholic drink, ru'o'on. Tea fungus preparations, kombucha, hongo, and medusen tee, and tee fungus are possibly related. Whether or not they are related to the Chinese red rice preparation angkak is questionable. Ang-kak, angkhak, ang-quac, anka, ankak, aga-koji or akakoji, beni-koji or benikoji are red rice preparations. Ang-kak is made by fermenting rice with certain strains of *Monascus purpureus* and is used in the Orient for coloring various foods. Sekihan is the Japanese term for ang-kak. These materials and/or inocula form the basis of many of the Oriental foods, only a few of which are well known in the United States.

MISO

Miso is a fermented food prepared in Japan, China, Taiwan, the Philippines, Indonesia, and undoubtedly other countries in the Orient. In Japan, it is second only to soy sauce in quantity produced. Miso might be compared with moromi, the mash from which soy sauce is prepared. Miso has the consistency and color of peanut butter. Hesseltine (1965) described the methods used in preparing miso. It is used as a soup and also as a flavoring substance for other foods. It is essentially a fermented blend of rice, soybeans, sometimes barley, and salt. A two-stage fermentation is used: the first, an aerobic fermentation is carried out by strains of *Aspergillus oryzae;* the second, an anaerobic fermentation is carried on by *Saccharomyces rouxii.*

To prepare this food, washed, soaked, and steamed rice is inoculated with a mixture of strains of *Aspergillus oryzae* and allowed to ferment 48 hr at 40°C in shallow pans. The rice must be moist enough to allow mold growth but dry enough to prevent bacterial growth. Strains of *A. oryzae* are used that will yield a proper amylolytic, peptolytic, and proteolytic enzyme action. Mycelial growth is arrested before spore formation occurs. This is the koji. At the same time, soybeans are washed, soaked, steamed, and cooled. These are blended with salt in the proportion of 4 parts molded rice or koji, 10.4 parts soybean, 2 parts salt, and 1 part old miso and water. This blend is fermented aerobically for 7 days at 28°C and then 2 months or longer at 35°C. It is then placed in a vat where the second fermentation, essentially anaerobic, continues. Since this is anaerobic, the mold mycelia die and strains of the osmophilic yeast *Saccharomyces rouxii* continue the fermentation. The yeasts are introduced in the old miso added in the blend. The high content of salt should prevent growth of bacteria. Fermentation may continue for 2 months after which it is allowed to age for 2 weeks at room temperature and then ground to a paste for consumption. Hesseltine (1965) stresses that there are weaknesses in the process, chances of contamination by bacteria, yeasts, and molds other than the desirable types. The level of 12% salt as used in the blend does not necessarily exclude the growth of other microorganisms. There are many variations of the process to be found in Japanese literature.

Several types of miso are made; kome miso is made with soybeans and milled rice, mugi miso is made with soybeans and barley, and mame miso, a reddish-brown product, is made from soybeans alone. Chiang and taotjung are Chinese names for miso. Hamanatto, produced in Japan, is a product somewhat similar to miso in flavor. Soybeans are soaked in water, steamed, cooled, and then inoculated with a koji prepared from roasted wheat and barley. The mold is *Aspergillus oryzae.* After fermentation for 20 hr, now covered with green mold, it is dried in the sun until the moisture is reduced to about 12%. The beans are placed in wooden baskets with strips of ginger, covered with salt brine, and allowed to age 6–12 months. The final product is red and turns black upon drying. It contains about 11% salt.

SUFU OR CHINESE CHEESE

In the Orient, soybeans have been consumed in one form or another for centuries. Chinese cheese, soybean cheese, sufu or su-fu is prepared by a mold fermentation of cakes of finely ground precipitated soybeans. It is assumed that there are variations in the method of preparation. As described by Hesseltine (1965), the beans are washed, soaked overnight, ground to a fine flour, blended with water, and filtered. The residue is discarded, the milky liquid is cooked to destroy the bean-like flavor, and then precipitated by addition of calcium sulfate. The white curd is pressed, cut into cubes of 1 × 1 × 2 in. called tofu, sprayed with an acid-saline solution, inoculated with a mold culture, usually a *Mucor*, and incubated at 12°–20°C for 3–7 days. The moldy cubes, known as pehtzes, are placed in a solution of 12% sodium chloride and 10% ethyl alcohol and then aged up to 2 months. The mold secretes enzymes that break down the soybean proteins to peptides and amino acids. This cheese is quite stable due to its high salt and alcohol content, but it is salty and could be improved both in body and texture. Apparently there are many variations in the process.

Hang and Jackson (1967) described a method for preparing the cheese in which the soybean milk was fermented by a culture of *Streptococcus thermophilus* to produce the acidity necessary to precipitate the cheese. The milk was incubated at 41°C, a temperature which would preclude growth of many other microorganisms. The resulting cheese was reported to exhibit a better body and texture than the cheeses produced by either calcium sulfate or by acetic acid precipitation, and it did not have the salty flavor of cheese produced by the conventional Chinese method. Presumably the cheese could be fermented by mold to produce the flavor associated with Chinese mold-fermented cheese.

Meitanza is produced by the fermentation of the solid waste material from manufacture of Chinese cheese. The residue, after separation of the milk, is pressed, cut into cakes, and fermented for 10–15 days. The cakes become covered with the white mycelia of *Mucor meitanza*. The cakes are dried and cooked with vegetables or in oil. Apparently this product is somewhat similar to tempeh.

Judging by the number of names applied to Chinese cheese, it seems possible there are great variations. Chao is a Viet Nam preparation. Chee-fan is a sufu cut into smaller cubes and placed in wine to age. Tsue-fan is also aged in wine. Hon-fan and red sufu are soaked in soy sauce and red rice is used to impart a red color.

Fuyo, fu-ju, and tosufu are other names applied to types of sufu. Tahuri, tuhuli, tojo, and tokua are Philippine names applied to soybean curd. Tahuri was formerly imported from China. The curd is allowed to cure for several months during which time it acquires a brownish yellow color. Taokoan of the East Indies is like sufu and meju is a Korean product.

Somewhat similar products are prepared from soybeans. Tao-cho is prepared by roasting the beans mixed with rice flour and fermented with an *Aspergillus,* sun drying, and then dipping in a brine of arengo, sugar, and glutinous rice. Taotjo, and taotji, are prepared from boiled soybeans mixed with roasted meal or glutinous rice. This is fermented with *Aspergillus oryzae,* and later the beans are placed in brine to which palm sugar is added at intervals.

MONOSODIUM GLUTAMATE

Monosodium glutamate, Aji-no-moto sauce, MSG, and Accent are used to enhance the flavor of many foods. Dr. Ikeda of Tokyo Imperial University devised a method for isolating glutamic acid from soybean hydrolysates in 1908. He had recognized that the salt of glutamic acid was the principle flavoring constituent of soy sauce. (For a more detailed discussion see Chap. 12.)

An interesting sidelight is that for many years it has been known that when beef was hung at 45°–60°F to cure, growth of fungi belonging to the genera *Thamnidium* and *Chaetostylum* occurred. Two patents have been obtained recently for the use of the mold *Thamnidium elegans* for tenderizing beef. The fungus apparently prevents growth of bacteria on the surface of the meat and imparts a desirable flavor to the meat.

ARROZ FERMENTADO OF ECUADOR

A fermented rice, arroz fermentado, arroz amarillo, or arroz requemado, is prepared and consumed in the Ecuadorean Andes (van Veen *et al.* 1968B). The moist unhusked rice is spread out on a cement or cane floor, covered with a tarpaulin, and allowed to ferment by molds and bacteria. Fermentation begins in about 3 days, allowed to continue for 4–5 days when it is turned over with a shovel and then allowed to ferment another week or more. Van Veen *et al.* (1968B) isolated *Aspergillus flavus, Aspergillus candidus, Actinomyces sp., Absidia corymbifera,* and the spore-forming bacteria *Bacillus subtilis* and closely related *B. cereus* and *B. pumilus.* During fermentation, considerable heat is produced. The rice is a golden to cinnamon brown color.

TEMPEH

In general, the fermented foods prepared in the Southeast are more attractive to the consumer than the raw ingredients.

Tempeh, tempe, or tempeh kedelee is an important food in Indonesia, New Guinea, and Surinam and is quite easily prepared (Fig. 10.1). It is eaten in several forms. The thin mat may be sliced, dipped in salt brine, and fried in vegetable oil to yield a golden brown, crisp product somewhat resembling bacon. It may be eaten in soups, and it is also eaten with soy sauce added as a dressing. The peoples of Indonesia do not care for soybeans as such, but find tempeh appetizing and apparently easily digested. Indonesian tempeh is made by fermenting dehulled partially cooked soybean cotyledons with molds, generally of the genus *Rhizopus* (Steinkraus *et al.* 1960; Djien and Hesseltine 1961). Tempeh is a very important source of protein in the Javanese diet. Tan-chey, prepared in Thailand is similar.

Courtesy of Dr. K. H. Steinkraus

FIG. 10.1. STEPS IN THE PREPARATION OF THE PROTEIN FOOD "TEMPEH"

Left to right: Raw dry soybeans; soaked, peeled soybeans; raw tempeh-beans covered with mold; and (bottom) tempeh deep fried.

Dry soybeans are washed, soaked overnight at about 25°C, and in the morning the seedcoats are removed and the soaking water is discarded. The beans are boiled for ½ hr. After cooling they are inoculated with spores of the mold *Rhizopus oryzae,* placed in shallow trays, and incubated at about 30°C for 20–24 hr. By this time the beans are completely covered and bound together by the pure white mycelia of the mold. At this stage the product may be consumed in one form or another (Steinkraus *et al.* 1960). Hesseltine (1965) lists six species of the genus *Rhizopus* which may produce acceptable tempeh. Steinkraus *et al.* observed that the addition of 1.5% acid to the beans when inoculated reduced the possibility of bacterial contamination. Steinkraus (1962) observed that in Malaya the beans are acidified by a lactic acid fermentation during the soaking. Steinkraus *et al.* (1960) reported that the soluble solids rose from 13 to 21% during normal fermentation and would increase to 27.5% if fermentation were allowed to continue. Soluble nitrogen normally rose from 0.5 to 2.0%, but if fermentation was allowed to continue, a distinct odor of ammonia developed. Hesseltine (1965) observed elaboration of proteolytic enzymes in an acid environment. Wagenknecht *et al.* (1961) studied the lipase activity. Steinkraus *et al.* (1965) adapted the tempeh fermentation to semicommercial production.

Bongkrek or tempeh bongkrek is made with copra presscake alone or blended with soybeans (van Veen 1967; van Veen *et al.* 1968A.). Copra or coconut meat is digestively distressing but apparently when digested in part by molds, it becomes a good source of protein, carbohydrate, and lipids.

Ontjom or Lontjom (Fig. 10.2) is a similar type of food prepared in Indonesia and utilizing peanut presscake (van Veen *et al.* 1968A). The presscake is fermented by the mold, *Neurospora sitophila* in much the same manner as tempeh and the product is consumed in the same way. The typical yellow-orange color of this mold yields a product of similar color. The tempeh mold, *Rhizopus oligosporus*, is used to produce a white ontjom.

Courtesy of Dr. K. H. Steinkraus

FIG. 10.2. ONTJOM PREPARED BY FERMENTATION OF PEANUT PRESS
CAKE

Tapé ketella, peujeum, pēyēm, tapai, and tapaj are Indonesian foods prepared with cassava. The cleaned cassava is cut into pieces, inoculated with ragi, a mold inoculum, and fermented for 5–7 days. The cassava becomes soft, has a sweet alcoholic taste, and may be consumed at this stage or after frying in oil.

Tape ketan is a similar product made with cooked glutinuous rice, wrapped, and fermented in banana leaves. The rice becomes soft, sweet, and alcoholic.

Minchin or wheat gluten is prepared in North China by fermentation using any of a series of molds. The fresh wheat gluten is placed in a container where it is allowed to mold for 2–3 weeks at room temperature. It will then be overgrown with molds and bacteria. About 10% salt is then added and it is allowed to stand a few more weeks to age. It is cut into thin strips and eaten with other food.

Tao-si is prepared in the Philippines from soybeans covered with wheat flour and fermented with *Aspergillus oryzae*. After 2–3 days' incubation, the mass is placed in earthen jars with salt brine and the fermentation and aging is continued for about 2 months.

Dawadawa is a preparation made in West Africa by fermentation of the African locust bean. The fermented mass is pounded into a paste, made into small balls, and sun dried. Platt (1964) states that the product contains a high level of riboflavin and protein.

FISH SAUCES

The Orientals learned the technique of hydrolyzing protein foods centuries ago. They have digested and enriched the flavors of many fish and meat preparations in the presence of high concentrations of salt (van Veen 1953, 1965). Today, there are a series of such products, some in which the product is enzymatically hydrolyzed by the enzymes native to the food and others hydrolyzed by enzymes of microorganisms. The fish liquids and pastes are used primarily as condiments such as in the use of soy sauces. They are naturally high in soluble proteins and also high in salt. Nuoc-mam, shottsuru, and patis are among the best known.

Nuoc-mam is a fish sauce prepared in Southeast Asia. Small fish are washed and packed without evisceration with high concentrations of salt in large earthenware containers and allowed to ferment. The fermentation is apparently due to the enzymes native to the digestive system of the fish. It will proceed at a more rapid rate if the temperature is somewhat higher than normal and will proceed very slowly, if at all, if the fish are eviscerated. After several months, a clear amber liquid rises. This is called nuoc-mam. It may be decanted or separated by pressing from the sediment. The clear liquid is highly prized. The sediment is used by the less affluent. A similar product, nuoc-mam-ruoc, is made from shrimp. In Burma the preparation is called ngapi or nappi and in Thailand, kapi. A similar preparation is made in the Philippines; the clear amber liquid patis is highly prized while the sediment bagoong is frequently eaten with freshly cooked fish.

Shottsuru is the fermented fish of Japan sometimes referred to as fish soy. Its origin may predate soy sauce. Prahoc is the name applied in Cambodia to the fish paste similar to bagoong. The fish are cleaned, packed into baskets, pressed between stones to remove water, using 2 lb of salt to 20 lb of fish, sun dried for a day, and then packed tightly into baskets. The fish are later pounded to a paste, placed in earthen jars in the sun, and allowed to ripen further. Since this product contains much less salt than that used in the preparation of patis and bagoong, fermentation by yeasts and bacteria may more likely occur.

Trassi-ikan is prepared from fish and trassi-udang is prepared from shrimp in Sumatra. Padec, made with fish, salt, and rice bran, is made in Laos, phaak or mamchao is made with rice in Cambodia, and mamcu-sak is a salt fish treated with roasted rice in Cambodia, and mamtom is a shrimp paste. Jotkal is a Korean food made from oysters and presumably fish, containing 15% of salt and fermented at 37°C.

Fish pastes containing 10–12% salt and having good keeping quality and a true fish-paste flavor have been produced by Tan et al. (1967). Containing less salt than the traditional fish paste, they can be consumed in greater quantities.

Burong dalag is a blend of rice and the fish dalag prepared by fermentation in the Philippines. In this blend, a comparatively low concentration of salt is used. Orillo and Pederson (1968) have isolated strains of the lactic acid-producing species *Leuconostoc mesenteroides, Pediococcus cerevisiae, Lactobacillus plantarum,* and *Streptococcus faecalis* as well as strains of the genus *Micrococcus* (Tables 10.4, 10.5). Yeasts were isolated after the sixth day of fermentation, and molds appeared on improperly-covered blends. This fermentation is conducted with lower concentrations of salt than is customary for so many fish and fish-rice fermentations, and the results indicate the contrast in fermentation in low salt concentrations (Fig. 10.3) to those using the upper extreme of salt concentration in which only the most tolerant osmophilic microorganisms could be expected to survive.

TABLE 10.4

CHANGES IN BACTERIAL FLORA DURING FERMENTATION OF THE FIRST PREPARATION
OF BURONG DALAG

		Estimated Number of Each Species of Bacteria $\times$ 10^6				
Time in Days	Total Count $\times$ 10^6	Leuco- nostoc mesen- teroides	Streptococcus and Micrococcus Species	Pedio- coccus cerevisiae	Lacto- bacillus plantarum	Yeast Species
0	7	...	...	...	...	...
1	2000	1680	320	...	...	...
2	2000	880	1120	...	...	...
3	1080	376	188	376	140	...
4	1400	112	...	560	728	...
5	3000	...	...	360	2640	...
7	1700	...	...	340	1156	132

TABLE 10.5

DEVELOPMENT OF ACID AND CHANGES IN BACTERIAL FLORA DURING FERMENTATION OF THE
SECOND PREPARATION OF BURONG DALAG

		Total Acid as Lactic Acid %	Total Bacterial Count $\times$ 10^6	Estimated Number of Each Species of Bacteria $\times$ 10^6			
Time in Days	pH			Aerobic Species	Strepto- coccus faecalis.	Pedio- coccus cerevisiae	Lacto- bacillus plantarum
1	6.72	0.01	0.32	0.32	...	...	...
2	4.50	0.37	670	...	670	...	...
3	5.10	0.34	650	...	364	286	...
4	4.40	0.63	880	...	220	555	105
5	4.00	0.89	760	...	...	544	116
6	4.05	0.90	600	...	...	108	492
8	4.10	0.92	450	...	...	80	370
10	3.95	0.91	550	...	...	...	...

A similar preparation, burong isda, is prepared with rice and mudfish while another, burong hipon, is prepared with rice and shrimp.

Studies in the fields of microbiology and chemistry could do much to clarify the numerous problems and underlying alterations in these food products. Such studies would lead to a better understanding and to more standardized methods of preservation of some of the foods of the Orient. Much has been accomplished in the various laboratories, particularly in Japan. The reviews of van Veen (1953), Platt (1964), Smith (1961), and Hesseltine (1965) have done much to acquaint scientists of the Western world with the products used in the Orient and elsewhere.

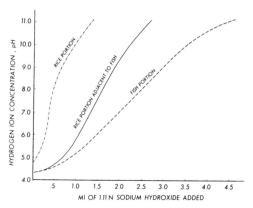

FIG. 10.3. TITRATION OF SEVERAL PORTIONS OF BURONG DALAG

A 10-gm sample, 8 days old. Acidity of rice portion at pH 7.0 is 0.4% as lactic acid and at pH 8.4 is 0.57% as lactic acid. Acidity of portion adjacent to fish at pH 7.0 is 1.4% as lactic acid and at pH 8.4 is 1.75% as lactic acid. Acidity of fish portion at pH 7.0 is 2.0% as lactic acid and at pH 8.4 is 2.72% as lactic acid.

TARO

Taro, a coarse plant of the arum family, also called eddo and dasheen, is native to many islands of the Pacific Ocean. Poi, prepared from the taro root, has occupied a unique place in the diet of the peoples of Hawaii. Poi is prepared by a lactic acid bacterial fermentation of a mash prepared from the taro tuber (Allen and Allen 1933) (Table 10.6). Although poi is considered disagreeable by some people because of its sourness, it is relished by the natives of the Hawaiian Islands. There are a number of names applied to taro and poi. Unfortunately, although taro tubers are high in carbohydrate, particularly starch, they are very low in protein. Although their caloric value is high, they are nutritionally poor.

CASSAVA

Cassava, a farinaceous plant of the spurge family, is an important source of food starches. It is consumed in great quantities in some societies. Unfortunately, like taro, it is low in protein. Cassava contains a glucoside

TABLE 10.6

MICROBIAL COUNTS ON SAMPLE OF POI DURING FERMENTATION

Age of Poi	Plate Counts × 10⁶ per Gram			Approximate Percentage	
	No. 2	No. 5	No. 10	Acid Producers	Yeast
Hours					
0	70	147	212	79	0
6	104	227	308	83	1
12	103	239	250	86	1
18	104	269	228	90	1
Days					
1	104	316	184	94	1
2	85	103	34	90	2
3	20	48	21	90	4
4	5	10	24	78	20
5	7	8	13	68	28
6	6	10	12	38	60

Adapted from Allen and Allen (1933).

that is readily hydrolyzed to yield hydrocyanic or prussic acid. In the process of heating such as occurs in bread making, some of this acid is volatilized. In some communities, a mash is prepared and then fermented to aid in removal of the hydrocyanic acid. In drying the fermented mash, most of the acid is dissipated. One such product is called gari in Nigeria. Tapioca and arrowroot are obtained from the Cassava root.

BIBLIOGRAPHY

ALLEN, O. N., and ALLEN, E. K. 1933. The manufacture of poi from taro in Hawaii with special emphasis upon its fermentation. Hawaii Agr. Expt. Sta. Bull. 70.

DJIEN, K. S., and HESSELTINE, C. W. 1961. Indonesian fermented foods. Soybean Dig. 22, No. 1, 14.

FUKUSHIMA, D. 1968. Personal communication. Central Research Institute, Kikkoman Shoyu Co., Noda-shi, Chiba-ken, Japan.

HANG, Y. D., and JACKSON, H. 1967. Preparation of soybean cheese using lactic starter organisms. I. General characteristics of the finished cheese. Food Technol. 21, No. 7, 95–96.

HESSELTINE, C. W. 1965. A millennium of fungi, food, and fermentation. Mycologia LVII, No. 2, 149–197.

LO, K. S. 1964. Pioneering soymilk in Southeast Asia. Soybean Dig. 24, No. 7, 18–20.

ORILLO, C. A., and PEDERSON, C. S. 1968. Lactic acid bacterial fermentation of burong dalag. Appl. Microbiol. 16, No. 11, 1669–1671.

PLATT, B. S. 1964. Biological ennoblement: improvement of the nutritive value of foods and dietary regiments by biological agencies. Food Technol. 18, No. 5, 68–76.

SHALLENBERGER, R. S., STEINKRAUS, K. H., and HAND, D. B. 1967. Changes in sucrose, raffinose, and stachyose during tempeh fermentation. Rept. 8th Dry Bean Res. Conf, Bellaire, Mich. Aug. 11–13. U.S. Dept. Agr.-ARS 74-41.

SMITH, A. R. 1961. Oriental methods of using soybeans as food. Agr. Res. Serv. 71, U.S. Dept. Agr.

STEINKRAUS, K. H. 1962. Personal communication. N.Y. S. Agr. Expt. Sta., Cornell Univ., Geneva, N.Y.

STEINKRAUS, K. H. et al. 1960. Studies on tempeh: An Indonesian fermented soybean food. Food Res. 25, 777–788.

STEINKRAUS, K. H., VAN BUREN, J. P., HACKLER, L. R., and HAND, D. B. 1965. A pilot-plant process for the production of dehydrated tempeh. Food Technol. *19*, 1 6.

TAN, T. H., VAN VEEN, A. G., GRAHAM, D. C. W., and STEINKRAUS, K. H. 1967. Studies on the manufacture of low salt fish paste. Philippine Agriculturist *LT*, No. 51, 626–635.

VAN VEEN, A. G. 1953. Fish preservation in Southeast Asia. Advan. Food Res. *4*, 209–232.

VAN VEEN, A. G. 1965. Fermented and dried seafood products in Southeast Asia. *In* Fish as Food, Vol. III, G. Borgstrom (Editor). Academic Press, New York.

VAN VEEN, A. G. 1967. The Bongkrek Toxins. *In* Biochemistry of Some Foodborne Microbial Toxins, R. I. Mateles and G. N. Wogam (Editors). M.I.T. Press, Cambridge, Mass.

VAN VEEN, A. G., GRAHAM, D. C. W., and STEINKRAUS, K. H. 1968A. Fermented peanut press cake. Cereal Sci. Today *13*, No. 3, 96–98.

VAN VEEN, A. G., GRAHAM, D. C. W., and STEINKRAUS, K. H. 1968B. Fermented rice, a food of Ecuador. Archivos Latinamericanos de nutrición *18*, No. 4, 363–373.

VAN VEEN, A. G., and STEINKRAUS, K. H. 1970. Nutritive value and wholesomeness of fermented foods. J. Agr. Food Chem. *18*, No. 4, 576–578.

WAGENKNECHT, A. V. *et al.* 1961. Changes in soybean lipids during tempeh fermentation. J. Food Sci. *26*, No. 4, 373–376.

YOKOTSUKA, TAMOTSU. 1960. Aroma and flavor of Japanese soy sauce. Advan. Food Res. *10*, 75–134.

CHAPTER 11

Coffee, Cacao, Vanilla, Tea, Citron, Ginger

COFFEE (COFFEA ARABICA)

The use of coffee in Abyssinia was recorded in the 15th Century. It was later introduced into Arabia. Advantage was taken of its physiological action in preventing drowsiness in connection with the prolonged religious services of the Mohammedans. Although coffee was known to be an intoxicating beverage and was frowned upon by the church, the coffee-drinking habit spread among the Arabians. In the early days, almost all of the world's supply came from the province of Yemen even though it was first found in Abyssinia. Brazil followed by Colombia, are the largest coffee-producing countries today. There are more than 30,000 plantations in Brazil (Fig. 11.1), and a plantation may consist of as many as 3 million trees. Colombia is reported to have more than 250 million trees.

Courtesy of ABC do Cafe, Departmento Nacional do Cafe, Brazil

FIG. 11.1. A COFFEE PLANTATION

Coffea arabica is an evergreen plant with large, smooth, and shiny oblong leaves and strikingly pure white jasmine flowers (Fig. 11.2) that grow in clusters and produce a rich fragrant coffee-like aroma. The fruit is a fleshy berry about the size of a small cherry (Fig. 11.3). As it ripens, it changes from a green to a cherry-red color. In Brazil it is harvested from May through October (Fig. 11.4). Each fruit contains two seeds or beans with the flat sides lying adjacent to each other contained in a thin, membranous, parchment-like endocarp embedded in a yellowish pulp. The fermentation of coffee involves digestion of this pulpy material to free the soft but tough greenish berries enclosed in the parchment coat and thin silver skin (Fig. 11.5).

247

FIG. 11.2. THE WHITE JASMINE-LIKE FLOWERS OF THE COFFEE PLANT

FIG. 11.3. CLOSE-UP OF COFFEE CHERRIES

The berries may be separated by a dry method or a more modern wet method. The dry method involves drying thin layers of fruit in the sun (Fig. 11.6), either as whole cherries or pulped cherries, and after drying, separating the berries from the parchment and remaining dried pulp. The fermentation that occurs during drying is manifested by a great increase of Gram-negative bacteria during the first 24 hr and an almost complete degradation of the pectic materials of the pulp in 1–2 days. In the wet method the cherries are placed in a tank of water from which they are later drawn off through pulping machines where the fleshy portion is pulped. They are then transferred to a second tank of water and later spread to dry. Fermentation occurs during the soaking period, leaving the coffee beans enclosed in the clean parchment and silver skin. When dry, the parchment and silver skin are broken mechanically and may be removed by winnowing. There are variations of these procedures.

    The fermentation of coffee cherries is spontaneous and involves a variety of microorganisms. Loew (1907) attributed the fermentation process to yeasts that formed alcohol that was later converted to acetic acid. Lilienfeld-Toal (1931) isolated five species of bacteria including strains of the coliform-

FIG. 11.4. HARVESTING COFFEE CHERRIES IN THE PHILIPPINES

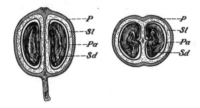

FIG. 11.5. STRUCTURE OF THE COFFEE CHERRY: PLUP (P), SLIMY LAYER
(SL), PARCHMENT ENVELOPE (PA), AND COFFEE BEAN WITH SILVER SKIN
(SD)

Courtesy of Dr. R. S. Breed

FIG. 11.6. COFFEE CHERRY CURING

like bacteria, the acetic acid bacteria, and aerobic spore-formers, in addition
to yeasts and molds. Vaughn *et al.* (1958) reported that during the first 12–24
hr of fermentation the pectic material in Brazilian coffee cherries was degraded by coliform-like bacteria, resembling species of the genera *Aerobacter* and

Escherichia. These organisms were abundant on the surface. The bacterial population increased to as high as 10^9 per gram. Pectinolytic species of the genus *Bacillus* and fungi of the genera *Aspergillus, Penicillium,* and *Fusarium* were found in depulped cherries. No yeasts were observed. de Menezes (1968) stated that he had obtained counts of 33 $\times$ 10^{12} organisms per gram.

Frank *et al.* (1965) observed that the predominant flora of fermenting Kona coffee cherries consisted of Gram-negative bacteria of the genera *Erwinia, Escherichia,* and *Paracolobactrum.* They observed that in pure culture study only the species *Erwinia dissolvens* was capable of decomposing the cherry mucilage layer and that 44 of 168 strains isolated were strains of this species. They observed that demucilaging could be completed in less than 5 hr when the pH was held at about 5.3. During continued fermentation, lactic acid bacteria, particularly the species *Leuconostoc mesenteroides,* developed in the liquid. Some lactic acid bacteria are capable of fermenting glucuronic acids obtained in hydrolysis of pectins. The coffee studies by Pederson and Breed (1946) in which lactic acid bacteria of the genera *Leuconostoc* and *Lactobacillus* were isolated, were conducted on samples shipped in closed vials from Mexico and Colombia (Table 11.1).

TABLE 11.1

GROUPING OF MICROORGANISMS ISOLATED FROM COFFEE CHERRIES AND BEANS

Cultures	Isolated from	No Growth	Yeasts	Mold	Contaminated Yeast and Bacteria	Microaerophilic Gram-positive Cocci and Rods	Aerobic Gram-positive Cocci	Aerobic Rods
1–40	Original coffee cherries	5	15	1	5	12	2	
41–51	Coffee beans— 1 hr	1				5	2	2
52–63	Coffee beans— 2½ hr	2				6	1	2
64–77	Coffee beans— 4 hr	2				13		3
78–93	Coffee beans— 24 hr	1				12		3

Source: Pederson and Breed (1946).

Mucilage removal can be accomplished by any one of several procedures (Frank and Dela Cruz 1964). Pereira (1957) reported that demucilaging resulted from the enzymatic activities originating in the cherry, but most investigators feel that bacteria are responsible. Frank and Dela Cruz (1964) and Frank *et al.* (1965) stated that Stern (1946), Vaughn *et al.* (1958), Coste (1959), and Carbonell and Vilanova (1952) concluded that bacteria were responsible for demucilaging. Johnston and Foote (1951) discussed the use of commercial enzyme preparations for dissolving the mucilage. They noted that digestion may be completed rapidly at pH 3.5–4.0 and further stated that unfavorable fermentations frequently occur in which putrefactive bacteria are numerous. It may be questioned whether the value of the natural lactic fermentation lies in its production of an acidity unfavorable to the growth of putrefactive bacteria. Frank *et al.* (1965) stated that unsatisfactory demucilaging can result from insufficient as well as excessive fermentation. Underfermentation interferes with the normal drying process, while over-fermentation results in adverse changes that affect the flavor and aroma.

CACAO *(THEOBROMA CACAO)*

The contribution of cocoa to the world's foodstuffs was made by the American Indians. Various conjectures have placed the origin of the cocoa tree in the Amazon Basin of Brazil, the Orinoco Basin in Venezuela, and in Central America. The cacao or cocoa tree is indigenous to the American tropics. Cacao cultivation was well established by the Indians long before the New World was discovered by Europeans. The Aztecs and Mayas cultivated the trees. Columbus first saw a cargo of cocoa beans in 1502. The Aztec cocoa beverage was a mixture of cocoa beans and corn and quite unlike the drink we know as cocoa today. Its bitterness made it unpalatable to the

FIG. 11.7. CACAO FRUIT ILLUSTRATING ITS GROWTH ON THE TRUNK OF THE TREE

Spanish. The Spanish eventually learned to add sugar, vanilla, and spices, and the beverage became a favorite in Spain. Although unknown in England and other European countries until 1579, chocolate became a favorite and fashionable beverage thereafter. Some time later, solid chocolate was prepared.

Cocoa is prepared from the seeds of the cacao tree. The small pink flowers and later the pods that develop are borne directly on the trunk (Fig. 11.7) or main branches of the tree rather than on the smaller branches as is the case of most tree products. The pods, when mature, are cut from the tree and cut open so that the 30 or 40 beans of each pod enmeshed in a white sweet pulp may be scooped out (Fig. 11.8). The curing of cacao beans by fermentation is a long-established practice and serves not only to remove the pulp but also is believed to produce desirable changes in aroma, flavor, and color (Knapp 1937). It was undoubtedly discovered by accident that the beans dry more readily if the pulp is removed. The beans with pulp are placed in suitable containers for fermentation (Fig. 11.9). The temperature may rise to 90°–115°F within a day or two but it should not be allowed to exceed 120°F. Fermentation may extend from 2 to 9 days before all of the fermented dissolved pulp drains away. Chemical changes occur during this period. The pulp contains about 80% water and 10% reducing sugars; the quantity of pulp present will govern the time necessary for fermentation. Humphries (1939) observed that a loss of dry matter occurred with an apparent, but not actual, increase in fat.

The typical yeast-like odor sometimes accompanied by an odor of acetic acid indicates a yeast followed by an acetic acid fermentation. Apparently, the fermentation depends to a considerable extent upon how tightly the

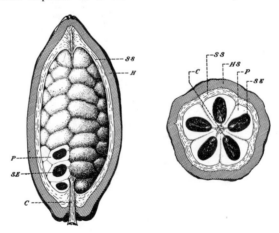

FIG. 11.8. STRUCTURE OF CACAO FRUIT: SEED WITH ENVELOPE (SE), PLACENTA (C), PULPY OR SLIME TISSUE (P), SOFT INNER LAYER OF FRUIT SHELL (SS), AND HARD OUTER SHELL OF FRUIT (HS)

FIG. 11.9. FERMENTATION OF COCOA BEANS

The anaerobic conditions in such a deep vat would favor a lactic acid
bacterial fermentation.

pulpy beans are packed. Well-aerated beans would favor fermentation by
yeasts and acetic acid bacteria while tightly-packed beans would favor a
lactic acid fermentation. According to Knapp (1937), yeast counts are high.
Species of not only *Saccharomyces ellipsoideus* and *S. apiculata* but also of
lactic acid, butyric acid, and acetic acid bacteria and several species of molds
may be present. Knapp stated that Lilienfeld-Toal found lactic acid and
acetic acid bacteria. Hoynak *et al.* (1941) were in general agreement and
note that apparently the environmental conditions have a marked influence
on the fermentation. During fermentation the pulp changes from a white to a
brown color and finally dissolves and runs off; the bean becomes more plump
and the shell becomes brown and crisp. The germ changes from dark brown
to brown to purple-brown, loses its power to germinate, and changes in physical structure. After drying, the crisp shell can be removed readily.

   Humphries (1939) has demonstrated decreases in tannins, theobromine,
and caffeine as well as reducing sugars and an increase in acids during fermentation. The tannins are thought to be altered by the natural enzymes
present in the bean, with the production of the desirable brown color.

   Forsyth and Quesnel (1963) reviewed the studies by Roilefsen and
Geisberger in Java and by Rombouts in Trinidad on the microbiological
changes in fermenting cacao. The seeds in healthy pods are sterile, but during
extraction and transport to sweat boxes, they become contaminated with a
variety of microorganisms. The changing environment determines the sequence that consists of yeasts, lactic acid bacteria, acetic acid bacteria, and
spore-forming bacteria. The yeasts, 15 or more species, and the acetic acid
bacteria play the major role. The lactic acid bacteria, usually heterofermen-

tative and related to *Lactobacillus fermenti,* play a minor role. The acetic acid bacteria were identified as strains of *Acetobacter rancens, A. aceti, A. oxydans,* and *A. melanogenus.* Forsyth and Quesnel state that the chocolate precursors are formed immediately after death of the seed. The death of the seed is caused by the acid and alcohol and the heat produced by the microbial fermentation. The chemical changes that occur will be effected by the fermentation and environment.

VANILLA

The Spaniards observed the Aztec emperor, Montezuma, drinking a chocolate beverage flavored with vanilla. In the 16th Century, Spain imported vanilla beans as well as cocoa beans.

Vanilla comes from the fermented and dried pods of several species of orchids belonging to the genus *Vanilla.* It thrives best in a hot, moist tropical climate. This perennial climbing orchid is the only orchid with fruits that are used for foods. It is one of the few aromatic vegetable products that cannot be used in its fresh state. The great bulk comes from the species *Vanilla planifolia* and *Vanilla fragrans.* Major producing countries are Madagascar, Mexico, the West Indies, Indonesia, and Oceania. About two million pounds are produced annually. The mature fruit is a pod from 6 to 10 in. long and about ½ in. in diameter, containing numerous small seeds. The best-cured vanilla pods are very dark chocolate-brown to nearly black and are covered with fine, needle-like crystals (Fig. 11.10). The major flavor substance is vanillin, an aldehyde of methyl-protocatechuic acid. Uncured pods are odorless. If left on the vine to ripen, they gradually turn from green to yellow and finally to dark brown or chocolate color. By this time the pods may have split open and the seeds may be exposed making them practically worthless. The characteristic flavor and aroma of the beans develop as a result of chemical changes that occur in the beans during curing (Fig. 11.11). The object of curing is to arrest the natural vegetative processes and hasten the changes that lead to the formation of the aromatic flavoring constituents.

Courtesy of L. George Rosskam, David Michael Co., Philadelphia

FIG. 11.10.   MEXICAN VANILLA BEANS

Courtesy of L. George Rosskam, David Michael Co., Philadelphia

FIG. 11.11.   CURING VANILLA BEANS

The beans are harvested when the green color begins to change to a dull green to yellow. During the curing of the fruits, the color changes from yellow-green to dark brown, and they lose 80% of their weight. During this curing, they develop their characteristic aroma and flavor. The beans are wilted either by spreading them out in the sun, by oven wilting at 60°C for 24–36 hr, by wilting in water at 60°–65°C for 2–3 min, or by such methods as scratching or freezing. The beans are allowed to sweat until they become flexible then dried to a suitable dryness after which they are conditioned. Sweating is conducted in sweating boxes or in blankets for about 24 hr during which a dark brown or chocolate color develops. The sweating process may be repeated several times or until a proper chocolate-brown color develops. There is considerable variation in practices, but the ultimate aim is to develop the color and typical aroma and flavor. After drying, the beans are tied into bundles for further conditioning. A number of such bundles are packed in wax papers and in tin-lined wooden boxes and conditioned for several months. Chemical changes occur that may be due entirely to the enzymes of the beans (Balls and Arana 1941; Arana 1943; Wild-Altamirano 1969). The writer observed a pronounced lactic acid bacterial fermentation involving strains of *Leuconostoc mesenteroides* and *Lactobacillus plantarum* during curing (Tables 11.2, 11.3). The chemical substance from which vanillin, the principle constituent, is formed occurs in the beans as a glucoside, glucovanillin. This glucoside is hydrolyzed during curing by the action of an enzyme beta-glucosidase to glucose and vanillin (Balls and Arana 1941; Arana 1943). Other glucosides, there are at least three, present in lesser amounts are also hydrolyzed to yield, with vanillin, the typical fragrance. The different varieties of vanilla beans yield slightly different flavor or aroma components. Well-cured vanilla beans are marked by the presence of numerous fine needle-like crystals on the surface of the bean. The vanillin content may

TABLE 11.2

CHANGES DURING PRELIMINARY CURING OF UNBLANCHED VANILLA BEANS

Time of Curing (Hr)	pH	Acidity as Lactic Acid %	Plate Count ($\times 10^6$/Gm)	Estimated Number of Each Type of Bacteria $\times 10^6$/Gm			
				Aerobic	Strepto-coccus Species	Leuco-nostoc mesen-teroides	Lacto-bacillus plan-tarum
0	5.71	0.25	0.66	0.66			
24	4.28	0.25	360		315	45	
32	3.98		390		310	80	
48	3.98	0.60	164		123	41	
72		0.81	54		29	11	14
120	3.47	1.15	390			390	
168	3.49	1.41	200			200	

Source: Pederson and Breed (1948).

TABLE 11.3

MICROBIOLOGICAL AND CHEMICAL CHANGES DURING PAN CURING OF BLANCHED VANILLA BEANS

Time of Curing (Hr)	pH	Acidity as Lactic Acid (%)	Plate Count ($\times 10^6$/Gm)	Estimated Number of Each Type of Bacteria $\times 10^6$/Gm			
				Strepto-coccus-Species	Leuconos-toc mesen-teroides	Lactobacil-lus plan-tarum	Lacto-bacil-lus brevis
24	4.42	0.29	1200	320	250	570	
33	4.06		800	160	640		
48	3.98	0.43	900	90	810		
72	3.93	0.53	95	29	66		
120	3.62	0.80	630	35	105	490	
168	3.61	1.08	628			597	31
240	3.62		600			570	

Source: Pederson and Breed (1948).

vary from 1–5%. Other constituents present include vanillyl alcohol, proto-catechuic aldehyde, heliotropin, and anisinic acid. Wild-Altamirano (1969) states that very little is known about the biochemical reactions taking place during ripening and curing of the fruit. He noted an increase in glucosidase and polyphenoloxidase with increased ripening of the fruit. Balls and Arana (1941) observed that the traditional methods of processing the beans to produce aroma causes a marked increase in rate of evolution of carbon dioxide and suggested that the curing might be an accelerated rate of oxidation. DeVries (1967) in his discussion of vanilla stated that a large and complex mixture of aldehydes, esters-aldehydes, and other chemical compounds are formed from the glucoside by the action of an enzyme during curing. Among the many studies on the curing of vanilla, little study has been given to the possible role of microorganisms. Many strains of lactic acid bacteria are capable of splitting glucosides and ferment the glucose. The heterofermenters produce carbon dioxide. The writer (Pederson 1948 observed the presence of considerable numbers of the species of *Leuconostoc mesenteroides* and *Lactobacillus plantarum* in curing vanilla beans but could not pursue the study.

GINGER

The underground stem or rhizome of *Zingiber officinale* is commonly used as a spice after drying. The rhizomes are sometimes collected in their raw state, washed, scraped, fermented, and then preserved in syrup as a sweet meat. Ginger preserved in a 10% brine for about 4 weeks yielded a superior product according to Cernuda (1948).

CITRON

A species of *Citrus* (*Citrus medica*), produces a large oblong fruit with a rough adhesive rind. The pulp is very acid and inedible and the seeds are bitter. The peel is very thick and when candied, has a very pleasing flavor and is highly esteemed as a confection. The citron peel is fermented to prepare it for candying. The fruit are immersed in a 3–3.5% salt brine solution shortly after harvesting. Ripening continues in the brine and there the yellow color tends to develop. A fermentation occurs during which the strong bitter principle is destroyed. The dense, opaque tissue expands considerably and becomes entirely translucent. There may be a vigorous growth of yeasts on the surface. This appears to be a typical vegetable fermentation brought about by lactic acid bacteria. The yeast growth may be essential in hydrolysis of the bitter principle.

TEA (*THEA SINENSIS*)

Tea was supposedly used in 2737 B.C. by the Emperor Shên-nung. The first authentic record traces to 350 A.D. in China and 593 A.D. in Japan. The word tea comes from the Amoy dialect *T'e*, pronounced tay. The leaves are obtained from the shrub called cha (Fig. 11.12).

The quality of tea depends upon the character of the plant and the methods dealing with the plucked leaf. The most important feature is regulation of the fermentation, an oxidation process to yield the well-known black tea. Green tea is not fermented but rather the leaves are heated shortly after picking in order to destroy the oxidizing enzymes.

The stages in manufacture of black tea are withering, rolling, fermentation, and drying. The plucked tea leaves are spread thinly on trays in the withering house where in 18–20 hr the leaves become soft and flaccid. The oxidizing enzyme apparently increases during this process. The leaves then

FIG. 11.12. TEA, A WHITE-FLOWERED EVERGREEN PLANT

pass through a rolling machine where the leaf cells are ruptured and the juice is expressed upon the surface of the leaves. Under the influence of the oxidizing enzymes, the constituents combine with oxygen. The leaves are transferred to darkened fermentation rooms where they are spread out in layers 1–2 in. thick. During subsequent fermentation, the leaves change color, the colorless tannin darkens, and the characteristic odor and flavor often develop. Although this appears to be essentially an enzymatic oxidation by enzymes natural to the tea, yeasts and bacteria are present on fermenting tea leaves. Improvement in flavor is associated with presence of yeasts. The fermentation reduces the astringency of the tea and develops the proper flavor and aroma. The principle constituents of tea are caffeine, tannin, and the essential oils that supposedly impart the stimulating effect, the body, the flavor, and aroma. When milk is added to tea it combines with tannins and the tea loses its astringency.

TABASCO SAUCE

The tabasco pepper is native to the humid soils of the state of Tabasco, bordering the Gulf of Mexico and Guatemala. Tabasco sauce is produced at Avery Island, Louisiana, from peppers originally obtained from Tabasco. The fiery red peppers are ground into a mash and steeped in their own juice in large oak barrels. The barrels are closed with perforated wooden covers, a thick layer of salt is spread on top which allows fermentation gases to escape without permitting air to enter. After a prolonged aging, vinegar is added to the mash and later the mash is filtered and bottled. Obviously, a fermentation occurs, but whether it is a mixed or pure culture fermentation by yeasts and lactic acid bacteria or whether it is a fermentation by the enzyme of the fruit has not been clarified.

BIBLIOGRAPHY

ARANA, F. E. 1943. Action of a β-glucosidase in the curing of vanilla. Food Res. 8, 345–351.

ARANA, F. E. 1945. Vanilla curing. Federal Expt. Sta. U.S. Dept. Agr. Circ. 25.

BALLS, A. K., and ARANA, F. E. 1941. The curing of vanilla. Ind. Eng. Chem. 33, 1075.

CARBONELL, R. J., and VILANOVA, T. M. 1952. A beneficial rapid and efficient method for liberation of coffee by means of the use of caustic soda. Bol. Técnico No. 13, El Centro Nacional de Agronomia, El Salvador. (Spanish)

CERNUDA, C. F. 1948. Ginger curing experiments. Rep. Federal Expt. Sta. Puerto Rico, U.S. Dept. Agr.

COSTE, R. 1959. The Coffee Tree and the Coffee in the World, Vol. 11. Les Cafes Part 1. Editions Larose, Paris. (French)

DE MENEZES, T. B. 1968. Personal communication. Geneva, N.Y.

DE VRIES, J. 1967. Vanilla. Food Technol. Australia 19, No. 12, 562–567.

FRANK, H. A., and DELA CRUZ, A. S. 1964. Incidental microflora in natural decomposition of mucilage layer in Kona coffee cherries. Food Sci. 29, No. 6, 850–853.

FRANK, H. A., LUM, N. A., and DELA CRUZ, A. S. 1965. Bacteria responsible for mucilage-layer decomposition in Kona coffee cherries. Appl. Microbiol. 13, No. 2, 201–207.

FORSYTH, W. G. C., and QUESNEL, W. C. 1963. The mechanism of cacao curing. Advan. Enzymol. 25, 457–492.

JACOBS, M. B. 1951. The Chemistry and Technology of Food and Food Products, 2nd Ed. Vol. 2. Interscience Publishers, New York.

HOYNAK, S., POLANSKY, F. S., and STONE, R. W. 1941. Microbiological studies of cacao fermentation. Food Res. 6, No. 5, 471–479.

HUMPHRIES, E. C. 1939. Some problems of cacao fermentation. Trop. Agr. 21, No. 9, 166–169.

JOHNSTON, W. R., and FOOTE, H. E. 1951. Development of a new process for curing coffee. Food Technol. 5, No. 11, 464–467.

KNAPP, A. W. 1937. Cacao Fermentation. John Bale, Sons and Curnow, London.

LILIENFELD-TOAL, O. A. VON 1931. Fermentation of Coffee. Estado de São Paulo, Directoria de Publicidade Agricola, São Paulo, Brazil. (Portuguese)

LOEW, O. 1907. The Fermentation of Coffee. In The Fermentation of Cacao. John Bale, Sons and Danielsson, London.

PEDERSON, C. S., and BREED, R. S. 1946. Fermentation of coffee. Food Res. 11, 99–106.

PEDERSON, C. S. and BREED, R. S. Feb. 1948. Unpublished. The lactic acid bacteria present in sweating of vanilla beans.

PEREIRA, J., JR. 1957. Rapid method for liberation of the mucilage of coffee pulp through the activity of appropriate enzymes. II. Rapid degumming of the coffee pulp in contrast to prolonged fermentation. Mucilage liberation. Arch. Inst. Biol. 24, 93. São Paulo. (Portuguese)

STERN, J. N. 1946. The fermentation of coffee. Proc. Trans. Texas Acad. Sci. 30, 103.

ROBERTS, E. A. H. 1942. Chemistry of tea fermentation. Advan. Enzymol. 11, 113.

VAUGHN, R. H. et al. 1958. Observations on the microbiology of the coffee fermentation in Brazil. Food Technol. 12, No. 4 suppl., 57.

WILD-ALTAMIRANO, C. 1969. Enzymic activity during growth of vanilla fruit. I. Proteinase, glucosidase, peroxidase, and polyphenoloxidase. J. Food Sci. 34, No. 3, 235–238.

CHAPTER 12

Organic Acids

In addition to the many foods altered by fermentation to improve their character, there are a number of products that are not consumed, per se, but added to foods in preparation. These include a number of organic acids, certain vitamins, and other additives.

Organic acids produced by fermentation processes have been used for centuries in one way or another. Vinegars produced from fruits and sour milk have been and are still used in food preparation to impart character and to preserve. When used, they impart not only the acidity due primarily to the acetic and lactic acids, but also flavors characteristic of the specific food sources and its products of fermentation. Some acids although present in only trace amounts are still very effective. The presence of certain components, e.g., proteins, lipids, vitamins, nucleic acids, and minerals, as well as carbohydrates, is necessary to promote fermentation. The various chemical entities may interact to impart distinctive flavors. The esters formed from alcohols and acids are examples. A trace of acids, such as caproic produced by the hydrolysis of lipids, may combine with alcohol to produce fruity flavors.

Fermentations are dependent upon the environment and may not necessarily remain static. It is well known that the fermentation of acetic acid continues to convert it completely to carbon dioxide and water. Certain substances such as the B vitamins and antibiotics may be synthesized by microorganisms. Others may accumulate in cell protoplasm. The great advances in microbial biochemistry during this century have included the isolation of many organic substances, the controlling of conditions for fermentation, and the accumulation by fermentation processes of a great variety of chemical entities. Many of these are being used in food products to improve their nutritive values, keeping qualities, and/or character.

VINEGAR (ACETIC ACID)

The manufacture of vinegar is of such ancient origin that the oldest literature records merely brief details of the methods and appliances used. Vinegar is mentioned in the Old Testament. Pliny relates that Cleopatra, to gain a wager, dissolved pearls in vinegar that she then drank. Hippocrates used vinegar as a medicine. It is still used by many individuals. An account of the methods used in making vinegar was given in 1616 by Oliver de Serres.

Vinegar is a word derived from the French vinaigre, meaning sour wine. Although it was originally applied to the product obtained by acetification of wine, it lost this original meaning long ago. It may be prepared from almost any watery substance that contains sugar and other nutrients to provide an alcoholic fermentation followed by an acetic fermentation. Malt vinegar, called "alegar," was prepared in England from ale or beer that had soured. Vinegar is primarily a dilute solution of acetic acid made by a fermentation process; however, it contains the unaltered soluble ingredients from which it is made as well as many fermentation products other than acetic acid.

The character of each vinegar is dependent upon the character of the substance fermented. Today, much of the vinegar used in the United States is prepared from fermented apple juice; however, wine vinegar is prepared in many of the wine-producing areas of the world. Pineapple vinegar is produced in fairly large quantities in some areas, and distinctive vinegars are produced from the juices of oranges, persimmons, plums, pears, peaches, dates, and other fruits. In fact, vinegar can be produced from numerous substances containing carbohydrates. Molasses, honey, and other sugar syrups are frequently used, and malt vinegars are popular in some countries, notably England. Distilled vinegars, used to a considerable extent in industry, are produced from grains. They lack the distinctive character of vinegars prepared from fruits or honey. Tea beverage vinegars are produced in Indonesia, Russia, and elsewhere by alcoholic fermentation followed by acetification.

Two microbiological processes are essential for the production of vinegar; the first is an alcoholic fermentation, the second, an oxidative fermentation of the alcohol to acetic acid. The first is a yeast fermentation and the second is a bacterial fermentation. Early man undoubtedly associated the clouding of wine or beers and the evolution of gas with the alcoholic fermentation, just as he associated the souring of wine or beer with the formation of a film or "mother" on the surface of his alcoholic beverage. He must have associated the first alteration with the absence of air and the second with the presence of air and established conditions so far as possible to promote these

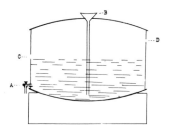

FIG. 12.1.   DESIGN OF A SIMPLE BARREL VINEGAR GENERATOR

Drawoff faucet (A), funnel for adding additional cider (B), 1- to 2-in.
hole for inlet and outlet of air (C and D).

changes. The practical conditions essential for vinegar production were known long before its scientific basis was established. It was observed centuries ago that barrelled wine or cider turned sour, particularly if the container was incompletely filled. Gradually, as it was realized that air was essential for this conversion, improvements were made that finally resulted in the basic processes now generally used: the Orleans or French method; the generator of Schüzenback or Boerhave process; and the recently developed submerged vinegar process. The Orleans or barrel process (Fig. 12.1) is the oldest and undoubtedly became established following the observation that when barrels of wine were partly full, the wine turned to vinegar much more rapidly than when the barrels were completely full. It was a simple additional step in the art to fill wine or cider ½–⅔ full into barrels tilted on their sides with air vents to provide air passage over the surface of the wine (Table 12.1). A spigot was placed at the bottom to withdraw vinegar when needed and an opening at the top was made so that fresh wine could be added. If the film or "mother" on the surface was not disturbed, acetification occurred more rapidly. This method has been and continues to be used in farm homes to produce vinegar. The method can be continuous if, as vinegar is withdrawn at intervals, more wine is added. When a barrel of stock is started in this manner, it may require several days, weeks, or months to obtain sufficient bacterial growth or "mother" to produce a rapid oxidation. Once established, conversion of wine or hard cider to vinegar takes place rapidly. The quality of vinegar produced by this process is considered superior by many. This may be true, since such vinegar is usually made in small barrels of 50 gal. capacity or less, and it naturally ages in the barrels before being withdrawn. Aging generally reduces the harsh flavor of freshly-made vinegar so that it becomes milder and more agreeable. The chemical changes that occur during aging are

TABLE 12.1

RATE OF GENERATION OF VINEGAR IN BARREL

| Time (Days) | Cider Added | | Acidity of Vinegar (%) | Vinegar Removed (Gal.) |
	Amount (Gal.)	Acidity (%)		
0	8	1.04	1.04	
7			1.23	
14			2.56	
21			4.94	3
21	4	1.00	3.30	
28			4.30	
35			5.35	5
35	5½	1.20	3.00	
42			4.10	
49			5.20	5
49	5	1.30	3.30	

similar to those that occur in wine. The best known chemical change is ester-ification of the ethyl alcohol and acetic acid to form the well-known, pleasing ester, ethyl acetate. Similar reactions occur between other acids and alcohol.

The Schüzenback, Boerhave, or generator process for rapid acetification of alcohol was introduced in Germany in 1832. This has become the estab-lished method for the production of commercial vinegars. The vinegar gener-ator is a large, cylindrical, tall tank designed to provide maximum surface exposure of vinegar stock to the air so that the acetic acid bacteria may oxi-dize the alcohol to acetic acid rapidly and efficiently. The generator consists of three parts. The middle part is filled with any type of material that will insure a large surface area; these materials include beechwood shavings, corn cobs, ceramic material, coke, or other materials. Above this is placed an apparatus to insure distribution of the vinegar stock over the generator ma-terial. The lower part serves as a chamber for collecting the vinegar. The simplest top is a cover with a number of holes through which the stock may flow, and the bottom consists of a similar arrangement to allow the vinegar to flow into the chamber below. More complex equipment is adapted to control the rate of flow of vinegar stock, introduction of air, and temperature. These modifications insure an even distribution of vinegar stock, the control of the rate of flow so that the bacteria can oxidize the alcohol efficiently, the introduction of sufficient air, and the provision for uniform temperature of 80°–85°F. Since this is an oxidation process, heat is generated; it is, there-fore, essential to keep the generator cool enough so that the microorganisms may act efficiently.

Aerated fermentations for the production of bakers' yeasts have been in use for a number of years. During the development of antibiotics on a large scale, the principle of aeration became a common practice. The principle of aeration was suggested by Haeseler in 1949 (Allgeier and Hildebrandt 1960). Hromatka and Ebner published a series of articles dealing with vinegar man-ufacture. Cohee and Steffen (1959) outlined the submerged method as used commercially.

Although there are many modifications of these processes, all are essen-tially similar to those described. The production of distilled vinegars from materials other than fruit products often requires the addition of certain nutrients for the growth of the bacteria, but wine and hard cider contain suf-ficient nutrient material. Needless to say, there are numerous patents de-signed to improve upon the general methods used.

Persoon, one of the first scientists to investigate the vinegar process, at-tributed the action to small living particles that he called *Mycodermi aceti.* This was disputed by Liebig, but the explanation was firmly established by Pasteur. The process is essentially an oxidation process carried out by the vinegar bacterium now known as *Acetobacter aceti.* Considerable variation occurs among the strains and species of the genus *Acetobacter.* Pure culture fermentations have been less successful than natural fermentation, and it is assumed that this may be due to the presence of variants of *Acetobacter*. Some strains and species of *Acetobacter* are capable of continuing the fermentation to carbon dioxide and water. Closely related species of the genus, such as *Acetobacter xylinum,* may be troublesome in a generator. *Acetobacter xylinum* often produces a thick, mat-like growth that may be difficult to remove from a generator. In an aerated system such as this, evaporation may result in yield losses. Continued oxidation of alcohol and acetic acid to carbon dioxide and water may reduce yields. Vinegar eels, tiny nematodes, vinegar flies, and vinegar mites or lice are frequently present in the aerobic surroundings. Sanitation, filtration of the vinegar, pasteurization, and subsequent elimination of air are essential for control of these pests. Blackening and excessive browning are other defects. The first is due to metallic contamination, notably by iron, while the second is the natural enzymatic browning that often occurs in the juices of fruits.

LACTIC ACID

The lactic acid produced in so many foods, particularly in milk and vegetables, has been used frequently as an acidifying agent in the preparation of other food products. Sour milks and creams are commonly used in baking, particularly in making cakes. They are also used in the preparation of meats and vegetables and for blending in soups. Lactic acid is used in certain infant feeding formulae, in confectioneries, sherbets, soft drinks, jams, and jellies. The salts of lactic acid are used in some baking powders.

Blondeau recognized the acid of fermented foods as lactic acid, and 30 yr later Lister isolated *Streptococcus lactis,* the species frequently responsible for acid production in milk. Avery produced lactic acid on a commercial scale 4 yr later. At the relatively low temperature of 30°C required for growth of this species, many other organisms are capable of growth and may contaminate the fermenting medium. During this period. Delbrück was investigating the lactic acid fermentation in distilleries and concluded that a fermentation at relatively high temperature favored the pure culture fermentation. The species *Lactobacillus delbrüeckii* was first described and named *Bacillus delbrüeckii* by Leichmann and later placed in the genus *Lactobacillus* by Beijerinck. This species is capable of active growth and fermentation at 45°–50°C, a temperature at which few other microorganisms can grow readily.

Lactic acid is now produced from mashes, molasses, sugar, and whey. The lactic acid formed in commercial fermentation is usually the racemic mixture of approximately equal amounts of the levorotatory and dextrorotatory forms. Either of the optical forms may be produced by selection of the proper cultures. The species of the genus *Streptococcus* when grown in pure culture, always form dextrorotatory acid while species of the genus *Leuco-*

nostoc always form levorotatory acid. Some species of homofermentative lactobacilli also produce levorotatory lactic acid. Racemization occurs readily in mixed fermentations.

Lactobacillus delbrüeckii is ordinarily used for the higher temperature fermentations, e.g., 45°–50°C in cereal mashes, while strains of *Lactobacillus bulgaricus* are more suitable for whey fermentation at the higher temperature. *Streptococcus lactis, Lactobacillus casei, Lactobacillus plantarum*, and related varieties are homofermentative with an optimum temperature of about 30°C. Heterofermentative species convert only about 45–50% of the sugar fermented to lactic acid. Certain spore-forming bacteria, particularly *Bacillus coagulans* var. *thermoacidurans*, produce large amounts of lactic acid. Lactic acid is also produced by species of various genera of yeasts and molds (Prescott and Dunn 1959). Prescott and Dunn present greater detail in regard to commercial production of lactic acid.

In general, the fermenting solution is neutralized by addition of calcium carbonate or calcium hydroxide. After fermentation is completed the calcium lactate solution is generally acidified with sulfuric acid, refined, and concentrated to the concentration and purity desired.

CITRIC ACID

Citric acid has become one of the most common acidulants used in the beverage and food industries. It is the predominant organic acid of the citrus fruits. The pure chemical has replaced lemon juice in many foods. It was first obtained in the solid state by Scheele in 1784.

Acid production from sugars by the species of molds is common, particularly for the genera *Aspergillus* and *Penicillium*. In general, any mold species grown upon a substrate containing fermentable carbohydrate will produce acid in the early stages of fermentation. The production of acids by a fungus was reported by Wehmer in 1893, and he used the generic name *Citromyces*. Later he reported the production of citric acid by two species of mold, *Penicillium luteum* and *Mucor piriformis*. Buchner and Wüstenfeld used a strain of *Citromyces citricus*. In 1917, Currie surveyed the possibilities with this mold and discarded it in favor of *Aspergillus niger*. Successful commercial development of citric acid production was accomplished by selection of strains of the genus *Aspergillus* for which the lag between total acid production and oxalic acid production was greatest. He found that oxalic acid production could be reduced to a minimum by using certain concentrations of sugar. Sufficient citric acid is now produced by these principles to fulfill industrial requirements.

GLUTAMIC ACID (MONOSODIUM GLUTAMATE)

Glutamic acid as monosodium glutamate enhances the natural flavor of many food substances. Monosodium glutamate, also known as ajinomoto,

Chinese seasoning, MSG, Accent, Vetsin, Zest, and other names, apparently was first recognized in China. Fermented soybean curd was observed to enhance flavor and zest to a limited and monotonous diet. Soybeans contain a very high amount of glutamic acid; in fact, 18% of the amino acid of soybean protein consists of this acid. Ikeda (1909) isolated the crystalline flavor compound and later a patent was granted for its production by chemical hydrolysis of vegetable protein. Considerable research followed and numerous patents have been granted. By chemical synthesis, the racemic form of glutamic acid is produced and it must be resolved into its two optically active isomers. This can be accomplished by use of a suitable enzyme preparation.

Considerable interest has been displayed in the production of glutamic acid and other amino acids by fermentation since by the microbiological production, the biologically active L-glutamic acid is produced. This is a distinct advantage. A great variety of microbial species have been shown to produce glutamic acid and they include species of the genera *Micrococcus, Brevibacterium, Bacillus, Pseudomonas, Escherichia, Serratia, Corynebacterium, Kluyveri, Cephalosporium,* and others. Soybean protein, wheat gluten, corn gluten, sweet potatoes, sugar beet molasses, molasses waste water, and other substances have been used in various studies (Prescott and Dunn 1959).

Two basic microbial methods have been developed: a 1-stage fermentation involving a single species; and a 2-stage process involving first the production of the intermediate α-ketoglutaric acid followed by conversion of this chemical to glutamic acid. These are discussed by Prescott and Dunn (1959) and Kinoshita (1959). Kinoshita states that to produce glutamic acid by a 1-stage fermentation, 2 contradictory conditions must be satisfied, e.g., the aerobic oxidation of sugar and the anaerobic reductive deamination of α-ketoglutarate to glutamic acid. He further states that *Micrococcus glutamicus,* a species that he had previously described, is an excellent glutamic acid producer. The necessity for careful control of biotin content in the medium is noted.

LYSINE

A number of foods, particularly cereal grains, are deficient in the amino acid, L-lysine. Kinoshita (1959) discussed the production of lysine by *Micrococcus glutamicus.* Prescott and Dunn (1959) stated that L-lysine is manufactured in the United States by a two-stage process in which diaminopimelic acid is first produced, followed by its decarboxylation to lysine. Nickerson and Brown (1965) have noted that several yeasts are able to produce high yields of lysine in the presence of suitable precursors, such as 5-formyl-2-ketovaleric acid.

OTHER ORGANIC ACIDS

Numerous organic acids other than those described are or may be produced by microorganisms under various environmental conditions. Several of these acids are present in fermented foods, ordinarily in very small quantities, and some are used in various nonfood industrial processes. The numerous acids of the fatty acid series, the acids of the citric acid cycle, the several sugar acids, keto acids, and others may be expected to be present in trace amounts in several fermented foods.

Propionic acid, produced with acetic acid and carbon dioxide in cheese by species of the genus *Propionibacterium,* contributes to the flavor of cheeses of the Swiss type. Although it may be produced from lactose, it is generally assumed that the associative action of lactic acid bacteria with the propionic acid bacteria results in conversion of the lactic to propionic acid. Ratios of propionic to acetic vary from about 5:1 to 3:1. Propionic acid could be produced readily if there were a greater demand. The sodium and calcium salts are used for their mycostatic properties in breads, cakes, fish, cheese, and other foods. Other acids of the fatty acids series are present in many fermented foods in trace amounts and impart distinctive flavors, particularly to cheeses.

Gluconic acid, produced by the oxidation of the aldehyde group of glucose, is used to a limited extent in the food and feed industries. First observed by Boutioux in 1878 in a fermentation by *Acetobacter* and believed to be lactic acid, it was later identified as gluconic acid. It is produced by *Aspergillus niger* in the citric acid fermentation and can be trapped by the use of calcium carbonate.

Guanylic and inosinic acid produced by *Micrococcus glutamicus* are used as food flavors.

Gluconic, fumaric, kojic, itaconic, and several other organic acids are produced by fermentation processes and are used in several industrial processes. Although food acids may be produced during the fermentation processes, they are not necessarily produced commercially by such fermentation processes. Tartaric acid, first isolated by Scheele from wines in 1769, may be produced by *Penicillium notatum, Aspergillus niger,* or *Aspergillus griseus.* Succinic acid is produced in yeast fermentations. Oxalic acid, used in the hydrolysis of starch to glucose, is produced by many molds. *N*-caproic and *N*-capric acid form esters to yield artificial fruit flavors. Although propionic acid, used as a mold inhibitor, is produced in small amounts in certain cheeses, the commercial product is prepared by chemical synthesis. Butyric acid, so common in well-ripened cheeses and in butter, may be produced by the fermentation of starch by *Bacillus subtilis.* Malic acid occurs in many fruits, and like itaconic, ketoglutaric, and many other acids that may be produced in fermentation, is not ordinarily added to foods even though it may be naturally present.

BIBLIOGRAPHY

ALLGEIER, R. J., and HILDEBRANDT, F. M. 1960. Newer developments in vinegar manufacture. Advan. Appl. Microbiol. 2, 163–182.

COHEE, R. F., and STEFFEN, G. 1959. Makes vinegar continuously. Food Eng. 31, 58–59.

FRAZIER, W. C. 1967. Food Microbiology, 2nd Edition. McGraw-Hill Book Co., New York.

IKEDA, K. 1909. On a new seasoning. J. Tokyo Chem. Soc. Tokyo Kagaku Kaishi 30, 820–826.

KINOSHITA, S. 1959. The production of amino acids by fermentation processes. Advan. Appl. Microbiol. 1, 201–214.

NICKERSON, W. J., AND BROWN, R. C. 1965. Products of yeast and yeast-like fungi. Advan. Appl. Microbiol. 7, 225–268.

PRESCOTT, S. C., and DUNN, C. G. 1959. Industrial Microbiology, 3rd Edition. McGraw-Hill Book Co., New York.

VAUGHN, R. H. 1954. Acetic acid vinegar. In Industrial Fermentations, Vol. 1, L. A. Underkofler and R. J. Hickey (Editors). Chemical Publishing Co. Div., Tudor Publishing Co., New York.

WÜSTENFELD, H. 1930. A textbook on vinegar manufacture. Paul Parey, Berlin. (German)

CHAPTER 13

Microbial Products Consumed in Foods

It is seldom realized that when we eat many of the fermented foods, we are also actually consuming with the basic food either the microbial cells themselves or their autolytic products. Even the mold so obvious in Roquefort type cheese is considered an integral part of the cheese and many people have developed a strong desire for it. In many mold-fermented foods, such as tempeh consumed by the peoples of the Orient, the molds remain as a part of the basic food. Westerners have not developed an appreciation for their flavor. The yeasts and bacteria in bread and the bacteria of sausage, sour milk, cheese, and other foods are consumed in either their living or dead state or as their autolytic products. In fact, yeast is sometimes added to bread for flavor and sour milk is consumed to introduce lactic acid bacteria in the intestinal tract. It is unfortunate that consumers have become so addicted to clarity that they have not appreciated the character lost by discarding the brines in which dill pickles and sauerkraut are fermented. These brines are consumed in other countries.

Food components required by microorganisms for their growth must be supplied in the food to be fermented; the organisms, however, are able to synthesize from the ingredients of the food certain important nutrients required by humans. These may accumulate in the protoplasm of the microbial cell or may be released into the food when the cell is autolyzed. They remain in the fermented food and improve its quality. In liquid foods, however, some cells may be autolyzed while others in suspension settle with the insolubles to form the sediment or lees. Cellular substance has a slightly bitter flavor which detracts from the flavor of some beverages. The cells from some such products are gathered and used as food or feed supplements.

The microbial cells constitute only a small portion of the total mass of a fermented food even when millions of organisms are present at some stage in fermentation. Nevertheless, nutritionally and flavorwise they are important. Dunn (1952) lists among nitrogen-containing compounds in yeasts the following: proteins, amino acids, the purines adenine and guanine, the pyrimidines cytosine and uracil, and the vitamins biotin, folic acid, nicotinic acid, para-aminobenzoic acid, pyridoxine, riboflavin, and thiamine. He includes adenosine triphosphate, adenylic acid, cephalin, and coenzymes I and II, and co-carboxylase, cytochrome, glucosamine, glutothione, and lecithin. All are important nutritionally even though some may exist in trace amounts. Carbohydrates, lipids, and minerals are also contained in the cell protoplasm. The protein content of some strains of yeast species may vary from 30 to 50%

of the dry weight. Dunn also presents the amino acid content of some strains of yeast. Molds and bacteria of the various species are equally rich in nutrients.

The need for protein- and vitamin-rich foods has been emphasized in many scientific and general papers and reviews. Among the more recent reviews are those by Lawrie (1970), Milner (1969), Pirie (1967), and Peterson and Tressler (1963). The fermented food industries provide an area in which microorganisms may supply some of the needs for protein-rich and vitamin-rich foods. During World War II, the Germans produced thousands of tons of yeasts and molds in quantity to supplement protein- and vitamin-deficient diets. Commercial production of food yeasts has been established in many countries.

VITAMINS

Vitamins synthesized or absorbed from the medium during the growth of microorganisms may be retained in the cells, may be degraded during metabolism, or may be excreted into the medium. They may thereby enrich the food in which they are produced or the microorganisms, such as masses of yeasts, may be used for food or pharmaceutical purposes. The value of yeasts fed to livestock was recognized during the latter part of the 19th Century. Although yeasts as a by-product of brewing and distillery operations are the principle sources, it is recognized that vitamins are synthesized also by bacteria.

YEASTS

Commercial brewer's or bakers' yeasts are rich sources of all of the B vitamins except B_{12}. Thiamine, riboflavin, pantothenic acid, niacin, pyridoxine, folic acid, biotin, para-aminobenzoic acid, inositol, and choline may be synthesized or accumulated in cells when substances such as grain extracts, malt sprouts, molasses, and other substances are fermented. The yeasts are capable of synthesis and absorption to a high degree. The degree of absorption and synthesis varies with strains and can be increased by selection of media. *Saccharomyces* strains are used for brewing and baking. Thiamine, niacin, and biotin are readily absorbed from the media by yeasts while pyridoxine and inositol are taken up to a lesser extent. Riboflavin occurs in fairly uniform amounts in yeasts because any excess is released into the medium. Yeasts are limited in their ability to synthesize vitamins but may accomplish this if certain precursors are present.

Yeasts for baking purposes may be obtained in several forms but most commonly as yeast cakes, as dry cakes or pellets. The white to dull yellow, soft yeast cakes consist of viable yeasts with a starchy absorbent binder and have a slightly sour yeasty odor. Although the dried yeasts may be obtained as a by-product of the brewing of beer, they can be produced also on any suitable medium. Crude molasses is an inexpensive source of sugar. Yeasts are used in baking, brewing, wine making, and other fermentation processes. Yeasts are also used for their nutritive value and as flavoring substances, particularly in bread.

Yeasts are one of the richest sources of the B vitamins and have provided the source for the isolation and biochemical investigation of great numbers of different enzymes.

Ordinary brewer's yeast is also an excellent source of proteins and other nitrogen-containing substances and makes a nutritious supplement to the diet of people living in areas where there is a shortage of proteins. Nutritional yeast in contrast to the yeast from breweries is propagated essentially for food. Yeast cultures are selected for their cultural and biochemical stability, their adaptability to environment, and their ability to reproduce themselves in high yield. Strains of the species *Candida utilis* are commonly used commercially since they have the ability to assimilate from a large number of nitrogen- and carbon-containing sources the nutrients required for growth. This species will grow rapidly, and it possesses high nutritive value, agreeable flavor, and good appearance. This species is often referred to as torula yeast and is sometimes classified as *Torulopsis utilis.*

A number of other species have been used experimentally and commercially (Prescott and Dunn 1959; Frazier 1967). These include other species of the genus *Candida* and species of the genera *Hansenula, Trichosporon, Mycotorula, Torulopsis, Saccharomyces,* and others. Some species such as *Candida utilis* are able to utilize some of the inexpensive waste products, such as sulfite liquor from paper pulp mills, wood hydrolysate, wood sugar stillage, as well as crude molasses, and produce from them yeast containing from 30 to 60% protein on a dry weight basis. Such yeasts find a great outlet in animal feeds. The complex flow sheet for production of food yeast from sulfite liquors is illustrated in the text by Prescott and Dunn (1959).

Certain species of molds and bacteria are also able to accumulate or assimilate proteins and vitamins. *Geotrichum candidum* has been grown for food use.

LIPIDS

During World War I, considerable research was conducted in Germany by Lindner and associates on the production of fat from yeast. Nägeli and Loew had demonstrated in 1878 that carbohydrates could be transformed into fats by yeasts. Lipid production by microorganisms is normal but high yields can be expected only under optimum conditions. Lipids are synthesized by microorganisms and they have been prepared commercially in times of great need. During World War I, the Germans used strains of *Geotrichum candidum* and *Trichosporon pullulans* for production of lipids. Lipid production has been demonstrated by species of the genera *Candida, Saccharomyces, Lipomyces, Rhodotorula,* and *Torulopsis* and by molds of the genera *Aspergillus, Penicillium, Geotrichum, Mucor,* and *Fusarium.*

The glycerides of fatty acids, phospholipids, and sterols are normal constituents of the protoplasm of microorganisms. The presence of more than 20 saturated and unsaturated fatty acids was noted by Vorbeck *et al.* (1963) in bacterial fermented foods in at least trace amounts, the C16 and C18 existing in greatest quantity.

Rhodotorula species are of interest because it has been shown that they produce high yields of fats up to 63% on a dry basis, and they can be grown in submerged cultures (Prescott and Dunn 1959). *Lipomyces* species also produce considerable amounts of fat. Among the sterols, ergosterol is most important although other sterols are formed. Ergosterol is important because it may be converted to vitamin D_2 by irradiation with ultraviolet rays. *Saccharomyces* species are the best producers of ergosterol among the yeasts while species of the mold genera *Penicillium, Aspergillus,* and *Neurospora* produce appreciable quantities. Most microbial ergosterol is synthesized commercially by yeasts.

ASCORBIC ACID

L -Ascorbic acid, vitamin C, is prepared from sorbose by first preparing an acetone derivative of sorbose which is then oxidized. Sorbose is produced by fermentation of the polyhydric alcohol, sorbital, by any one of several species of the genus *Acetobacter. Acetobacter suboxydans* is generally used. The common vinegar bacteria *Acetobacter aceti* is inefficient in this conversion.

Strains of *Pseudomonas fluorescens* produces 2-ketogluconic acid from glucose or gluconic acid. This is used in the production of D -araboascorbic acid which is used as an antioxidant in prevention of rancidity in oils and fats. This can be used as a substitute for L -ascorbic acid for such purposes, but it has much less potency as an antiscorbutic.

BUTANEDIOL

The natural occurrence of butanediol found in a food product, wine, was that of Henninger in 1882. The first oxidation product of 2,3-butanediol is acetylmethylcarbonol or acetoin. In 1906 Harden and Walpole proved that *Aerobacter aerogenes* produced appreciable quantities of 2,3-butanediol. Acetoin, 2,3-butanediol, and diacetyl occur naturally in many fermented food products such as butter, bread, cheese, coffee, vinegar, wines, and beers. Diacetyl has a strong odor and it is considered to be the chief component of butter aroma as well as a component of the aroma of these other foods. 2,3-Butanediol is produced by yeasts and a number of species of bacteria.

NATA

Several food-like products are produced by fermentation which have little nutritive value. Nata is the cellulosic mat-like substance that develops on the surface of nutrient liquids (Fig. 13.1), notably pineapple juice, nata de pina, and coconut water, nata de coco. It is sometimes confused with dextran. The thick, leathery membranes that formed on the surface of liquids by *Aceto-bacter xylinum* (Brown, Holland), were identified by Khouvine *et al.* (1932) as cellulosic and related to cotton cellulose. The organisms have been identified on numerous occasions since then.

FIG. 13.1. EXPERIMENTAL NATA DE COCO

Note the profuse growth in the flask on the left to which was added 0.1% alcohol.

Any kind of pineapple, trimmings from canning, culled pineapple, or over-ripe pineapple can be used for making nata. The pineapples are peeled, sliced, and ground. Add 50–100% volume of water and 5% sugar. This blend is placed in a glass jar to a thickness of 3–5 in. and then covered.

Pineapple juice is surface inoculated with a small amount of growth from another fermenting jar of nata. Within a few days a thin film will form which, if undisturbed, will form into a mat 1–3 in. in thickness after 2–4 weeks. This mat may be removed and a second mat, or even more, may form as long as there is sufficient nutrient. A trace of alcohol will increase the rate of mat formation, but too much alcohol will result in production of a soft mat. The nata is washed several times, once with hot water, and bleached in the sun. The white nata is cut into pieces, cooked in a syrup with the desired flavor, and eaten as a dessert. It is primarily cellulose and therefore has little food value.

DEXTRANS

Other microorganisms produce polysaccharide capsular material during their growth. Dextrans are microbiologically produced polysaccharides commonly found in sugar factory equipment and sugar solutions. They are also found in the sap exudates of many vegetables and in the early stages of many vegetable fermentations. Pasteur and many others before him observed these gummy substances. They are ordinarily considered products of the fermentation of sucrose by *Leuconostoc mesenteroides*. Different strains of *Leuconostoc mesenteroides* vary in the consistency of the dextrans they produce, some producing a semiliquid mucus dextran while others produce all gradations to an almost rubber-like dextran. Dextran is formed more readily in crude sugar solutions. The addition of plant extracts to purer sugar solutions stimulate their production. Various dextrans differ in their chemical configuration and chain lengths. They are readily precipitated as relatively pure products by saturating the medium with alcohol.

Dextrans are used as stabilizers in food products, syrups, soft center confections, and ice creams. They are, however, more valuable as blood plasma extenders.

Bacterium vermiforme, a heterofermentative *Lactobacillus* resembling *Lactobacillus brevis* and many strains of heterofermentative lactobacilli can be induced to produce dextran.

Other microbial polysaccharides are being studied with the possibility of their use in controlling the consistency of foods. Certain phosphomannans that are produced by species of the yeast genus *Hansenula* are promising. These polymers consist primarily of mannose units.

ANTIBIOTICS

The antibiotics have become extremely important as pharmaceuticals, and considerable experimentation is underway in utilizing some preparations in preservation of foods. Nisin is an antibiotic produced by *Streptococcus lactis* which prevents growth of *Clostridium* species in cheese. Kojic acid is an antibiotic produced by *Aspergillus* species. Bacillin and subtilin produced by *Bacillus subtilis* are effective in preventing growth in foods of certain spore-forming bacteria.

BIBLIOGRAPHY

DUNN, C. G.   1952.   Food yeast.   Wallerstein Lab. Comm. *15*, No. 48, 61–83.
FRAZIER, W. C.   1967.   Food Microbiology, 2nd Edition.   McGraw-Hill Book Co., New York.
KHOUVINE, Y., CHAMPETIER, G., and SUTRA, R.   1932.   X-ray study of the cellulose of *Acetobacter xylinum*.   Compt. Rend. Acad. Sci. (Paris) *194*, 208–209.   (French)
LAWRIE, R. A.   1970.   Proteins as Human Food.   Avi Publishing Co., Westport, Conn.
MILNER, M.   1969.   Protein-enriched cereal foods for world needs.   Amer. Assoc. Cereal Chemists, St. Paul, Minn.
PIRIE, N. W.   1967.   Orthodox and unorthodox methods of meeting world food needs.   Sci. Am. *216*, 27.
PETERSON, M. S., and TRESSLER, D. K.   1963.   Food Technology the World Over, Vols. 1 and 2.   Avi Publishing Co., Westport, Conn.
PRESCOTT, S. C., and DUNN, C. G.   1959.   Industrial microbiology, 3rd Edition.   McGraw-Hill Book Co., New York.
VORBECK, M. L. *et al.*   1963.   Lipid alterations during the fermentation of vegetables by the lactic acid bacteria.   J. Food Sci. *28*, No. 5, 495–502.

CHAPTER 14

Food Fermentation as a Approach to
World Feeding Problems

When one studies the development of fermented foods, one is confronted with certain basic truths. Fermented foods are prepared by processes that have been developed in homes during a period of centuries by accidental observation of natural phenomena. These food processes were developed without knowledge of the basic fundamental changes that occurred in the food transformation. The methods were passed on from generation to generation. From time to time certain improvements were made in the processes, based upon the careful observations of certain individuals. Until the last century, fermentation was an art based upon centuries of experience.

For thousands of years, the peoples of the world have relied upon foods preserved by fermentation processes. Practically every civilization has developed some type of fermented food and some type of fermented beverage.

Although fermentation is one of the oldest methods of food preservation, it is still the least understood among methods of preservation. Even in many industrial operations in which fermentation is the essential step, the operators are often unaware of the role of the microorganisms involved. Relatively few individuals know why foods ferment, acquire their acceptable qualities, and remain safe for consumption. They are unaware of the fact that microorganisms, e.g., bacteria, yeasts, and molds of certain species, are responsible for the changes that occur during fermentation.

The essential basic facts in regard to the role of microorganisms and the chemical and physical changes that occur in the food consist of knowledge that has been gained in the past 100 yr. Industrialization followed. The application of this knowledge has influenced practices to yield superior foods of more uniform quality.

Fermented foods are the result of the activity of a few specialized types of microorganisms among the thousands of species of bacteria, yeasts, and molds known to mankind. Fermentation processes are dependent for their success upon the control of environmental conditions that insures the growth of the desired species and discourages all other microorganisms. Environmental conditions must be established to provide these suitable conditions. The microorganisms that cause the desirable changes in various food products can be distinguished from one another and from those that are responsible for undesirable changes resulting in bad flavors, spoilages, illness, and even deaths.

Growth and fermentation by several species of microorganisms, usually developing in sequence, occur in many food fermentations. In each food product, certain species nearly always initiate fermentation. The growth products and alterations brought about by one microbial species are frequently essential to the nutrition of and growth of succeeding species in a sequence.

With knowledge available concerning the effects of physical and chemical environment necessary for growth of various species, desired changes in food products of almost any type may be established. The control, inhibition, or elimination of undesirable species concommitant with favoring growth and fermentation by desired species should be the ultimate goal.

Even though great advances have been made during the past century, advances in the industry have involved improvement in methodology with resulting standardization and quality improvement rather than development of new products. Many fermented foods have never been investigated. Studies of such foods may result in improvement of their quality and constancy.

The need today is not only for production of more food, particularly protein- and vitamin-rich food, but also for preservation of the available food. It is useless to produce more food unless it can be harvested, preserved, and made available to the people. Many countries are rich in natural resources with excellent soil and abundant rainfall. These lands of abundance should have no problem in feeding their people; however, the difficulties of preservation, transportation, and the economic level of families are a deterrent to nutritious feeding. One of the most important means of aiding underdeveloped countries would be for the more affluent countries to devote more of their resources to applied and basic research concerning food problems. Fermentation has proved itself as a cheap and effective means of food preservation.

Index

Absidia, 31
  corymbifera, 239
Acetobacter, 27, 225, 263–264, 267, 272
  aceti, 27, 254, 263, 272
  melanogenus, 254
  oxydans, 254
  rancens, 254
  suboxydans, 272
  xylinum, 28, 220, 264, 273
Achromobacter, 56, 113, 184, 225
Actinomyces, 31, 239
Aerobacter, 68, 74, 93, 119, 137–138, 184, 249
  aerogenes, 183–184, 272
Aji-no-moto. *See* Glutamic acid
Alcaligenes, 68
Alcoholic, beverages, 38, 199–231
  spirits, 228–230
Algae, 17
Alternaria, 31
Antibiotics, 274
Arrag, 85, 87
Arroz, amarillo, 239
  fermentado, 239
  requimando, 239
Ascomycetes, 17, 31–33
Ascorbic acid, 272
Aspergillus, 31, 34, 180, 184, 205, 219, 224,
    239, 250, 265, 272, 274
  candidus, 239
  flavus, 221, 239
  griseus, 267
  niger, 265, 267
  oryzae, 218, 229, 233, 236–237, 241
  soyae, 231, 233, 235
Atsumandie, 236

Bacillus, 68, 74, 113, 126, 137, 184, 225, 250,
    266
  brassicae fermentati, 71, 113
  caucasicus, 71, 84
  cereus, 239
  coagulans var. *thermoacidurans,* 265
  cucumeris fermentatae, 113–114, 125
  delbrueckii, 264
  mesentericus, 184
    var. *fuscus,* 144
  natto, 236
  panis fermentati, 182
  pumilus, 239
  subtilis, 236, 239, 267, 274
Bacterium
  brassicae, 114
  busae asiaticae, 76, 221
  gracile, 207
  lactis, 70
    longi, 70
  vermiforme, 274
Bagoong, 7, 242
Bakhar, 236

Basidiomycetes, 17, 31
Beer-like beverages, 218–223
  arrack, 219, 223
  awamori, 222
  balche, 222
  binuburan, 219
  braga, 221
  busa, 221
  chang, 221
  chica or chichara, 220
  chicha, 220
  chonta, 222
  corn beer, 220
  kaffir, 221
  kalja, 220
  kanji, 219
  kaschiri, 222
  konya, 221
  kuva, 222
  kvas, 220
  machewa, 221
  magon, 221
  masata, 222
  mezcal, 222
  millet beer, 221
  napoho, 221
  nohelick, 220
  okelehao, 219
  pachwai, 219
  pogotour, 220
  pulque, 221–222, 230
  rakshi, 220
  rice, 209
  ru'o'on, 219
  saké, 218–219
  shaw-shing-chu, 218
  shima, 222
  sonti, 219
  sool, 209
  sora, 220
  teekvass, 220
  tepache, 221
  tesquino, 220
  thumba, 221
  tiswin, 220
  toddy, 219
  toowak or tuwak, 222
  torani, 219
  tulpi, 220
  yagjoo, 209
  zaca, 220
  zythum, 220
Beers, 1, 9–10, 46, 64, 85, 111, 199, 211–218,
    272
  barley, 6
  bior or pior, 212
  boozah or bousa, 211, 221
  dark beer, 217
  ginger, 223,
  heather beer, 223

hek or heqa, 211
hopped, 211
hung-chu, 212
kin, 212
machmetzeth, 212
near, 218
pale, 216–217
porter, 218
rice, 218–219
samshu, 212
spiced, 211
sprossenbeer, 223
stout, 218
tchoo, 212
tibi, 223
Betabacterium, 71
  brevi, 71
  caucasicum, 84
  longum, 71
Betacoccus, 70, 91
Bongkrek, 231, 240
Botrytis, 231, 240
  cinerea, 205
Bread, 4–6, 9–10, 64, 173–198, 212, 222–224,
    269, 272
  acorn, 196
  beer, øllebrød, 183, 191
  bisquits, 1
  black, 86, 195
  breadroot, 196
  breadsticks, 190
  buckwheat cakes, 1, 192
  chemically-leavened, 187
  chufa, 196
  cinnamon rolls, 190–191
  classification, 186
  crackers, 190, 193, 196–197
  Danish pastry, 191–193
  dessert breads, 189
  doughnuts, 190, 197
  fermentation, 185
  flat breads, 175, 190
  flavor, 188
  French, 190–191
  garlic, 190
  graham, 191
  grapenuts, 197
  griddle cakes, 192
  idli, 6, 181, 188, 192–193, 196
  ingredients, 176
  Italian straight dough, 190
  jamin-bang, 196
  kanom-tan, 193
  knäckebröd, 86, 175, 195
  macaroni, 187
  microbiologically leavened, 188
  microorganisms, 179–183
  Munich rye, 195
  oatmeal, 191
  pampooshke, 6
  panary fermentation, 177, 185
  pancakes, 192
  pasha, 191
  piñon, 196
  prairie potato, 196

prairie turnip, 196
pretzels, 190, 197
pumpernickel, 181, 187, 191, 194–195
puto, 6, 181, 183, 188, 193, 196, 222
raisin, 191
rice, 191
rogbröd, 190
rye, 190–191
rye-krisp, 195
rye rusk, 190
salt-rising, 193, 196
sour dough, 6, 174, 187–189, 193, 195–196
sour dough pancakes, 192
sour milk pancakes, 192
sour rye, 7, 188
spaghetti, 187
sunflower, 195
Swedish health, 195
tein-mein chang, 222
texture, 188
turnip, 196
types, 190–193
unleavened, 187
waffles, 192
whole wheat, 191
wild rice, 191, 196
zweibach, 191
Brettanomyces, 32
  lambicus, 224
Brevibacterium, 266
  linens, 105
Brucella, 74
Burong, dalag, 8, 242–244
  hipon, 8, 243
  isda, 243
Butanediol, diacetyl, 272
Butter, 1, 4, 6, 9–10, 64, 66, 76, 87–93, 111,
    272
  brown butter, 88
  clarified, 6
  ghee, 88
Butyric acid, 267

Cacao and cocoa, 251–254
Candida, 32, 59, 219, 271–272
  utilis, 33, 59, 182, 271
Casava, 244
Cephalosporium, 31, 266
Cereal foods, 173–198
Chaetostylum, 239
Chao, 238
Ch'au-yan, 235
Chee-fan, 238
Cheese, 4, 6, 9–10, 34, 64, 66–67, 72, 93–106,
    111, 231, 267, 269, 272
  albuminoid, 100
  American, 97, 102
  brick, 105
  camembert, 34, 64, 95, 97, 106
  cheddar, 1, 9, 64, 95, 97, 102
  cottage, 1, 7–8, 76, 95, 99
  cured cheese, 100
  devonshire cream, 10, 86
  grana, 101, 103
  hard grating, 96

hard plastic, 96
liederkranz, 105
limburger, 43, 95, 105
mold fermented, 105
mysost, 95, 100
neufchâtel, 100, 106
parmesan, 103
plastic curd, 103
provolone, 95, 101, 103
ricotta, 95, 100
roquefort, 9, 11, 34, 43, 46, 64, 95, 99, 106, 267
soft curd, 95
soft curd plastic, 95
surface ripened, 105
Swiss, 1, 13, 39, 43, 70, 95, 97, 103–105
uncured, 95, 98
whey, 1, 100
Chemical alterations, 36–44
Chiang, 239
Citric acid, 265
Citromyces, 265
Citron, 257
Cladosporium, 31
Classification, 12–35
Clostridium, 138, 274
botulinum, 165
Coffee, 247–251, 272
Corynebacterium, 266
Cryptococcaceae, 32
Cryptococcales, 31–32
Cryptococcoideae, 32–33
Cryptococcus, 32

Dawadawa, 241
Debaryomyces, 32
Dematiaceae, 31
Dextran, 40, 115, 273
Diplococcus, 19
Dispora caucasica, 84

Endomyces, 32
fibuliger, 184, 221
Endomycetoideae, 32
Entomophthorales, 31
Enzyme-induced reactions, 36, 44
Erwinia, 250
dissolvens, 250
Escherichia, 68, 74, 113, 137, 250, 266
coli, 183
Eubacteriales, 18
Eumycetes, 17, 31–33

Family, 17–18
Fermentation, complexity, 40–41
Fermented foods, origin, 2
Flavobacterium, 56, 93, 113, 119, 184, 225
Fish, paste, 241
sauce, 241
soy, 241
Fumaric acid, 267
Fungi, 31
Fungi imperfecti, 17, 31–32, 34
Fusarium, 31, 250, 272

Genera, 17–18
Geotrichum, 31, 75, 219, 271–272

Ginger, 257
Gluconic acid, 267
Glutamic acid, 239, 265–266
monosodium glutamate, 239
Growth, associations, 63–64
curve, 60–63
factors, 49–50
acidity, 52
carbohydrates, 58
lipids, 60
minerals, 59–60
moisture, 50
nucleic acids, 60
nutrition, 57
oxygen relationship, 53
presence of microorganisms, 56
proteins, 58–59
purines, 60
pyrimidines, 60
starters or cultures, 56–57, 75, 89, 91, 158, 160, 213
temperatures, 54
vitamins, 59
microorganisms, 45–65
population, 49
products, 45–46
sequences, 64

Hanseniaspora, 32
Hansenula, 32, 59, 219, 233, 271, 274
anomala, 59
Helminthosporium, 31
Hon-fan, 238

Itaconic acid, 267

Jotkal, 242

Kapi, 242
Kapoosnyck, 7
Ketjap, 235
Ketoglutaric acid, 267
Kholodnyk, 6
Kimchi, 71, 131, see also Vegetable products
Kloeckera, 32, 204
apiculata, 204
Kluyveri, 266
Koji and Ragi, 236–237
Kojic acid, 267, 274
Kombucha, 237
Kvass, 6

Lactic acid, 264–265
Lactobacillaceae, 18
Lactobacilleae, 18–19, 24
Lactobacillus, 18, 24–25, 54, 68, 70, 74, 98, 122, 126, 149, 158, 161, 163, 171, 182, 210, 221–222, 224, 227, 233, 250, 264, 274
acidophilus, 26, 71–72, 81, 87
bifidus, 26
brevis, 22–23, 25, 27, 63, 113–116, 119, 122, 129, 132–133, 135, 137–138, 163, 182–183, 195, 206, 256, 274
var. vermiforme, 223
buchneri, 27, 182, 206

bulgaricus, 22, 26, 63, 70–72, 74–75, 78, 80–82, 85, 8/, 101, 103–104, 265
   var. *yoghurtii,* 71, 74, 82, 84
casei, 23, 26, 63, 68, 70–71, 74, 80, 82, 85, 99, 101, 182, 265
caucasicus, 25, 71
delbrueckii, 26, 182, 206, 215, 221, 226, 228, 233, 264–265
fermenti, 27, 71, 182, 206, 254
frigidus, 227
gracile, 227
helviticus, 26, 71
heterohiochi, 218, 227
hilgardii, 206, 227
homohiochi, 218, 227
lactis, 26, 71, 104
leichmanni, 26, 182
malofermentum, 227
parvus, 227
pastorianus, 26, 182, 206, 226
plantarum, 22, 24, 26, 62–63, 113–116, 119, 122, 127, 129, 132–133, 135, 137–138, 148, 157, 161, 163, 182–183, 195, 206, 210, 221, 242–243, 255–257, 265
thermophilus, 26
trichodes, 206, 227
viscosus bruxellensis, 224
yoghurtii, 71, 82
Leuconostoc, 18, 22, 24, 40, 70, 91–92, 98, 126, 161, 182, 207, 210, 222, 227, 250, 264
citrovorum, 23, 70, 76–77, 80, 91, 98, 206
dextranicum, 23, 77, 91, 206
mesenteroides, 21, 23–24, 27, 40, 47, 54, 62–63, 70, 112–117, 119, 121–122, 126–127, 129–138, 149, 157, 163, 171, 182–183, 188, 193, 196, 206, 210, 222, 242–243, 250, 255–257, 274
Lipids, 271
Lipomyces, 272
Lontjom, 241
Lysine, 266

Malic acid, 267
Mamchao, 242
Mamcu-sak, 242
Mamtom, 242
Medusen tea, 237
Meitanza, 238
Meju, 238
Melanconiales, 31
Microbial products consumed, 259–274
Microbiology, development, 15
Micrococcus, 68, 157, 163–164, 184, 242–243, 266
aurantiacus, 161
glutamicus, 266–267
malolacticus, 206
variococcus, 206
Microorganisms, activity, 15
application of names, 16
early observation, 14
of fermented foods, 12–35
size, 47
small plants, 14

Milk, fermented, 72–87
acidophilus, 72, 79–80
arrag, 76
busa, 73, 76
butter milk, 1, 6, 9, 72–73, 76–80, 85, 87, 89
cieddu, 73, 76
cream, sour, 6, 79, 86
curd, 5, 7, 73
dahi, 73, 76, 79
filbunke, 73, 79–80, 86
gioddu, 79
kefir, 7, 9, 79, 83–85, 222–224, 230, grains, 70, 73, 76
kissélo mleko, 73, 79
kuban, 76
kumiss, 7, 73, 76, 79, 85, 87, 191, 222–224, 228, 230
langmjölk, 70, 79–80
leben, leben-raid, 7, 9, 72–73, 76, 79
mast, 79
mazun, matzoon, 73, 76, 79
oxygala, 79
skyr, 73, 80, 83
sour, 5–6, 76, 86–87, 269
soy milk, 231
taettemjölk, taette, 70, 73, 76, 79–80, 86
yoghurt, 6–7, 9, 22, 70, 72–74, 76, 78–79, 81–83, 86–87, 228
zhum, 86
Milk products, 2, 66–107
Minchen, 241
Miso, moromi, 231, 237
kome miso, 237
mame miso, 237
mugi miso, 237
Monascus purpureus, 236
Monilia, 31, 93, 184
sitophila, 184
Moniliaceae, 31, 34
Moniliales, 31, 34
Monosodium glutamate, 239
Mucor, 31, 184, 205, 219, 224, 236–237, 272
meitanza, 238
piriformis, 265
rouxii, 221
Mucorales, 31
Murcha, 236
Mycoderma casei, 106, 263
Mycotorula, 271

Nadsonia, 32
Nappi, ngapi, 242
Nata, 220, 273
nata de coco, 8, 272
nata de pina, 8, 272
Natto, 236
Neurospora, 272
sitophila, 241
Nuoc mam, 242
Nuoc-mam-ruoc, 242

Olives. See Vegetable products
Ontjom or lontjom, 231, 241
Oomyces, 31
Oriental foods, 231–245

Organic acids, 260–268
Oxalic acid, 267

Pachwai, 236
Padic, 242
Paracolobactrum, 250
Patis, 7, 242
Pediococcus, 18, 21, 55, 122, 126, 132, 149, 158, 182, 207, 210, 226
acidilactici, 21
cerevisiae, 20–21, 115–116, 119, 122, 127, 129, 135, 157, 161–163, 193, 206, 226, 242–243
soyae, 232–233
Penicillium, 31, 34, 180, 184, 205, 219, 250, 265, 272
camemberti, 106
candidum, 106
luteum, 265
notatum, 267
roqueforti, 33, 106
Peptostreptococcus, 19
Peujeum, 241
Pēyēm, 241
Phaak, 242
Phycomycetes, 17, 31
Phylum, 17
Pichia, 32
Pickles. See Vegetable products
Pityrosporum, 32
Plant, 17
Podvarak, 7, 135
Poi, 244–245
Prahoc, 242
Propionibacteriaceae, 18, 27
Propionibacterium, 18, 27, 138, 267
shermanii, 104
Propionic acid, 267
Proteus, 74
Pseudomonadaceae, 27
Pseudomonadales, 18, 27
Pseudomonas, 56, 68, 74, 93, 113, 119, 184, 266
fluorescens, 93, 272
fragi, 93
mephiticus, 93
nigrificans, 93
putrefaciens, 93

Raitas, 86
Ran, 5
Ranu, 236
Red sufu, 238
Rhizopus, 31, 34, 184, 205, 219, 224, 229, 236, 240
oligosporus, 241
oryzae, 34, 240
sonti, 219
Rhodotorula, 32, 272
Rhodotoruloideae, 32
Roghan josh, 5
Rømmegrot, 86

Saccharobacillus pastorianus, 226
Saccharomyces, 32–33, 59, 76, 204, 217, 219, 224, 270–272

apiculata, 253
beticus, 204
bruxellensis, 224
busa asiaticae, 76, 221
carbajali, 210, 222
carlsbergensis, 30, 33, 216
cerevisiae, 33–34, 59, 180, 182, 202–203, 216, 224, 229
var. ellipsoideus, 28–30, 33, 203, 225
cheresienses, 203
chevalieri, 204
ellipsoideus, 203, 225, 253
intermedius, 223, 225
kefir, 84
oviformis, 204
pastorianus, 225
rouxii, 232–233, 237
saké, 218
turbidans, 225
validus, 225
Saccharomycetaceae, 32
Saccharomycetales, 32
Saccharomycetoideae, 32
Saccharomycodes, 32
ludwigii, 220
Saraimandie, 236
Sarma, 135
Sauerkraut. See Vegetable products
Sausage, 1, 4, 7, 9–10, 64, 153, 172, 269
abnormal, 164–166
Alessandro, 156
Alpino, 156
Arles, 156
beerwurst, 154
cappicola, 156
cervelats, 156, 158, 161, 167, 169, 171
chorizos, 156–157, 171
dry, 169
farmer, 158
Genoa, 9, 154, 156, 158, 163
Göteborg, 158
Holsteiner, 158
kosher salami, 171
landjaeger, 158
Lebanon bologna, 9, 154, 158, 161–162
Lyons, 156
medwurst, 158
Milano, 156
mortadello, 156
oryae, 154
pepperoni, 156
pork roll, 161
salami, 9, 154, 156, 158, 167, 169, 171
semi dry, 156, 171
Siciliano, 156
smoking, 171
summer, 1, 9, 154–155, 158, 161, 163, 167
Thuringer, 9, 158, 161, 171
Sausage cultures, 160–161
Schizomycetes, 17–18
Schizosaccharomyces, 32
pombe, 220
Scopulariopsis, 31
Sequence of growth, 5
Serratia, 68, 74, 266

marcescens, 183–184
Shigella, 74
Shottsuru, 242
Soy, 235
  sauce, 9, 34, 231–235, 237, 239
Soybean cheese, 231, 233, 238
Soybean milk, 235
Sphaeropsidales, 31
Species, 17–18
Spirits, apple-jack, 229
  aquavit, 230
  arrack, 223, 228–229, 230
  brandy, 223, 228
  cognac, 229
  gin, 230
  kirsch, 229
  lao-rong, 219
  liqueur, 209
  mirabelle, 229
  rum, 229
  rusiviina, 209
  sautchoo, 223, 228
  skhow, 223, 230
  slivovitz, 229
  sool, 209
  southern comfort, 229
  tequila, 222, 230
  vishnyovoka, 209
  vodka, 209, 223, 236
  whiskey, 214, 223, 228
Sporotrichum, 31
Stemphylium, 31
Streptobacterium casei, 71
Streptococceae, 18–19, 226
Streptococcus, 18–19, 24, 70, 93, 98, 114,
    126, 132, 158, 163, 171, 182, 226, 243,
    256, 264
  citrovorus, 70, 77, 91
  cremoris, 21, 68, 70, 74, 77, 98
  faecalis, 20, 129–130, 163, 193, 242
  lactis, 16, 18, 20, 63, 68, 70, 74, 76–77,
    80, 85, 91–92, 98–99, 101–102, 114,
    264–265, 274
    var. *diacetilactis*, 70
    var. *hollandicus*, 70, 79
    var. *liquifaciens*, 74
    var. *maltigenes*, 70
    var. *tardus*, 70
  paracitrovorus, 70, 77, 91
  thermophilus, 19, 70, 74, 78, 80, 82, 101,
    103–104, 238
Succinic acid, 267

Tabasco sauce, 258
Tahuri, 238
Tamari, 238
Tan-chey, 240
Tang-chang, 240
Tao cho, 239
Taokoan, 238
Tao-si, 241
Taotjo, 239
Tao-tjung, 237
Tapai, 241
Tape ketan, 241

Tape ketella, 241
Tarhana, 86
Taro, 244
Tartaric acid, 267
Tea, 257–258
Tempeh, 34, 231, 239–240
Thallophyta, 17
Thamnidium, 31, 239
  elegans, 239
Thermobacterium, 71
  helviticus 71
Torula, 76, 84, 93
Torulopsis, 32, 204, 233, 271–272
  utilis, 271
Toyo, 235
Trasse-udang, 242
Trassi-ikan 242
Trichosporoideae, 32
Trichosporon, 32, 271
  pullulans, 271
  variable, 184
Trichothecium, 31
Tricoderma, 31

U-t-rat, 236

Vanilla, 254–257
Vegetable products, 108–152
  brine solutions, 123–130
  beets, 6
  Brussels sprouts, 133
  carrots, 133–134
  cauliflower, 110, 133, 136
  celery, 133
  chard, 133
  Chinese cabbage, 133
  green beans, 133
  lettuce, 133
  mustard, mostasa, 133–134
  okra, 133
  olives, 9, 110, 113, 116, 123, 136–139
    brined, Greek, 137
    California, green, 137
    ripe, 137
    French, 136
    Sicilian, 137, 139
    Spanish green, 136–137
  onions, 110, 133
  pak-gard dong, 134
  peas, 133
  peppers, 110
  pickles (cucumber), 6, 9, 110, 137, 147
    blackening, 147
    bread and butter, 124
    dill, 1, 123, 125, 127, 129–130, 269
    salt stock, 123, 129–130, 140–143
    softening, 143–145
    sour, 7, 125, 130
    spiced, 125
  radishes, 134
  sarma, 7
  sauerkraut, 1, 7, 9, 64, 110, 112–113, 116–
    123, 137, 148, 269
    choucroute, 117, 135
    red, 7

tomato and juice, 130, 134
turnips, 134
vegetable blends, 115, 130–133
  burong gulay, 8, 132
  dua chua, 131
  karanyshe dong, 131
  kimchi, 7, 64, 110, 115, 131
  nukamiso, 131
  pawtsay, 64, 115, 131–132
  sajur asin, 131
  so chican, 131
  tang-chai, 134
  tarhana, 131
  yentsai, 131
whole head cabbage, kiseo kupus, 7
wild plants, 135
Verjuice, 117
Vinegar, 1, 4, 7, 117, 208–209, 260–264, 272
  alegar, 261
  distilled, 261
  fruit, 261
  honey, 261
  malt, 261
  pineapple, 261
  tea, 261
  wine, 261
Vitamins, role of, 38–39, 59, 270–271
Wine-like beverages, 208–211
  cider, 208
  mead, 211, 223
  metheglin, 211

pulque, 210
sparkling cider, 209
Wines, 1, 9–10, 46, 64, 111, 200–208, 211, 231, 272
  Bordeaux, 204
  champagne, 85, 207, 209
  Charmat, 208–209
  dessert, 201
  dry, 204
  espumante, 207
  flavored, 201
  port, 1, 201
  red table, 201, 212
  red seibel, 206
  sauterne, 204–205
  schaumwein, 207
  sherry, 201, 203
  sparkling, 201, 207–208
  spumante, 207
  vermouth, 201
  vins mousseaux, 207
  white, 201

Yeasts as food, 270

Zygomycetes, 31
Zygorrhynchus, 31
Zygosaccharomyces, 233
  major, 233
  soya, 233